Water Baby

The Story of *Alvin*

Water Baby

Victoria A. Kaharl

New York   Oxford   Oxford University Press   1990

Oxford University Press

Oxford New York Toronto
Delhi Bombay Calcutta Madras Karachi
Petaling Jaya Singapore Hong Kong Tokyo
Nairobi Dar es Salaam Cape Town
Melbourne Auckland

and associated companies in
Berlin Ibadan

Copyright © 1990 by Victoria A. Kaharl

Published by Oxford University Press, Inc.,
200 Madison Avenue, New York, New York 10016

Oxford is a registered trademark of Oxford University Press

All rights reserved. No part of this publication may be reproduced,
stored in a retrieval system, or transmitted, in any form or by any means,
electronic, mechanical, photocopying, recording, or otherwise,
without the prior permission of Victoria A. Kaharl.

Library of Congress Cataloging-in-Publication Data
Kaharl, Victoria.
Water baby : the story of Alvin / Victoria A. Kaharl.
p. cm. ISBN 0-19-506191-8
1. Alvin (Submarine)—History.
2. Underwater exploration—History.
I. Title.
VM453.K34 1990 623.8'205—dc20 89-48504

9 8 7 6 5 4 3 2 1

Printed in the United States of America
on acid-free paper

for yimkin
December 9, 1928–October 26, 1988

and for johnnymac
October 17, 1934–June 7, 1988

Contents

Prologue / xi

I Awakening, 1956–1964 / 3
II Hard Times, 1964–1971 / 49
III Wonderland, 1971–1982 / 133
IV The Submarine Blues, 1982–1989 / 251

Epilogue / 345

Index / 349

... *the fairies had turned him into a water-baby.*

A water-baby? You never heard of a water-baby?... There are a great many things in the world which you never heard of; and a great many more which nobody ever heard of....

Here begins the never-to-be-too-much-studied account of the ... wonderful things which Tom saw, on his journey to the other-end-of-Nowhere....

Tom came to the white lap of the great Sea-mother, ten thousand fathoms deep ... as he walked along in the silence of the sea-twilight, on the soft white ocean floor, he was aware of a hissing, and a roaring, and a thumping, and a pumping, as of all the steam engines in the world at once. And, when he came near, the water grew boiling hot ... [and] as foul as gruel; and every moment he stumbled over dead shells....

... he was on the edge of a vast hole in the bottom of the sea, up which was rushing and roaring clear steam enough to work all the engines in the world at once; so clear, indeed, that it was quite light at moments....

... as the vapors came up out of the hole ... they were changed into showers and streams of metal ... gold-dust ... silver ... copper ... tin ... lead, and so on, and sank into the soft mud, into veins and cracks and hardened there....

Tom ... is now a great man of science, and can plan railroads, and steam-engines ... and knows everything about everything, except why a hen's egg don't turn into a crocodile, and two or three other little things which no one will know.... And all this from what he learnt when he was a water-baby, underneath the sea....

Charles Kingsley, *The Water Babies*, 1863

This tale is true.

Prologue

The sea at the equator off the western bulge of South America is like bathwater, ideal for launching the midget submarine *Alvin*. All of the various oils in the sub are topped off, hoses capped, battery fluids checked. Men in T-shirts and shorts, each with a specific task, huddle around the craft. One man—tall, with a goatee and straw hat—makes checks on a clipboard.

The pilot climbs into the perfectly round hull that can fit only three people. All around are switches, like an airplane cockpit. The round windows seem incredibly small. He throws a switch, the mechanical arm stretches, and the man with the clipboard pencils another check.

Finally the two passengers throw their legs over the conning tower or sail, as they call it, and join the pilot. They sit on the floor, resigned to the discomfort—but they are young.

The round hatch above their heads shuts. If there is to be any fear on a dive, it usually comes now, making new passengers grow pale and clammy. In the dozen years that *Alvin* has been diving, nobody has fainted; the potential fainters don't stay inside long enough to hear the hatch close.

It is the first time for the chemist; he's cool. Permission to dive is given from the surface and *Alvin*, about the size of a UPS delivery truck, disappears into the Pacific amid a froth of bubbles. It takes about two hours to reach the bottom at 8000 feet.

The scientists take turns gathering the goods they have come for—water for the chemist, rocks for the geologist. It is black except for what *Alvin*'s powerful lights illuminate. And it is cold; the temperature always hovers just above freezing everywhere in the deep sea.

The geologist spots an interesting rock and directs the pilot to move *Alvin*'s long stainless steel arm toward it.

The chemist sees a large purple anemone as he gazes absent-mindedly out the porthole. The water seems to be shimmering like the air above a sidewalk in summer. Of course it cannot be; it must be an optical illusion. He blinks, looks again. It *is* shimmering. The geologist, too, has seen. They shout together.

Alvin moves up slope toward the shimmering water, and its spotlights capture a sight so sensational and bizarre that the scientists are near frenzy.

What they and their colleagues saw off Ecuador that sultry winter of 1977 were tall red-tipped worms undulating slowly, blind white crabs scampering across the rocks, clams and oysters as big as dinner plates, messes of tangled goo that looked more like day-old spaghetti than life, pretty pale-orange things that resembled dandelions gone to seed, and pink blue-eyed fish and more. Here in the frigid blackness of the deep ocean was a whole community of weird creatures thriving in warmth without sunlight and photosynthesis, the food-making process fundamental to life. Or so we were all taught.

The spaghetti, it turned out, was a conglomeration of worms. The dandelions were not plant but animal. The oysters were not oysters. The red-tipped tube worms, some of them eight feet long, proved strange indeed. They had no gut, no mouth, no anus. How did they eat?

There would be more surprises to come, as there were surprises before this. For more than a quarter century *Alvin* has been taking scientists into the deep ocean to see a new world, one that is still new and will continue to be for many decades to come. *Alvin* has a long waiting list of people eager to explore Planet Earth's last great frontier. But the story of *Alvin* begins when almost no one cared to venture into that vast blackness, when getting inside the deep sea was less a need than a hankering of but a few people who had the wisdom to dream and the courage to realize their dreams.

Water Baby

I Awakening, 1956–1964

So many questions, so many mysteries. It is only by going down ourselves to the depths of the sea that we can hope to clear them up.
    Auguste Piccard

1

Since very early times, at least three hundred years before Christ, humankind has used all kinds of contraptions—suits of armor and animal skins, kettles or diving bells, leather hoods, snorkels, nose clips—to get beneath the waves. It was in Leonardo da Vinci's time when the first practical-looking underwater boats began to appear, at least on paper. For most of their early history, submarines were not entirely trusted, and to a world that fancied itself civilized, sneaking up on someone under cover of water was a dirty way to fight. Da Vinci designed a "submarine boat" but this, he said, "I do not divulge on account of the evil nature of men." By the First World War, Germany was capitalizing on that evil and winning. With few exceptions, war was the motivation behind the evolution of the submarine.

In the sixteenth century a Swedish historian reported on two sealskin submarines rowed off Greenland, and in 1620 the Dutchman Cornelius van Drebble built one of wood and greased leather with oars that stuck out the sides. It was so successful he built another and rowed it up and down and slightly beneath the Thames. Despite his salesmanship, the British Admiralty was not interested.

The submarine was first used as a weapon in the American Revolutionary War. David Bushnell, a thirty-five-year-old American colonist, had an idea to sneak up at night on English ships in a submarine and drill holes in the keels to insert explosives and blow them up. Bushnell called his ingenious wooden submarine *Turtle,* apparently because it reminded him of two upright tortoise shells belly to belly. Instead of oars, the *Turtle* had a revolutionary hand-driven screw propeller in front and a rudder aft. It carried lead for ballast, took in water to descend and let out water to ascend. A fungus (foxfire) lit up the compass, and an abbreviated conning tower kept out water while the craft was at the surface.

General George Washington decided to use this secret weapon when a British fleet appeared off New York. The *Turtle* operated perfectly, but her pilot could not drill through the metal the British had placed over their hulls to keep out wood borers. Bushnell designed other submarines but his country wasn't interested in them.

Robert Fulton, who knew about Bushnell's craft, designed the *Nautilus*. Like the *Turtle*, it used the hand-turned screw for propulsion and was meant to sneak up on enemy ships to affix explosives. He couldn't get his funds in America, so he went to France and convinced Napoleon Bonaparte to put up the construction money in 1800.

Fulton's greed got him into trouble. He asked the French for more money, including a bonus for every British ship he sank with the *Nautilus*. When the French declined, he offered to blow up French ships for the English. He took the British Admiralty's disinterest and Napoleon's rage as his cue and left for America, where in the final years of his life, he developed what he is best remembered for, the steamboat.

By the mid-nineteenth century, as the Civil War rent the young United States, the South built hand-powered submarines from cylindrical steam boilers used on Mississippi River boats. The *H. L. Hunley* has gone down in history as the first submarine to sink an enemy ship. Despite the distinction, the *Hunley* was a singularly calamitous vessel. On the *Hunley*'s maiden cruise, a big paddle-wheel vessel passed, and before the hatches could be closed, the submarine was swamped and sank. Only the man standing in an open hatch escaped. On her second foray, the *Hunley* got tangled in a ship's line and sank in Charleston Harbor. Three people escaped. Again she was salvaged and outfitted with a new crew, and again she sank, this time taking the entire crew.

Finally the Confederate fathers decided to send her on a suicide mission to sink the Union corvette *Housatonic*. Only six volunteers could be found and they insisted that both hatches remain open. On the night of February 17, 1864, the *Hunley* set off on the water with her partial crew. The Yanks spotted her about a hundred yards away. The *Hunley* rammed its single torpedo against the *Housatonic* in an area where the gunpowder was stored. The explosion sank the *Housatonic* and the *Hunley*. Most of the crew aboard the corvette escaped, but not so the men aboard the *Hunley*.

After that bittersweet victory, the Confederates never used another submarine. "This disaster," an early historian wrote, "was due to the excellence of the use of torpedos which had been arrived at by the Confederates... it was a practice rather abhorrent to the minds of trained fighting men and owed its development by the naval officers of the South to necessity rather than desire."

Submarines could still be civilized. In honor of Tsar Alexander II's coronation, a band went to the bottom of Kronstadt Harbor in 1855 in the Bavarian-designed *Diable-Marin* (Marine Devil) and played the Russian national anthem. It must have been very close inside with the musicians puffing and the four crew members huffing as they ran on the treadmill which drove the screw that powered the submarine. The craft became mired in the mud and they narrowly escaped.

In 1862 the French built the first submarine with an engine. Compressed air turned the *Plongeur*'s propellers. Her first dive, however, ended in near disaster; her crew had great difficulty making the submarine surface. The French had none of the Confederates' grit. They immediately declared the *Plongeur* unsafe and turned her into a water storage tank.

But a new era of engines rather than muscle power had begun. A British steam-powered submarine followed in 1879 (although it sank and took all hands), and a half dozen years later, the first electric submarines.

The new technologies brought hard lessons. Submarines began to operate farther from coastal waters and were run over and sunk by surface ships. Carbon monoxide from gasoline engines killed sailors; hydrogen gas from leaky batteries caused explosions. Sub crews took aboard mice, which squeaked or panted and collapsed at gasoline vapors. And submarines continued to sink because of hatches left open.

The United States Navy commissioned its first submarine in 1895 from John P. Holland. The project was most unpleasant for Holland, who argued that the navy's specifications were impossible and that the supervising officers didn't know anything about submarines. The craft that was built was a failure, as he predicted. The following year he built another according to his own specifications and with his own financing. In 1900 the navy swallowed its pride and accepted the *Holland* from the Holland Torpedo Boat Company, which soon became the Electric Boat Company.

Holland's contemporary was Simon Lake, who adapted his submarine for exploration by adding wheels to the bottom and a manipulator for taking samples. In the 1890s he recovered sunken cargoes of ore and coal in Long Island Sound in his own submarine, took underwater photographs and scooped up clams.

The submarines of World War I used diesel on the surface and electric motors when submerged. They were crewed by about thirty men who called them "pig boats." Quarters were tight, there was no refrigeration, no shower, one toilet, no heat and a constant stench. By World War II the evil of underwater warfare was widely accepted.

All these various underwater boats were shallow; until the early 1930s, no one had gone much deeper than 600 feet and lived to tell about it. New depth records were set by Otis Barton and

William Beebe cramped in a two and a half ton steel ball attached to a cable. The contraption was crude and dangerous, but it did descend in 1930 to 1426 feet, and in 1934, to 3028 feet, the deepest anyone had ever been.

The first untethered vessel that could go to the deep sea, and one with far greater depth capability than the Barton-Beebe steel ball, was the bathyscaph, invented by Auguste Piccard, a soft-spoken Swiss physicist and engineer. Piccard had already built an airtight steel sphere suspended from a balloon, which in 1932 took him up to 53,400 feet, the very edge of space where no human had ever been. Working from the principles of his air cabin, he built an "ocean balloon." It resembled a blimp. Under its belly hung a steel ball for two passengers. Piccard knew the small glass portholes on the Barton-Beebe steel ball had cracked and leaked. Glass was very strong, stronger than even steel, but it was brittle. A tiny crack could be catastrophic under the ocean's pressure. He chose the new material methylmethacrylate being manufactured as Plexiglas for his windows and made them slightly conical with their tapered ends facing inside for a wide-angle view. Instead of the lighter-than-air helium or hydrogen used in balloons, Piccard filled his blimplike float with thousands of gallons of gasoline, which is about 30 percent lighter than water. Several tons of iron shot made it heavy enough to sink. To ascend, it dropped the steel pellets. He called it a bathyscaph, from the Greek *bathos*, for deep, and *scaphos*, boat.

In 1948 the world's first bathyscaph dived to 84 feet and without passengers went to 4544 feet. But under tow in search of calmer waters, it fell apart.

The sixty-nine-year-old Piccard unveiled his second bathyscaph, *Trieste*, in 1953, and with his thirty-one-year-old son Jacques descended to a record 10,393 feet, almost two miles. Once the highest-flying man of his time, Auguste Piccard had become the deepest diving.

2

The ocean. Really it is four oceans and more than twenty seas splayed contiguously over about 71 percent of the Earth. The average depth of the sea is about two and a half miles or 13,000 feet—not much compared with the globe's bulk; if the water planet were the size of a basketball, only a damp film would cover it. But by volume, the ocean comprises about 99 percent of the planet's inhabitable living area, and from its waters comes about a third of all plant life.

On February 29, 1956, 103 ocean scientists gathered for a symposium in Washington, D.C., to address the difficult challenge of probing the deep sea. Oceanographers dropped instruments to measure temperature, dredged the bottom for what it would give up, and dangled cameras into the blackness. One of their most useful tools was sound; they used to it gauge depth and the seafloor topography, which offered all the variety of land—canyons, plateaus, rolling plains. They analyzed seawater for its constituents and tried to determine where it came from and where it was going. From temperature, salinity and pressure they deduced the general direction and speed of currents. But they had data from only clement areas, sampled only when wind and wave permitted. The deepest realms of the sea were the most mysterious. Trying to study the depths from a ship was like trying to study the land at night from an airplane flying at 13,000 feet.

On the morning of their second and final day, one of the twenty-five symposium speakers suggested going down in person, far beneath the waves through which they had spent all their professional lives dangling one thing or another from the end of a line.

"I'm sure if the airplane had been the only craft invented to date and we didn't know about ships, we'd still be doing oceanography, but we'd have a different point of view about the ocean and we'd be learning different things about it," Allyn Vine, a young oceanographer, said. "If we'd had only a deep-diving submarine and not the surface craft, I presume that we'd still have had this meeting and that we'd still be doing oceanography, but I feel sure we'd have an entirely different concept and different perspective of the ocean."

Vine said he was ashamed that ocean scientists had virtually ignored the submarine for their research. Yes, he knew you couldn't see anything out of a modern submarine. The glass peepholes were removed in the war, he said,

> for the perfectly good reason of not wanting water to come in after a depth charge, but we oceanographers have not put any of those glass ports back in.... An experiment of this type would involve far less money and accumulated time than this meeting has.... I believe firmly that a good instrument can measure almost anything better than a person can if you know what you want to measure...but people are so versatile, they can sense things to be done and can instigate problems. I find it difficult to imagine what kind of instrument should have been put on the *Beagle* instead of Charles Darwin.

Vine was forty-one years old, a geophysicist with a knack for tinkering and a curiosity that wouldn't quit. His wife understood early on. She didn't worry the day he got home twenty hours late. Vine was in the Midwest awaiting a flight to Boston when he got talking with a fellow at the airport. The discussion so captivated Vine that he got on the stranger's plane to continue it. The plane landed in California.

The second of four boys, Allyn Collins Vine was born to Lulu Collins and Elmer Vine, a second-generation butcher, in Garrettsville, Ohio. Allyn left his home state to do his graduate studies in geophysics at Lehigh University in Bethlehem, Pennsylvania. Still a thesis shy of his Ph.D., he went in 1940 to the decade-old Woods Hole Oceanographic Institution on Cape Cod. From his new base at the Massachusetts laboratory, Vine and his colleagues used sound to learn what was on and under the ocean floor. They threw TNT and dynamite overboard, a technique borrowed from the oil companies. By knowing how quickly the sound waves from the explosives traveled, they could infer characteristics about the material sound penetrated, whether it was mud or rock, and what kind of rock. Their recorders captured the sound waves graphically in gentle peaks and valleys that looked like the landscapes of Chinese ink paintings.

Like other American oceanographers, Vine put aside his pet research projects in World War II to help the navy better understand one of its principal battlegrounds. He rode subs and taught the battle-bound submariners about a layer of temperature change in the ocean which deflected the pulses of sound. A submarine could hide under this layer, called the thermocline, from enemy sonars.

Vine couldn't shake the idea that submarines were perfect for oceanographic research, and he said so repeatedly to anyone who might make a difference, especially to the navy. Some of his letters were cosigned by Bill Schevill who shared an office with Vine at Woods Hole.

Schevill, who also taught submariners about the thermocline, was intrigued with "fish noise." Oceanographers had identified many of the sounds. Porpoises had a distinctive gurgle. The maddening cacophonous crackling that had taxed the best imaginations in wartime came from hundreds of shrimp snapping their little claws. But there were still plenty of mysteries. Some thought a peculiar low sound was a whale's heartbeat or an underwater earthquake. Someone else said it was part of a Communist ploy. Schevill wondered what in Sam Hill it was. One memo he cosigned with Vine proposed putting a clear plastic dome over a sub's conning tower so they could see what they had spent all those years hearing. What better way to study the ocean than by being inside it? It was a shocking thought back then, Schevill knew. Scuba was fairly new. The only time oceanographers got wet, he said, was when they fell overboard.

A conventional submarine wasn't the only way to get down there, Vine continued in his talk at the Washington symposium. He suggested a craft designed for research, not war, something that could be driven around like a submarine but go much deeper. Such a thing did not exist. Another possibility was the recently developed bathyscaph; its enormous bulk made it difficult to maneuver, but still, it could get them down there.

Vine always had an idea. He passed them out like a gentleman farmer with a bumper tomato crop, needing no credit for the sauces they went into. He was just glad to grow them, the strikingly perfect ones and the bruisers. A naval officer once said about a third of the research the navy sponsored after the war sprang from a Vine idea. One Vine peddled with special zeal was to hire the blind to work in submarines. It was something he realized suddenly, perhaps from spending so much time in submarines watching the acoustic operators hunkered down by the sonar gear, eyes shut in concentration, trying to interpret the melodies of the sea. But when Vine took his idea to the blind, they protested politely that they didn't know how to get around on a submarine. Incredulous, he asked them how they got around in the dark. The navy wouldn't even listen. It was one of those good ideas, Vine knew, whose time had to come. He was patient and busy enough thinking in that supersonic way of his. Following his presentation in Washington, there was no discussion, not a single question or remark related to any of his ideas for getting scientists down inside the deep ocean.

The navy, having learned from the war how little it knew about the environment beneath the waves, was hungry for knowledge of the ocean. It had established an Office of Naval Research which funded most of the oceanographic research in the United States. But few at ONR cared about getting inside the deep sea.

Bob Dietz, a marine geologist at ONR's London branch, did care. He had persuaded Jacques Piccard to accompany him to the Washington symposium. The Piccards had tried and failed to interest the United States Navy and the recently formed National Science Foundation in the *Trieste*. A single dive cost at least $1000, money the Piccards didn't have. In 1955 the big ocean blimp hadn't been in the water.

The young Piccard and Dietz gave presentations on the bathyscaph. They said it could be used to investigate deep sea currents, about which little was known, even whether powerful currents existed in the deep ocean; they could see firsthand what caused those strange signals on the sonar, and because they could see what they were doing, they could selectively choose what to sample. But the greatest rewards of getting inside the ocean, Dietz said, were unpredictable. "Without doubt," he said, "the main result from the use of a bathyscaph...would be the discovery of processes operating in the ocean of which we know nothing at present."

It was difficult to imagine those unknowns; they couldn't anticipate the surprises awaiting them in that faraway blackness, the creatures too fantastic to imagine, the discoveries that would solve the mystery of seawater's chemistry and make the theorists change their ideas about how earliest life on our planet evolved.

Opinion was divided about what was most needed. There were not enough federal research dollars to go around, and they needed so much. Oceanographers were still navigating by the stars. They had no instrument that could measure a current's speed, direction and depth simultaneously from a moving ship. "Whoever accomplishes this," one scientist declared, "will find himself immortalized by the oceanographic community." All of them had their own ideas for what constituted worthy research; going in person to the deep sea offered only potential.

Piccard and Dietz stirred interest enough to solicit questions and friendly exchanges. A scientist politely pointed out that there were lots of things they would not learn in the deep sea just by looking.

But Willard Bascom, thirty-nine, could hardly stay in his seat.

An ocean scientist, but trained as a mining engineer, he hadn't thought much about the deep ocean until he heard reports of large sharks seen from the bathyscaph. There weren't supposed to be big animals in the deep sea. Bascom wanted to know what else was down there, and he knew that most of his colleagues did not. Worse, it was almost lunchtime. The symposium would break up that afternoon.

Suddenly Bascom was on his feet.

"I think oceanographers, and I mean the people in this hall today, should declare themselves for some sort of deep-sea vehicle or bathyscaph for the United States," he said. "What can a person do better with instruments?... It seems to me, Al Vine's remark this morning that the best possible instrument to send around on the *Beagle* was Charles Darwin is pertinent. This thought might readily apply to the bathyscaph because by doing so we know what we should be looking for when we send instruments down. I think it's time for us to stand up and be counted."

His outburst took the eminent geologist Maurice "Doc" Ewing aback. "Do I understand that that suggestion was in the form of a motion for consideration here?" Ewing asked.

Bascom hadn't thought it through. "I would like to make it," he replied slowly, "in the form of a motion in the broadest sense, that oceanographers form themselves behind an organization which will be in favor of producing a deep-sea vehicle as soon as possible."

Ewing suggested Bascom form a committee to write a resolution. In response to Bascom's three choices, Ewing pointed a long finger at Admiral Edward "Iceberg" Smith, the director of the Woods Hole Oceanographic Institution. Smith got his nickname from directing the Coast Guard's International Ice Patrol, established to chart the icebergs in the North Atlantic after the *Titanic* sank in 1912.

Smith said he was honored to be among such esteemed company, but this committee needed balancing. Everyone knew, he said, that Vine was an extremist. Ewing volunteered two others for balance.

Before adjourning that afternoon Bascom read aloud the committee's statement, drafted over lunch:

> The careful design and repeated testing of the bathyscaph has clearly demonstrated the technical feasibility of operating manned vehicles safely at great depths in the ocean. The scientific implications of this capability have enormous potential. We as individuals interested in the scientific exploration of the deep sea, wish to go on record as favoring the immediate initiation of a national program aimed at

obtaining for the United States undersea vehicles capable of transporting men and their instruments to the great depths of the ocean.

There was a question from the floor. "I am concerned by the words 'enormous potential.' That is quite a big statement. What is meant by 'enormous' in this particular application?"

"The last several days have been spent discussing the possibilities still open to us in oceanography," Bascom replied impatiently. "One promising way to determine which way to go would be to go down. That is what we mean by 'enormous' because it is a very big place to look around."

"You mean geographically?"

"Can anyone think of a better word than 'enormous?' I think the point was well taken," the symposium chairman agreed.

"Tremendous."

"I think 'very great.' "

"Far-reaching," someone else shouted.

"I believe that would be a good one," the chairman said. "Is it agreed that we change this to 'far-reaching'?"

"Is it intended," a voice came from the back of the room, "that this be a number of vehicles? Was the word 'vehicles' plural?"

"It was meant to be plural because it was meant to indicate a trend," Vine explained. "If we are to be successful at all, we should have more than one."

"If we vote for this, does this commit us in any way to go down in it?" another scientist wanted to know.

"May I add," Bascom addressed the chair, "it was our feeling that you as chairman of this symposium would see to it that it at least reached the President of the Academy [of Sciences] and the Chief of Naval Research."

"Would it be fair to regard this resolution as indicating one direction at least in which our facilities were inadequate?" another scientist asked.

"That, I think, is the point," the chairman said, and asked all in favor to say aye. There was general response, according to the conference record. And to his "Contrary?" there was no response.

3

It was an extraordinary collective resolve to develop deep-sea vehicles because most oceanographers had not the slightest desire to go in person to the deep ocean. Yet the few of them who did were able to convince their colleagues, if not to share their vision and longing, to trust in it.

The Office of Naval Research sent Allyn Vine and a few others from ONR to Italy to see the *Trieste*. The visitors were struck by what they saw. The unique craft was maintained entirely by a crew of three: the lanky Jacques Piccard, an Italian mechanic, Guiseppe Buono, and a twelve-year-old called John*nee* (Gianni). The Americans did not dive in the bathyscaph but climbed in and around it and negotiated with Piccard to lease it the summer of 1957 for fifteen dives in the Tyrrhenian Sea off Naples.

In 1958 the United States Navy bought the *Trieste* for $250,000, and Piccard and Buono accompanied their beloved bathyscaph to the Naval Electronics Laboratory in San Diego to help train American pilots.

The Americans immediately ordered from the Krupp steel works in Germany a new passenger sphere with thicker walls. Six minutes after one o'clock on the afternoon of January 23, 1960, the refurbished *Trieste* descended to the deepest known spot on Earth, a mammoth gash in the floor of the Pacific about 200 miles southwest of Guam. Piccard and U.S. Navy Lieutenant Don Walsh dropped 35,800 feet, nearly seven miles, to the floor of the Mariana Trench.

Running the navy's bathyscaph program was one of the most thrilling times of Andreas Rechnitzer's life. It was the biologist Rechnitzer who, in the heat of excitement during the Mediterranean *Trieste* dives ONR chartered, hung a hunk of provolone and a piece of beef from the bathyscaph to see what would happen. (Not much; some tiny unspectacular crustaceans nibbled.)

Rechnitzer relished every dive, despite *Trieste*'s severe limitations. Moving it in and out of the water was a big production. After every dive, the 30,000 gallons of aviation gasoline had to be emptied. Moving it horizontally in the water was virtually impossible. "She was really an elevator," Rechnitzer said. "We never got more than a few hundred yards in her, always just going a fraction of a knot, but most of the time we would just drift with the current."

What Rechnitzer, Walsh and *Trieste*'s other pilot, Larry Shumaker, really wanted was a new, more perfect deep-diving vessel, one that could *do* something. They wanted a much smaller craft that could be easily moved, something that could be driven and didn't use gasoline for buoyancy. Bigger windows would be nice. *Trieste*'s two portholes were big enough for only one eye. They reckoned that given present technology, the deepest this more perfect submersible could go would be between 3000 and 5000 feet. The three men had long talks about it with Harold "Bud" Froehlich, an engineer from General Mills, which had made a mechanical arm for *Trieste*.

On one plane ride back home to Minneapolis, Froehlich sketched a small submarine with the features the San Diego team dreamed about. On his next trip to California, he showed the *Trieste* men his design of a more perfect deep-diving vessel. Rechnitzer encouraged Froehlich, who in turn convinced his superiors at General Mills to spend money on the design. Froehlich said they had a buyer, the man in charge of the Navy's Deep Submergence Program.

And Rechnitzer really thought so for awhile. "I found out that I was on the wrong side of the fence," he said. "The navy said it was not in our mission to go deep. [Admiral Hyman] Rickover said anything deeper than 2500 feet was a waste of money. I was fed up. My management wasn't moving ahead to buy the General Mills–designed submersible or anything else."

So, Rechnitzer quit. And General Mills wondered what to do with the design. Of course the company would try to sell it; but who would want it?

A different scenario had been under way on the East Coast. Charles B. "Swede" Momsen, Jr., the Chief of Undersea Warfare at the Office of Naval Research in Washington, D.C., found himself in an odd position for a decorated submarine commander and career navy man. He was trying to convince oceanographers of the worth of getting inside the deep ocean.

J. Louis Reynolds of the Reynolds Metals Company had gone to Momsen with the design of an aluminum deep-diving submarine, hoping the navy would put up the money to build it. The *Aluminaut* looked like a small conventional sub, but it had two small portholes, outside lights and a depth limit of 15,000 feet. It was designed for Reynolds by Ed Wenk, who had quit his job as a navy engineer because he could not persuade the navy to let him design a deep-diving submarine. The admirals didn't believe that it was possible without making unhappy tradeoffs in weight because the deeper a submarine went, the thicker its walls had to be to withstand the greater pressure.

Momsen thought the *Aluminaut* was a good alternative to the *Trieste*. But the navy's Bureau of Ships, not ONR, built submarines. However, if there were enough scientists who wanted to use such a submarine in their work, Momsen thought, they could rent it from Reynolds with ONR funds. Momsen called the country's two largest establishments for marine science—the Woods Hole Oceanographic Institution (WHOI) in Woods Hole, Massachusetts, and the Scripps Institution of Oceanography in San Diego.

The Scripps scientists weren't interested. At almost 50 feet, they reasoned, it would be too big to poke around the nooks and crannies at the seafloor. Sponsoring it would require a large initial investment with no guarantee that interesting results would emerge. And there would be the risk to life. No, they were not interested.

But at Woods Hole, Allyn Vine lobbied his colleagues and, with Momsen, lobbied WHOI's new director, Paul Fye. Six months after taking charge of the Cape Cod laboratory of four hundred employees, Fye wrote J. Louis Reynolds on December 15, 1958:

> Rather glowing reports of a deep diving submarine...have been reaching me....I wish we...could help with the construction costs. ...However, as you know, this is a nonprofit, privately endowed Institution with only limited funds. However...if the submarine should become available to the Institution, we would be prepared to operate and employ it in the best interests of science and the navy.

Reynolds agreed to rent.

In the spring of 1960 Woods Hole sent to ONR Vine's proposal for funding to operate the world's first deep-diving submarine. Only ten WHOI scientists were willing to put their names on the proposal; counting Vine, made eleven.

"Many WHOI scientists want to use a deep manned submersible in their own research work," the proposal began. Following justification, Vine discussed safety, which he knew was Fye's biggest concern. "The leading advocates believe that a properly designed and operated deep submarine will be as safe as a normal submarine, a surface ship, an airplane or an automobile...." The proposal asked for $3,135,120 to cover a year of preparation while the craft was being built, and $1 million for two years' rental.

Momsen was shocked. ONR approved one-year research contracts with no guarantee of continued funding for the same project, as WHOI knew. He understood Fye's concern of being stuck with an unfunded submarine, but WHOI was just going to have to take the risk. He asked Woods Hole for another proposal—for $1 million for one year.

The second proposal arrived at ONR four months later, and that December, the executive committee of WHOI's trustees formally agreed to rent the "exotic craft." Electric Boat of General Dynamics was under contract to Reynolds to review the design and ultimately build it.

What did WHOI or Electric Boat or Reynolds know about the strength and elasticity of aluminum alloy 7079 at 15,000 feet? Not much. But neither did anyone else. Electric Boat was a builder of welded-steel combat submarines designed to withstand pressure to only about 1000 feet. The *Aluminaut* was to be bolted and pasted together in ten large donut sections because the aluminum was not weldable. The executive committee insisted that the submarine had to make a dive without any people aboard to 17,000 feet, well above the expected crush depth of 23,000 feet.

On January 30, 1961, Momsen called Woods Hole with good news: approval of the million-dollar proposal would soon be in the mail, and he could see the way clear to budgeting another million dollars for each of the next two years. A rental contract was drawn up by WHOI's counsel.

But ONR and WHOI could not pin down Reynolds, who kept changing his mind—about the rental fee, the charter period, liability and who ultimately would own *Aluminaut*. At $500,000 annual rent, ONR insisted that the title of the craft be passed on to the navy after a three-year charter; the navy figured it would have paid for it by then. When Reynolds changed his mind, he seemed to do it just before he got on an airplane, forcing Momsen and Fye to await his return or try to find him, which usually proved impossible.

The *Aluminaut* design review was several weeks behind schedule in the spring of 1961. Vine asked the "garage gang" at the dockside WHOI machine shop (formerly a garage) to build a full-scale model of the *Aluminaut*'s front half. His officemate Bill Schevill climbed into the "Pineknot," as the garage gang called the plastic and wooden model, and, pretending that he was inside the ocean, called out to Vine what he could and couldn't see. Passersby climbed in, too. Vine had plenty of opinions, all aligned with his. "More windows. Gonna miss the mermaids with just two windows," Schevill said. "Gonna miss the *whales*."

Besides more windows, Vine wanted the submarine to hover

like a helicopter and use a twin rather than a single propeller for greater maneuverability. For two years, he had been repeating his arguments like a litany, but nobody budged. More windows would cost more, reduce payload, and according to another argument, more holes in the hull, no matter that they were for windows, would weaken the craft's structure, about which not enough was known anyway.

Finally, Vine and Schevill took matters into their own hands. They were a study in contrasts, the tall bony Schevill with a long face that tapered in a neatly trimmed goatee; he moved in long easy strides and his voice resonated in a baritone. Vine was short, chunky and staccato; his glasses framed a round face. Both men laughed easily. The driver who picked them up at the train station for a conference at Lakehurst, New Jersey, a few years back was appalled at their conversation. It wasn't that he meant to eavesdrop; he couldn't help it. They were talking about dynamiting the northeast coast.

"The fat little gangster didn't bother me too much," the driver told his next passenger, who was a scientist headed for the same conference, "but I was really surprised at that rabbi." The scientist knew it had to be Vine and Schevill. They had been discussing the oceanographic seismic work in progress.

Their thoughts were on windows when they drove to the Groton shipyard. Although they couldn't change anyone's mind, they convinced the Electric Boat and Reynolds representatives to come see the Pineknot.

At Woods Hole Schevill got another biologist to help with the demonstration in a WHOI parking lot. The younger biologist took the pilot's position before a porthole, while Schevill peered out the other window and spotted a giant squid. He ordered the pilot to steer toward it, but the pilot couldn't see the big squid and finally, the Pineknot stopped, jammed under a cliff overhang that neither window took in. Now what? Now, gentlemen, they were stuck for good, far beyond the reach of rescue. Reynolds and Electric Boat agreed to investigate the feasibility of adding two more windows.

Paul Fye was determined that none of his scientists was going to die at the bottom of the sea in an aluminum submarine. He had an abiding skepticism typical of a scientist and an attention to administrative and political detail that was not characteristic of his peers. Fye, forty-nine, held a Ph.D. in physical chemistry, and before becoming WHOI's director held various administrative posts at the Naval Ordnance Laboratory. He had a reputation for being a perfectionist and made it a point to learn first hand about everything that went on at the institution he led.

Fye hired two mechanical engineers to be his eyes and ears during *Aluminaut*'s construction—at the mill where the aluminum parts were to be forged, at Electric Boat where the craft would be assembled, and at private and naval laboratories, learning the classified and nonclassified state of the art of deep-diving craft and, hopefully, aluminum alloy whatever-it-was.

The first engineer was Jim Mavor, thirty-eight, who had a master's degree and had worked as a naval architect for three years at a navy engineering lab and shipyard; he was teaching college engineering and working part-time at WHOI. The second engineer was Joe Walsh who was eight years Mavor's junior. After earning his Ph.D., Walsh worked eighteen months for an engineering consulting firm and then took a year off to travel. He had just returned from his round-the-world trip and now, he decided, it was time for a "real job."

Woods Hole was almost desperate to find the administrators and pilots for this venture. The project director would be a full-time position, which to a scientist meant giving up research. Vine was not a candidate. "They wanted a stick-to-one-thing person," Vine said. "I was already working one and a half full-times." And WHOI's salaries were low, about half the amount industry offered.

In search of pilots, Vine asked the navy to peruse its personnel rosters for an ex-submarine officer with training in geology, biology, physics or meteorology. An "aptitude and interest in trying new things" also would be nice, he wrote. Vine eventually reviewed 105 ex-submariners, none of whom was hired. WHOI's ideal pilot was a dream. They never found their man; he found them.

William Ogg Rainnie, thirty-six, was a 1946 U.S. Naval Academy graduate. After eight years in the navy, about half that time spent on submarines, Rainnie resigned his commission as senior lieutenant because he was tired of the rootlessness of moving his family every year. In 1960 he joined the National Academy of Sciences as a liaison to research labs, a job related mostly to underwater mines and acoustics. But he was still looking for his special niche in life. Over lunch in Washington with one of Fye's aides, Rainnie asked if WHOI might have anything to offer him. "You don't happen to like riding in submarines, do you?" the aide asked.

He was hired as "alternate pilot." Vine noted in his journal that Rainnie's first day on the job, September 11, 1961, was his mother Lulu's birthday. Perhaps it was a good omen.

It wasn't. After three years of changing his mind, Reynolds did it again, informing Momsen and Fye by messenger that he would not relinquish his submarine to the navy after the charter period or pay for the tethered test dive to 17,000 feet. Now the rent stood at $667,000. But Reynolds signed a $2 million contract with Electric Boat, and the construction of *Aluminaut* finally began. The blueprint included two more windows and a twin screw.

WHOI's "alternate" and only pilot wondered if he had made a wise career move. Momsen worried that if the funding he held in abeyance was not committed soon, it would be taken for other research. Rainnie, Vine and the two engineers were located in different offices; they weren't a team.

Rainnie had a good listener in Earl Hays, a WHOI physicist. They had gotten to know each other at their sons' Boy Scout meetings and camp-outs. It was at a Boy Scout troop meeting that Rainnie was sounding off. What they needed most, he said, was a full-time leader, and as he looked at the balding, avuncular Hays, the man who kept a bigger-than-life-size poster of Albert Einstein on an easel in his office, it occurred to Rainnie that he was looking at a solution.

Fye fully approved Rainnie's suggestion, and greatly relieved, announced that the project finally had a full-time manager. Now, Fye thought, with a respected oceanographer in charge, the opposition to the submarine project among WHOI's scientists would dissipate. At the same time Fye created WHOI's first formal departments and made Hays chairman of the Applied Oceanography Department, which embraced thirty-one employees: the engineers and technicians who made moorings for scientific instruments, an embryo computer group and the submarine project.

WHOI bought an old three-story tenement at 38 Water Street on the corner of the same dockside block that held the institution's other two buildings. A drugstore leased the first floor and the top story still had tenants. Hays, Rainnie and the engineers made their headquarters on the second floor and hired a secretary, May Reese, a former Wave and the wife of a Coast Guardsman. They christened themselves the "Deep Submergence Group" and called their headquarters "the Drugstore," a name that is still used. Hays gave Rainnie a new title: Project Coordinator and Chief Prospective Pilot.

In April 1962, three years and four months after Fye's first letter to Louis Reynolds, the aluminum company mogul had another change of heart. Reynolds offered three new options; each ignored the title transfer to the navy.

"Dear Swede," Fye wrote Momsen, "here is the last chapter of the sad story of negotiations with Louis Reynolds...."

4

Bud Froehlich, the young General Mills engineer, brought his design of a more perfect deep-diving vessel to Washington. He called it the *Seapup*. It had that chubby baby look of a Volkswagen bug. It was eight feet wide, eighteen feet long, and tapered like a fish, with room for two inside its steel passenger sphere. In an emergency, the sphere could be separated from the rest of the craft and floated to the surface like a balloon. It had a large aft propeller that pivoted and doubled as a rudder, two smaller props on either side, portholes in front. Sticking out the nose was the same model mechanical arm that had been made for the *Trieste*. The craft could hover and rotate and reach 6000 feet, the four-page General Mills brochure said.

ONR had received several designs for space-age underwater craft from other companies that hoped the navy would be forthcoming with the construction money. Industry expectantly perceived a "wet NASA," an infusion of the same kind of energy, resolve and government money behind the recently created National Aeronautics and Space Administration.

Of all the designs that crossed his desk, Swede Momsen liked Froehlich's best. The *Seapup* also had something none of the other designs had—a price tag. It was on sale for $225,000 and not firm; General Mills had offered it to the *Trieste* team for a rock-bottom $98,000.

Woods Hole's Deep Submergence Group liked it, too. The men at the Drugstore started talking about alternatives to *Aluminaut* almost as soon as they were given a name and a place to call home. At first, their thoughts were for an "interim" craft, something small and inexpensive that could be used while the *Aluminaut* was being built. But as they talked, they realized that this other craft could be better than the Reynolds submarine. Small meant easier to move out of water and in. They had been to San Diego and knew about the *Trieste* problems. In fact, they had been to more than two hundred labs, naval and private, learning the state of the art of everything and anything that was being experimented with or just thought of in the context of deep submergence. They knew enough to know what was possible and they knew what they wanted in their own deep-sea research submarine.

But General Mills was not about to build the *Seapup* with its own money and rent it to oceanographers. The company needed a buyer.

In those days, Swede Momsen was heard to say more than once: "Hell. Most oceanographers are so damn narrow-minded. If the scientists don't know what they need, I do, and damn it, I'm going to get them a deep-diving submarine...." He was a renegade, a can-do bureaucrat, and to the men at the Drugstore, a hero. This was not a simple rental to conduct research, this was building a vessel, a prototype, and doing it with ONR money. The Bureau of Ships, builder of all navy vessels, would not like having its authority usurped. Momsen had to figure that one. Justification, that's what he needed.

Momsen advised the Woods Hole team to write the specifications for their underwater machine as quickly as possible. He was not sure what would happen to the funds he stowed for the *Aluminaut* project if it were known that the Reynolds deal was dead. Others within the federal bureaucracy were already asking him about that money.

While Bill Rainnie and Earl Hays flew to Minneapolis to get help from Froehlich in writing the specifications, Momsen put his head to work. He found his cause in ARTEMIS, the navy's secret underwater listening station off Bermuda; it happened to be at 6000 feet. It was perfect. He was about to call Woods Hole with his plan when an official memo arrived from the navy ordering him to cancel the funding for Woods Hole's deep submergence project.

But Momsen had been protecting those million dollars for a long time. And *hell*. He tossed the memo in a desk drawer, dialed the Cape Cod laboratory, and authorized the Deep Submergence Group to send out the specifications to bid. On second thought, there wasn't time to write the bidders—call them. He suggested inviting them to a pre-bid conference. The crux of his plan was having a fully tested vehicle delivered next summer when the navy would install new hydrophones at ARTEMIS. The *Seapup* could assist with the installation and inspect the array. After Woods Hole chose the bidder, Momsen explained, ONR would formally authorize WHOI to buy it with *Aluminaut* funds. But they had to move quickly.

In addition to General Mills, Momsen suggested six other companies that might be interested in bidding: Electric Boat, North American Aviation, General Motors, Lockheed, Philco and Pratt and Whitney Aircraft. Momsen and Woods Hole knew some would not bid; they were coasting on the euphoric anticipation of a wet NASA, but at least three had spent money on designs. None of the seven had ever built a deep-sea submarine, although Electric Boat was constructing the *Aluminaut,* and only EB was a builder of any kind of underwater craft. But the untried and impatient team at the Drugstore confidently went with what it had and called all seven firms.

At least two representatives from each company came to the May 28, 1962, pre-bid conference at a Boston hotel to review the twenty-four pages of specifications, which with few exceptions were copied from Froehlich's *Seapup*. This was a fixed-price contract with deadlines and penalties for late delivery and "unsatisfactory performance of the craft," which had to operate at 6000 feet and go once to 7500 feet without passengers.

To guarantee the craft's performance, the bidders had to have their homework already done or be willing to take an enormous risk. It was obvious from the questions that some were there only to learn. Someone asked what the mechanical arm was for. It was the kind of query that made Allyn Vine take a quick breath and squint. How could it not be perfectly clear that under the kinds of pressure that could mash a person to talc, they couldn't very well open a window and reach out for a fish or a rock?

Another question: What did passenger hull separable from supporting structure mean? It was the craft's failsafe feature. With the turn of a large screw, the passenger sphere would release and float up. Submarines had no such thing.

For the most part the specifications addressed performance, that is, what the submarine had to be able to do; little was said about how to achieve the desired capabilities. The windows were the most detailed items. The specifications asked for four, each precisely positioned (e.g., "starboard looking downward 15 degrees and 54 degrees off centerline").

Bud Froehlich and Andreas Rechnitzer sat across the table from each other at the pre-bid conference, each confident that his company would soon be building Woods Hole's underwater machine. Rechnitzer, who was now the director of North American Aviation's new ocean sciences department, was so confident that he had already chosen a name for the submarine: *Andrea*. But by the June deadline, only their companies had submitted bids and neither conformed to the specifications. General Mills bid a fixed price of $472,517; North American Aviation bid $723,600 for a cost-plus fixed fee. Neither would guarantee the craft's performance.

The other companies wrote to say that while they were interested, it was impossible to build such a prototype in ten months and folly to attempt it under a fixed-price contract. Philco called it a crash program. Electric Boat privately told Woods Hole it was crazy.

Of course it was reasonable for a company to want to make money, but the firm that would build this submarine would not participate to make money, but to be first, to get into the business, and because it was a nonexistent business with promise, it was reasonable to expect few companies to take that risk.

But the bid response didn't bode well to Fye. He wondered if they were being unrealistic in their contractual demands or in thinking that WHOI could operate a submarine. Momsen was adamant about the fixed price; he thought too many companies were wasting taxpayers' dollars with open-ended contracts. But seven big, reputable companies had said no. Was it lack of courage to take the risk, or had he, Paul McDonald Fye, erred in judgment? Fye wasn't the kind of person who needed a majority to approve what he thought or did. WHOI scientists who had argued against a research submarine in 1959 continued to argue with him in 1962. "They thought it was a waste of money," Fye said. "Of course they always thought it was their money."

For an objective opinion, Woods Hole turned to one of the basic tools of the science trade, peer review. They found three willing peers: Bob Frosch, the director of Hudson Laboratories of Columbia University; Fred Spiess of Scripps, who wasn't interested in the *Aluminaut;* and Ralph Kissinger, a retired navy captain who had worked as a consultant with Ed Wenk on the *Aluminaut* design. They were asked to make a case, if they could, for building a deep-sea research submarine to be operated by WHOI.

The three quickly reached a unanimous conclusion. Frosch said they knew more about the moon's backside than the ocean's bottom; someone had to take the risk. Spiess said he found himself in an awkward position since he could not imagine why a navy would need to be convinced of the need for a deep-sea submarine. To make a case for it, he said, seemed rather like having to make a case for learning to walk.

The two companies that had sent nonconforming bids agreed to re-bid and gamble on guaranteeing performance. Woods Hole admired Bud Froehlich but thought North American Aviation had more in-house capability for this job. But both bids were judged to be technically equal and General Mills' bid of $498,500 was $96,500 lower than North American's. So the maker of Wheaties, Breakfast of Champions, was awarded the contract.

5

"National Defense Begins at Breakfast"
  General Mills ad for Wheaties

Cadwallader C. Washburn and John Crosby couldn't have known in 1877 when they merged their Minneapolis mills that their firm would become a multi-billion-dollar conglomerate that sold not only flour, but also cookbooks, golf shoes, polo shirts embossed with a little alligator, luggage, wallpaper and a submarine.

In addition to the bakers, General Mills employed people to repair and build machinery, such as an automatic flour sack sealer and a contraption that mixed and cooked and packed and sealed Kix and Cheeri-oats. In 1940 the firm proudly volunteered its Mechanical Division to the United States government to build torpedo and gun parts. After the war the government contracts kept coming, and the Mechanical Division grew to several thousand employees. One of them was Bud Froehlich.

Froehlich's dream was to be an aeronautical engineer and to fly the aircraft he designed. The prospect of college was remote; the only money that came in was the $12.96 a week from his mother's job washing dishes at a restaurant. After high school he joined the navy, and after the war, the GI bill sent him to college. By 1946 he had a bachelor's degree in aeronautical engineering from the University of Washington and his first job as an aircraft designer.

The progress of Froehlich's dream was interrupted in 1951 with a telephone call from home. His mother was gravely ill. He and his wife Avanelle took out all their savings and hitched up their house trailer to their 1940 Ford convertible for the long drive from the West Coast to Minneapolis. While there, Froehlich happened to see an ad for an opening at General Mills. He hired on that October.

Unlike the airplane companies, where Froehlich had thick manuals that specified how to do everything, he had the freedom at General Mills to follow his curiosity and do things his way. Most of their contracts came from ONR for projects involving stratospheric balloons. One of the food maker's engineers had studied aeronautical engineering under Auguste Piccard's twin brother Jean, who was a professor at the University of Minnesota and an avid balloonist himself. It was Jean Piccard, the engineer said, who had convinced General Mills management to take on the ONR research contracts.

Froehlich worked on a balloon to circumnavigate the globe like a satellite to get the CIA behind the Iron Curtain. (The upper winds were stronger than expected. The balloon landed in a back yard in New Jersey.) He designed helium gas valves and breathing systems for balloons that took navy pilots to record heights. Other balloons lofted rockets to the stratosphere to measure radiation for James Van Allen, who would discover a radiation belt around the planet. In 1958 Froehlich was made section head of the General Mills Nuclear Equipment Department which designed and built manipulators for the *Trieste* and the nuclear power industry.

It was a great company for Froehlich and one that appeared to be thriving in 1962. Annual sales were $546,400,880. More than a hundred new products, like Toffee Swirl Cake Mix and Sugar Sparkled Twinkies, went out to tempt the consumer that year, and General Mills engineers also built a digital computer for NASA's Goddard Space Flight Center and a rocket for the navy. It was also the year John Crosby died at age ninety-four and the year his flour company was awarded a contract from WHOI to build the world's first deep-diving submarine.

Woods Hole's upper management contact with General Mills was with Jim Summer, a tall, handsome West Point graduate and former air force pilot. Summer had joined General Mills two years before with a secret mandate from the top to clean house in the electronics division. He knew exactly what he was going to do. The company never should have invested in the design of a submarine in the first place, but it was futile to argue the point now. He would close Froehlich's division but first build this submarine and try to recoup their investment. Froehlich had convinced him to go with the project. Summer liked Froehlich; he thought he was probably a genius. Of course he couldn't tell him what was going to happen to his department.

General Mills was legally bound by its bid, but Summer was loath to guarantee a hull to 6000 feet, and he sure as hell didn't want anything to do with the tethered 7500-foot dive. It would be expensive to stage and difficult, perhaps impossible, to predict what would happen. Summer knew there would be problems with corrosion. He wasn't sure machinery could run in the pressure of the deep sea. WHOI's seven-foot steel ball would be welded, not clamped or glued together like the *Trieste* spheres.

Contract negotiations took some creative turns. They discussed a sliding-scale payment in which General Mills would accrue a bonus or penalty for every 100 feet of variation from 6000 feet, up or down. Instead of one passenger sphere, Summer suggested building two, and Woods Hole would take the one that went deepest. He offered to guarantee one sphere to 4500 feet, or if WHOI agreed to the two spheres, he would guarantee both to 5000 feet. If the spheres could go to 6000 feet, he would buy back one for $37,500.

Fye thought Summer's idea to build two spheres was clever. But Woods Hole wanted 6000 feet, guaranteed. Negotiations stretched over the summer of 1962, with each side parrying to cover itself by trying to prevent the ultimate disaster, death or financial hemorrhage or a nightmare of liability.

Bob DeHart, chief of structural research at the Southwest Research Institute in San Antonio, kept track of the negotiations. For nearly five years, beginning in 1958 with his old boss Ed Wenk, DeHart had been investigating the feasibility of deep-diving craft by testing different materials and shapes under pressure. He thought it was the dawn of the era of deep diving. DeHart wanted to test the pressure-worthiness of forthcoming deep-sea craft, and had in mind a super tank, say 80 inches across, able to simulate the pressure at a depth of about 10,000 feet. There was nothing like it in the country, maybe the world. Southwest was nonprofit and poor, and DeHart knew his management would demand better justification than his gut sense that someday a huge pressure chamber would be needed. Now WHOI provided a real cause. If WHOI had access to a super pressure chamber, there would be no need to pressure-test small models of the submarine or conduct the extra-deep tethered dive without passengers. They could test the real thing. It made sense to everyone; Fye and Summer shook hands.

Knowing it would be months before the contract was actually processed and signed, Momsen authorized Woods Hole to send General Mills a purchase order of $10,000 which WHOI did not have yet from ONR. They needed to get started. The delivery date had already slipped. It was August; the submarine was to be finished in exactly a year.

Momsen was in a hurry for another reason. The Bureau of Ships was about to issue a directive specifically to remind him that all navy-funded vessels were designed and built under its jurisdiction. Momsen would use the phony purchase order to show that the ONR-WHOI project began before the directive became effective.

The contract, signed on September 4, 1962, would be amended three more times. But that fall the agreement looked like this: for $590,743, General Mills would build three passenger spheres, each to be tested in Southwest's super pressure tank which DeHart agreed to make bigger to fit the WHOI spheres. The sphere with the most flaws would be sacrificed for knowledge by being pressure-tested to collapse all the way to whatever that number was; no one was sure. The contract called for a bonus/penalty of $27.78 per foot of collapse depth between 9900 feet and 11,700 feet. General Mills stood to gain an extra $50,004 if the passenger sphere was good to 11,700 feet. Or the company could lose it.

Summer was not too sanguine about the project; he figured it was probably foolish, but worth the risk, and he secretly admired Momsen, not least for being wilier than he. "That Momsen knew just what he was doing," Summer later said. "He had a mission and it took a certain amount of something, guts, I guess, to take this contract with a food company. I could just imagine his colleagues asking him who in the world he got to sign the contract."

Momsen had his hands full with the Bureau of Ships which recommended making the passenger sphere of HY100, a high yield strength steel that had never been used before in a submarine. WHOI knew the vessel would be a prototype, but guinea pig was not one of Fye's selling points to his trustees, nor was it a concept that rested comfortably in his mind. HY100 got mixed reviews. Woods Hole heard it was especially difficult to weld. But WHOI and ONR felt they had no choice but to go with the Bureau's recommendation.

BuShips also provided "Special Provisions for Design and Construction," voluminous specifications printed in the font size used for telephone directories. Froehlich paled as he flipped through them, and General Mills added $112,000 to the contract.

The Special Provisions weren't really special. They contained mostly the standard American Society of Mechanical Engineers (ASME) code governing the construction of pressure vessels, which included submarines and boilers. Face to face with BuShips' officers, Froehlich pointed out that 95 percent of the code obviously didn't apply to this one-of-a-kind submarine. The submarines of the world's navies were not meant to withstand the pressure of the deep ocean; they were cylindrical, not spherical; and they were huge, not small.

All of the code applied to all of the submarine, BuShips said.

But according to the code, if you made a hole in a pressure

vessel, you had to put back in the same amount of material taken out to reinforce the area around the opening. Froehlich said if he obeyed the code for the hatch, for instance, the submarine would be unstable, because the weight of the reinforcing steel would throw off the center of gravity.

The entire code applied, BuShips repeated.

But it wouldn't float upright, Froehlich pressed, it would sit in the water *upside down.*

The faces from BuShips stayed stony, and one of them intoned: "The Bureau of Ships is under no obligation to relax the specifications for the convenience of the manufacturer."

Deeply distressed, Froehlich put his confession of premeditated disobedience in writing and sent the memo to Summer. It said he was about to violate the BuShips specifications in order to proceed with the project because he could not in good conscience build an unsafe submarine. Summer never answered the memo.

Froehlich's design awaited many details and there were many companies General Mills needed to build a submarine. He had lined up only one firm, Lukens Steel Company; but suddenly the Pennsylvania steel mill had second thoughts and decided that HY100 was too experimental for a fixed-price contract. Froehlich tried twenty other vendors, but each refused for the same reason. Then he tried Hahn & Clay in Houston. Larry Megow, the plant manager at the Texas shop, said yes. Yes, they could meet the time constraints, stay within a fixed price, follow any specifications. No problem.

Megow's alacrity made Froehlich so skeptical that he asked the superintendent of the General Mills machine shop to go to Houston. The superintendent said he liked what he saw. For one more good measure, Earl Hays, Bill Rainnie and Allyn Vine went unannounced to Houston, making as great an impression on Hahn & Clay as the fabricator made on them. The firm's president, Gene Clay, would remember them as the most unpretentious men of learning he had ever met. Here they were, in suits and ties and sandals with socks; and Hays, a physicist, talked about his beloved fruit trees and how he made his own cider every year and looked forward to retiring so he could sit in his yard and watch the grass grow.

Froehlich subcontracted to Hahn & Clay to weld the hemispheres together into three spheres and machine, sandblast and paint them. The Houston shop would also be responsible for finding a company to make the steel and shape it into hemispheres. Larry Megow wasn't worried; he planned to go to his principal supplier Lukens Steel.

Lukens, more comfortable without the responsibility of welding, agreed. The steel mill rolled out nine tons of HY100 steel plate and cut out six discs. In December of 1962 Hays and Rainnie watched the giant circles being spun into hemispheres. The discs, pulsating with the brilliant orange of 2100° F, were placed onto a spindle that turned like a potter's wheel. To the side was a large cylinder, which like a potter's hands moved up and down and in and out, making each pancake a hemisphere. Before Christmas, the six hemispheres were on their way to Houston to be welded together into three perfect spheres. Finally, *Alvin* was in sight.

The Deep Submergence Group began calling the submarine *Alvin* before the contract with General Mills was signed. Engineer Joe Walsh had caught Hays giggling to himself as he taped a cartoon of Alvin the chipmunk on his door.

"I thought it looked a little like Al Vine," explained the giggling Hays.

"Yeah," Walsh agreed. "In the cheeks."

Alvin the chipmunk blared over America's radios in the top ten in 1958 with his mimicker Ross Bagdasarian. It was during those frustrating *Aluminaut* days that the Deep Submergence Group on the second floor of the Drugstore would hear Bagdasarian

screaming "A-a-a-alvin!" and the chipmunk's obedient castrato crooning. Perhaps it was enough edge to interrupt their brooding and remind them that there was greater inanity in other worlds.

Hays and Rainnie told Froehlich they wanted to call their submarine *Alvin* after Al Vine because it was he who had the vision. Froehlich was quick to respect their wishes. He didn't know that Hays had a chipmunk on his door.

Some of the contractors Froehlich hired for the submarine project didn't initially believe that this one-of-a-kind craft was really going to be named *Alvin*. Not that *Seapup* was any more distinguished. In its first technical report on pressure tests, the Southwest Research Institute felt obliged before using the name to explain that the submarine "has been referred to as 'ALVIN' by General Mills and WHOI." And General Mills public relations explained to an inquiring reporter that *Seapup* was "dropped by the navy after a contract was obtained and the vehicle became *Alvin*—named for the Navy project head.... As I recall the man's name was Alvin Pine, tho maybe Vine...." Lukens called it "*Alvin* the *Seapup*."

It was *Alvin* on the homemade holiday card the Deep Submergence Group sent out that Christmas of 1962. And it was the Alvin Group the folks in the Drugstore began to call themselves.

6

We had a lot of curiosity in here on the part of the experts and not just from Woods Hole, but the Navy and whatnot, sayin' it was impossible to get some of these tolerances.
        Gene Clay, president of Hahn & Clay

    It was the most nerve-racking period in the entire career of Hahn & Clay's quality control man, Gene Woodruff. A short, slight young man with thick black hair, he was not at all comfortable with the numbers for Woods Hole's spheres; they were about as low as they could get. The passenger spheres had to be 1.33 inches thick, plus or minus .03 inches, the width of eight sheets of paper.

    Hahn & Clay had followed extremely close tolerances before, but for small things that would endure internal, not external, pressure. Internal pressure in a slightly imperfect sphere would push outward, compensating for out-of-roundness, and if not checked, the ball would eventually rupture. But external pressure gave no warning of impending catastrophe. All it took to cause an instantaneous implosion was an invisible imperfection in the chemistry of the steel or a deviation of only a fraction of an inch in the geometric purity of the sphere. When steel was welded, it puckered, and puckering changed the shape of a vessel, however slightly. If they misjudged, they couldn't take the spheres apart and remachine them.

    General Mills promised Hahn & Clay a $10,000 bonus for finishing the spheres with the maximum collapse depth under deadline, and Larry Megow had every intention of getting it. Megow, who dressed Texan but spoke Wisconsin, had never worked as a production-line welder, but he had supervised welders and developed welding procedures. For the experimental HY100, he added a few more steps to the navy's specifications. In the shop they called it "Larry's system," which, Megow said, was no big deal.

    But welding any vessel designed for human occupancy under external pressure is a very big deal. To weld the HY100, the air temperature at the plant had to be 125° F. The welders had to keep their torches a specified distance away from the steel, and they had only a specified amount of time to lay in each line of weld, otherwise the material would get too hot and change its molecular structure. The welding torches had to be baked to ensure that they held no drops of invisible moisture. A droplet of dew in a weld line became a pocket of steam, and that weakened the weld. Woodruff made a large chart with all the specifics for the welders.

When the first two hemispheres were welded together and it was time to take sphericity measurements, Woodruff held his breath. It was a big moment in the plant. "Believe me," Gene Clay said, "their guardian angels was settin' on their shoulders because they hit it within two to three thousandths of an inch. That's less than the thickness of a strand of your hair. We were all just as proud as a peacock."

By the spring of 1963, Hahn & Clay was waiting for WHOI to make up its mind so they could take the next step of cutting out holes for the windows.

Joe Walsh had recalculated and confirmed Froehlich's equations on which the submarine design was based, but his boss Jim Mavor wasn't satisfied. Allyn Vine thought Walsh's calculations were unnecessary, and he told him so; they should trust Froehlich's numbers.

Although Vine was spending more time on his own research and less on the submarine project, he continued to drop in at the Drugstore and offer ideas and opinions. Vine put a lot of stock in common sense. He liked to say, "My father used to tell us, 'For thirty years your grandfather cut porkchops and he's still got his own fingers. Now, that's credibility.' " But his differences of opinion on the craft's engineering irked Mavor and Walsh.

It was Vine's way; it puzzled Bobby Weeks who piloted WHOI's small seaplane, frequently taking Vine up. On one trip Vine asked Weeks if he could land the plane. "Have you ever landed a plane before?" Weeks asked. "No," Vine said, "but it can't be that difficult; after all it's just a matter of applied physics."

Weeks did not let him land the plane. Mavor and Walsh would not let him horn in on their engineering.

Mavor accepted the 1.33-inch thickness of the sphere walls. He knew the classical engineering theory used to arrive at that figure was unproved, but it was the only criterion available. What bothered him most were the hatch and windows. When they cut out those holes, about 25 percent of the sphere's total surface area would be removed. To restrengthen the passenger ball, each hole would be surrounded by a thickened steel donut that would gradually thicken from 1.33 inches to 3.5 inches. The calculations for the donuts were based on Auguste Piccard's data, but expanded. The *Trieste*'s windows measured two and a half inches across from inside and there were only two of them; *Alvin*'s would be five inches in diameter and there were four of them. And according to classical theory, they would dangerously weaken the sphere. Froehlich didn't believe it; theory was not gospel. He was as confident as Mavor was distrustful.

Even the Bureau of Ships kept quiet. A BuShips officer admitted that the Bureau had no experience with spheres and he for one was very interested in seeing how Froehlich would handle this "formidable problem."

In April Mavor finally gave in and agreed to go with Froehlich's design for the reinforcements. The August 1963 delivery was pushed to October, and an impatient Larry Megow gave his subcontractor the okay to start forging the reinforcing donuts.

• • •

Woods Hole and everybody else figured when we started cutting those big holes in that sphere it would literally start collapsing. There was lots of speculation. No one knew what was going to happen. You know, those holes were so close together. I mean, you had the one on the front and the one on the bottom and the two on the sides and those things all came so close. I held my breath.
    Gene Woodruff

The big overhead fans weren't doing a good job of pushing around the 125° F air in Hahn & Clay's machine shop, which was as long and as wide as a football field. Despite the goggles and face mask, Rooster Myers felt every speck of ground-up HY100 steel—up his nose, in his eyes, his mouth, his ears, everywhere. He had been sanding by hand the inner walls of a sphere to thin out the areas that were thirty-second heavy, that is, a thirty-second of an inch—the width of a pencil lead—too thick. The only way to get it down with the hemispheres welded together was to do it by hand from inside. He had crawled in through a top hole cut for the hatch.

Not that Rooster was complaining; he was proud to work on those big balls. They had machined the insides of the six hemispheres, welded them in a single seam at the equator, and placed them on a rotating table so two razor-sharp tool bits could shave off the steel. HY100 was hard stuff. Sometimes a tool bit lasted only five minutes, and sometimes, when the bit first hit that center welded seam, it tore off or just stuck there and melted.

At each step, Woodruff climbed in to check the thickness of

the walls, and the engineers from Southwest Research brought up their newfangled equipment to measure it again. Southwest's ultrasound scope found in the first sphere ten laminations, concentrations of impurities present when the steel was poured. Two laminations each covered 12 square inches, which seemed unacceptable, but nobody knew for sure. There were laminations to some extent in all steel. If they cut out the laminations and welded in a new hunk of steel, they'd have more opportunity for trouble. The thickness of the sphere was just within tolerance—at the equator, it was at the maximum 1.360 inches and thinned up and down to the minimum 1.30 inches.

It seemed a shame to ruin a perfect sphere by punching five holes in it. But once all the holes in the first sphere were cut and the ball did not collapse or flatten, Woodruff didn't have to hold his breath again. Dress rehearsal was over. The next two spheres were closer to the ideal numbers.

By mid-July the reinforcing donuts had been welded around the ports in the first sphere; the second hull had its welds X-rayed; and number three was being machined. Megow sent WHOI and Froehlich photographs, proof of the promised can-do. Each picture showed a big naked steel ball with round holes. The variation was his captions.

> ALVIN No. 1 after being sandblasted in and outside and being sprayed with a sticky coat of paint. That sand swirling around inside that perfect sphere sure kicked up a storm.

> ALVIN No. 2 outside in the Texas sun getting her picture taken.

To test the purity of the steel, samples from each batch of HY100 were sent to Southwest and three naval laboratories. Elsewhere small-scale models were tested for surface stability in wave-making tanks. Surprise came not from any tests or calculations, but from management. That summer of 1963, Litton Systems, Inc., took over the engineering and aerospace departments of the General Mills Electronics Division.

The Minneapolis engineers, with the Woods Hole team looking over their shoulders, were still trying to make the motors and the valves and pumps work, still learning what would not work in high external pressure. WHOI had revised portions of Froehlich's design by adding steps on the conning tower and a one-eyeball porthole in the hatch. Instead of Froehlich's inflatable conning tower, designed to collapse like a balloon, *Alvin*'s sail would be permanent. WHOI decided against using steel-shot ballast because it was irreversible. Once all the shot was gone, that was the end of a dive. Froehlich had to come up with another system.

Almost every change forced a trade-off because every item added poundage and took up space and there was a finite amount of pounds and space. All the lessons and second thoughts were reflected in the changing design and the new company's nervousness. Litton had cause for concern over its inherited $590,743 contract. The purchase orders alone totalled $468,807.48.

Alvin's delivery date was pushed up to March 1964. The first two spheres went to Minneapolis to have their holes filled with Plexiglas windows, each a flattened cone three and a half inches thick that flared 45 degrees, from five inches on the inside to a foot on the outside. And the big balls returned to Texas in December for pressure tests. Froehlich chose to sacrifice sphere No. 1 and use No. 2 for the submarine.

Southwest's super tank, essentially a cylindrical pressure cooker with a hemispherical bottom, was a real bargain at $100,000, built of a low-cost, experimental steel called T-1. Its 40,000-pound lid screwed deeply into the tank, which was designed to withstand the pressure at 13,501 feet.

About half the chamber's fifteen-foot length was sunk in the ground and filled with oil. Sphere No. 1 was gently lowered into the tank and subjected 536 times to the brutal pressures at 5626 feet, 6076 feet and 7426 feet. The engineers measured how much the steel moved with strain gauges, bits of looped wire taped all over the sphere. By measuring the electrical resistance of the tiny loops, they could tell how far to within a whisker the wires stretched or contracted, and this told them how infinitesimally much the steel moved, small but important measurements for calculating the sphere's strength.

Despite the laminations, sphere No. 1 held up just fine. In January 1964 sphere No. 2 performed as well at the same pressures and was shipped to Minneapolis for assembly with the rest of the parts that would transform it into a deep-diving submarine. Then No. 1 went back into the oil bath for the big test to collapse. It would also be the super tank's big test, the first time it would be taken beyond 4000 psi (pounds of pressure per square inch).

DeHart scheduled the test for a Saturday in February 1964 and tied to the lid a steel cable attached to a pole stuck in the ground. Joe Walsh and a handful of Southwest engineers gathered around the instruments and dials in the back alley building and started to take it up.

At 4300 psi there was an explosion. In the next second the engineers calculated the probable trajectory of the tank's lid, concluding that 40,000 pounds of steel were headed for the tin roof above them. They ran, all of them headed at once to the only other door at the far end of the building. "I remember the instantaneous transport of myself, like a Tibetan monk using the mind to will myself out of that building," Walsh said.

Outside they were enveloped in a yellow fog mist. The oil in the tank had vaporized, going from 4300 psi (equivalent to 9676 feet) to a vacuum and doing it in less time than it takes to blink. Oil was everywhere. It dripped from the building, the parked cars and the lone tree in the alley, and it covered the engineers. A 500-pound chunk of steel rested about two inches from a new Pontiac, once bright red, now dripping in grease. A bigger shard had landed on the bed of a pickup truck and bent it to the ground.

The super tank looked oddly untouched and its 40,000-pound lid, undamaged, was in place. The deadman was broken and its 40-foot cable was gone. When they removed the cover, they saw that the shrapnel had come from the upper threaded portion of the tank. The lid had shot up at least 40 feet and dropped back onto the tank, driving it about three feet deeper into the ground.

In the tank sat sacrificial sphere No. 1, undamaged. "It had one little ding in it," DeHart's assistant Ed Briggs said. "How that happened I don't know. The vacuum had sucked off the bolts in the hatch. It was fine. It was really miraculous, as a matter of fact, almost unbelievable."

The one-of-a-kind pressure chamber was no longer an option, but the chosen sphere was already tested and Woods Hole was satisfied, almost smug, as were Hahn & Clay and Froehlich. Their sphere, the supposed bad ball, had endured virtually unscathed the explosive release of 4300 psi. And sphere No. 2 was even better.

The March delivery date came, and before it went Litton advised Woods Hole that it owed $442,752 more for "scope increases," that is, for the features that did not appear in Froehlich's design submitted with the General Mills bid in 1962, and for the additional effort required by such things as WHOI's stringent quality control and tardiness in approving design drawings.

Rainnie wrote Litton:

> [You] took a rather wild excursion into the realm of irresponsibility on your part... the furor raised over this question has, as is usual in such cases, two sides.

It would be some time before that furor was settled. In the meantime, they had a submarine to put together.

Alvin was assembled in the same room General Mills built its specialized equipment for making Wheaties. Into the front half of an aluminum frame skeleton went sphere No. 2, now freshly painted white. The rest of the craft was tucked beneath, behind and above the perfect ball, which was 6 feet, 10 inches in diameter. A large circular piece of fiberglass with a tall stack in the center went over the sphere. The five-foot-high sail, which had three windows, was tailored to fit the six-foot-four-inch Rainnie. There were five more windows: four in the sphere, just big enough for both eyes, and a two-inch plastic circle in the hatch. Sleek white fiberglass skins tapered around the passenger ball to a pivoting propeller more than four feet across. Froehlich got the idea for it while boating; it would behave just like an outboard motor. Amidships were two smaller rotating props. At the port side was the mechanical arm, the last General Mills would build.

In *Alvin*'s belly were three compartments that held lead acid golf-cart batteries, a recent technology. The electrical wiring was wrapped in long encasements or penetrators which were piped through holes around the center and bottom portholes.

Instead of steel shot, *Alvin* carried 600 pounds of oil in six aluminum balls. To ascend, the oil would be pumped into two rubber-sided compartments. The inflated membranes would increase the craft's displacement, pushing it up with about 350 pounds of force. To dive, oil would be pumped back into the balls.

Another ballast system, the same kind used on modern combat submarines and the eighteenth-century *Turtle*, took in seawater to descend and discharged it to ascend. On *Alvin* this ballast system was meant to provide more freeboard.

To vary *Alvin*'s trim, Froehlich's design called for a system which used 600 pounds of mercury in three fiberglass balls the size of basketballs. Very little mercury weighed a lot and it flowed. It was ideal for putting a lot of weight in a small area, and by moving it back and forth from stern to bow, changing trim. That was the system used in Jacques Cousteau's prototypical *Diving Saucer*, launched in 1959, and that's where Froehlich got the idea. But he never discussed it with the French. Mercury had disadvantages; it corroded most metals, and as the General Mills engineers learned, it refused to move under high pressure. The *Diving Saucer* was designed for the pressures at only about 1000 feet. This had to work at 6000 feet.

Froehlich left the details to his assistant Forest Grimm, whose background was in building tractors and most recently, mechanical arms. Grimm tried all kinds of pumps but could not find one capable of pumping even saltwater under high pressure. So, he added oil to the mercury. The pump moved the oil which floated to the top of the ball, and the pressure of the rising oil forced out the mercury, which could be directed aft or forward to vary *Alvin*'s pitch by about 25 degrees.

There were still more balls, the shape so perfectly suited to external pressure. Nine aluminum spheres, each two feet in diameter, added 4000 pounds of buoyancy. These were encased in a new material composed of millions of hollow microscopic glass balls held together with epoxy. This syntactic foam, as the material would later be named, looked and felt like flour, soft and white without the grit of even refined sugar. Mixed with resin, it turned a light peach color, and before hardening, it was as flexible as clay.

Nine electric motors were in pressure-compensated boxes filled with oil. Inside the hub of *Alvin*'s tail Grimm packed a ten-horsepower motor, the same kind he had used so many times in tractors. And into every remaining bit of space were packed more of the glass bubbles and epoxy. These were taking the place of the buoyant gasoline of the bathyscaph and the helium of stratospheric balloons.

From nose to tail *Alvin* measured 22 feet and at its widest, eight feet, the width Froehlich chose because it was the legal width limit of any object that could be transported on a highway without special permits or an escort.

Alvin was decidedly mongrel, a cross between aircraft, spacecraft and submarine. It would go under water like a submarine but spend the night out of water to be recharged like a golf cart. That meant it had to be small and light enough to be regularly moved. The engineers used aircraft terms, calling the bottom skids landing gear. They spoke of flying *Alvin*. Unlike a submarine, it was meant to land, something a submariner was trained to avoid or face the humiliation of going aground.

"I soon learned," Mavor recalled, "that because of the very tight weight requirements, it was necessary to apply aircraft technology with marine technology. Throughout her design and construction the Bureau of Ships and other marine people expressed alarm at the selection of materials and the frame strength. Some of the design factors were in direct conflict with the most established criteria for submarine design. When you designed a ship or a submarine, most of it was inside the hull. For *Alvin*, the only thing that needed to be in the hull were people and a few instruments. She was more like a spacecraft."

Nor did submarines have *Alvin*'s built-in safety provisions. The arm could be detached from inside the passenger hull in case it became entangled. To gain buoyancy quickly in an emergency, the batteries could be dropped and all 600 pounds of mercury could be leaked. With several turns of a large shaft at the bottom of the sphere, the passenger hull and the fiberglass sail section could be freed from the rest of the craft and floated to the surface.

Alvin's safety factor was much higher than planned. Safety factor is a guide for designing products engineered for human use. At one extreme are combat submarines and test fighter planes with safety factors of 1.5; for commercial aircraft it is 4; and for elevators, 11. The number does not consider a maverick pilot or a bleary-eyed air traffic controller. Safety factor for a submarine is arrived at by dividing the collapse depth by the designed operating depth. A submarine with a high factor of safety would not necessarily be safer—its walls might be extra thick to better withstand pressure, but that might make it too heavy and prevent it from surfacing. Clearly, not a safe craft.

According to the engineer's calculations, the collapse depths of the three spheres fell between 15,000 and 16,100 feet. Hull No. 2 was indeed the best. With the intended 1.8 safety factor which was set by WHOI, it could operate as deep as 8950 feet. But *Alvin*'s depth limit remained unchanged. At 6000 feet the passenger sphere's safety factor was 2.7. For the other critical components, it was even higher: the portholes were tested to the pressure equivalent of 16,877 feet, and 22,502 feet for the penetrators which threaded all the electrical wiring into the sphere.

It would be fifteen years before the standard ASME boiler and pressure vessel code was revised for deep-diving craft. The new code would finally address a sphere and conclude that the craft's safety relied on its ability to bring its passengers to the surface in an emergency without power.

Alvin's first pilots would pack into the passenger sphere scuba gear and life vests, but it would not be possible to open the hatch against the crushing pressures of the ocean, not even in fairly shallow water. Rainnie would carry a hacksaw. He naively figured if the shaft to release the passenger hull ever got stuck, he could cut it free.

On May 23 *Alvin* was carefully tied down in two sections on two flatbed trucks and driven out of Minneapolis. The submarine arrived in Woods Hole in the early morning darkness three days later.

7

Everybody was there. It was as if they were commissioning the battleship Missouri.
    John Cooper, WHOI scientist

There is nothing I would enjoy more than being present with you on 5 June, but I am in a bight and cannot possibly make it. You may rest assured, however, that we here in the Bureau are mindful of the importance of Alvin *and are continuing our efforts to support you.*
    Captain S. Ries Heller, Jr., Bureau of Ships

For some time now Paul Fye had been thinking about names for the submarine. He was leaning toward *Deep Sea Explorer*. The Deep Submergence Group did not tell him they had a name, and while Fye had heard the craft referred to as *Alvin*, he didn't take it seriously. Neither did Allyn Vine. Choosing a name for a naval or research vessel involved a standard ritual and not a little politicking. Besides, it was ridiculous; *Alvin* was the name of a chipmunk. But when Fye realized that the Deep Submergence Group was serious, he climbed the stairway of the Drugstore for a rare visit.

The group anticipated his objections. Joe Walsh prepared several arguments, beginning with the dictionary. Alvin derived from the old high German *Alwin*, which meant noble friend. Walsh invented acronyms: Admirable Little Vehicle for Inner space Navigation, A Little Visit to Interrogate Neptune, how about A Last Voyage Into Nowhere?

Fye was not amused.

They got serious. Alvin was a contraction of Allyn Vine; they couldn't help it if there was a chipmunk with the same name.

"They were a pretty independent bunch, almost aggressive in supporting this idea of the submarine," Fye said. "I said to them, 'Come on, fellas, let's not be silly about this.' But they said it was too late to change. The name was in all the drawings."

Fye let them have their way but he insisted on a full-blown commissioning organized under his direction. He asked Jim Wakelin, Assistant Secretary of the Navy for Research and Development, to deliver the main speech, and Vine to say a few words. Vine would not be in town, but his wife Adelaide agreed to swing the champagne. Rainnie instructed his team, which had grown to about a dozen technicians, engineers and pilots-to-be: "Friday uniform is with ties—no sneakers."

Despite Vine's protests that it was bad luck to commission a submarine on a Friday and no Navy submarine was ever commissioned on a Friday, this one was. On June 5, 1964, several hundred people sat on the Laboratory of Oceanography rooftop, hung out the windows, and crammed into the WHOI parking lot before the world's first deep-diving submarine. On its glistening white fiberglass skins was written: "BUILT BY LITTON" and "RESEARCH SUBMARINE" and "ALVIN."

It was a brilliant sunny day and you could see across the sound to Martha's Vineyard and the Elizabeth Islands. *Alvin* was draped like a big combat submarine in red, white and blue scalloped bunting. Its single claw grasped a rock. Two airplane ladders were at the sides so that everyone could climb up and peer down into the empty passenger hull.

It looked more like a giant white bathtub toy than a submarine. The metaphors were loose. A weird fishing lure, the press said. Science fiction–like. A pregnant guppy. The Drugstore team was incensed when another reporter likened *Alvin* to a washing machine. It was rather confusing. As a vessel, the traditional pronoun was she; but the submarine had a male name.

Vine was in the deep Atlantic Ocean in the French navy's new bathyscaph *Archimede* examining the walls of the Puerto Rican Trench. He sent his regards via telegram which Fye read to the audience.

Bud Froehlich was half listening. His concentration had been interrupted by two men in naval uniform. "Did you see that?" one of them said, pumping his thumb at *Alvin*. "There's no shaft holding the aft propeller. That thing's a fake!"

It may well have appeared so. *Alvin*'s tail wagged in the stiff breeze because the hydraulic lines were empty, as were the battery compartments and air ballast chambers. Had *Alvin* been launched now, like any submarine at a commissioning, it would have sunk. *Alvin* had been rushed out of Minneapolis by a near frantic Litton management which kept telling Froehlich: "We gotta get that thing outta here!"

Fye thanked ONR for being the impetus of the program and providing strong moral support. "We were fortunate," he said, "in having Captain C. B. Momsen in the right spot at just the right time."

Before approaching the speaker's podium, Momsen asked Jim

Water Baby

Wakelin how he'd like to be introduced. Should I tell them, Momsen said with a grin, that you tried to cancel this project? Momsen didn't mention the memo ordering him to divert no further funds. He said: "Much of our support was moral and some of it was real."

Wakelin had supported the *Trieste* and even talked to Louis Reynolds once during the *Aluminaut* negotiations. Now that he thought about it, maybe that memo was a kind of bureaucratic hiccup. Wasn't it Reynolds who couldn't make up his mind? It didn't matter now. Wakelin had no doubt of the Navy's considerable need for *Alvin* and the other deep-diving craft that were sure to follow.

The sinking of the U.S. nuclear submarine *Thresher* a few hundred miles off Cape Cod the previous spring heightened the government's awareness of its lack of capability in the deep sea. The tragedy was played out in the world press and a naive public learned some shocking truths, that there was no craft on Earth except the elevator-like bathyscaph that could go to the deep sea, and anyone stuck in the deep sea was doomed to stay there.

Alvin was going to make a difference, Wakelin said. It didn't matter that the bathyscaph could go deeper. With a 6000-foot-depth capability, *Alvin* would have access to nearly half the water volume of the ocean, the portion in which almost all marine life resided, he said, and about one-sixth of the ocean bottom, an area almost equal to the surface of the moon.

"I understand," Wakelin said, "there is a saying among biologists that towing a sampling net through the ocean is like running madly through an open field blindfolded holding a butterfly net over one's head. *Alvin* will put the biologist where he belongs—in the center of the marine environment with his net and eyes open."

There were lots of analogies to describe what it was like to sample the ocean blindly from the undulating deck of a ship. It could be Herculean and dangerous, not to mention sickening. The younger computer generation of ocean scientists would call it "puke-over-the-side oceanography." The paradox of the metaphors was that the majority of scientists still did not believe that *Alvin* would be an important tool.

"To all the men who have helped plan, design, and construct *Alvin*," Wakelin said, "we can say to you that you have done your job magnificently. To the men who will operate this pioneer submarine and conduct its research program, we wish you well and note with a touch of envy the exciting challenges that await you in the ocean's deep frontier."

Adelaide Vine, with a small bouquet of spring flowers in one hand, lifted a green bottle of Parma California champagne. Vine had coached her to whack it against *Alvin*'s mechanical arm, the sturdiest exterior part of the submarine.

"I christen thee *Alvin!*" she said, delivering on the first swing, and a solo bugler blew imperfect notes, followed by a band of sailors playing a robust "Anchors Aweigh."

And with that, the DSRV or Deep Sea Research Vessel *Alvin,* noble friend, was formally brought into the world. It was a new species of craft whose time had come. Kindred vessels were under development at several firms. Later in the year the Reynolds submarine, which cost $4 million, would finally get wet, but flaws in its hull would prevent it from reaching even half its intended 15,000-foot depth.

The WHOI-ONR sub would lead an extraordinarily successful and famous life. Woods Hole would take credit for its design and so would the navy. But in 1964, *Alvin* inspired little confidence or interest in any but a very small group of men. It was at WHOI because no other marine lab in the country wanted it.

II Hard Times, 1964–1971

They played hard, they drank hard and they worked damn hard.
    Marge Stern, Alvin Group secretary

8

Now that Alvin *has finally tasted saltwater, we are reasonably close to our goal of taking some of you to the environment that has been somewhat inaccessible. We hope to do this efficiently and safely and intend to proceed . . . without committees involved, priority ratings, psychological and physical exams, etc. . . .*
            Earl Hays, Chairman of the Applied Oceanography
            Department, to WHOI scientists

Bud Froehlich, who spent the summer in Woods Hole, was not quite finished. According to his tallies of poundage, *Alvin* was too heavy. After the commissioning, the Woods Hole team added blocks of the tiny glass bubbles and epoxy high onto the sides of the plump submarine. Now Froehlich faced the critical proof. If *Alvin*'s centers of gravity and buoyancy were aligned properly, the submarine would sit straight up in the water. The difference between the two invisible points in space was only 3.24 inches, and it was a difference that couldn't be measured with a ruler.

Alvin was lowered by crane from the dock into the water.

It was a special moment for Froehlich. Not that he was surprised, he said, but seeing was believing and he could not help but think that if he had obeyed the Bureau of Ships, the sail would not have reached heavenward.

Now people could dive in *Alvin*. After Froehlich, Allyn Vine was the second guest to dive. He dropped 25 feet in Buzzards Bay and at bottom shook Bill Rainnie's extended hand. "We made it," Rainnie said, and they laughed for a long time. A leak detector for a ballast tank read positive. The speedometer didn't work. Visibility was almost zero. But who cared?

The first 77 dives in Cape Cod waters were brief and shallow, the deepest to 70 feet, and were intended to uncover the bugs and allow the pilots to get to know the craft. Those who wanted to dive during the trials had only to stand at the WHOI dock or hang around the beach and look eager as the submarine went through its trials. But the biggest trial was not in the water.

Because of the *Thresher* sinking, which was blamed on a faulty pipe joint, the navy promised an outraged Congress to make submarines safer. Now, subs would be built with more stringent specifications and would be required to pass a final comprehensive inspection to gain something called certification. Without it, a submarine would not be allowed to swim.

DSRV *Alvin* was not immune. The letter saying so arrived at WHOI July 7, 1964, followed twenty days later by six men from BuShips. They said *Alvin*'s welding did not conform to navy specifications; they were unhappy with the oil that bathed *Alvin*'s components because Mr. Hoover of the Hoover Electric Company in California refused to disclose what was in his special brew, claiming it was proprietary. He wouldn't tell even WHOI. (It looked and smelled like kerosene.) And they were appalled at the Ford automobile relays in the electrical system. *Alvin*, they concluded, should not be certified:

> WHOI does not have the technical ability to properly access [assess?] the product that they are receiving.
>
> Documentation and remaining planned tests will not give complete assurance of personal safety.
>
> Metals and fabrication procedures indicate a lack of sophistication in design and construction practices for deep sea environment. The same is true for piping systems and components.
>
> Further audit would be too time consuming for BUSHIPS personnel to undertake....

Woods Hole was enraged.

"WHOI, Rainnie, that whole gang, were very unhappy with us," Richard "Skee" Dzikowski, the head man at BuShips, recalled. "We had a set of guidelines prepared by BuShips for *Alvin*. They were stringent perhaps in light of *Thresher* and the state of the art. I was very skeptical at the time. Here was this taxicab and we didn't know how it was built. We didn't have anything to go on. Nobody in the world had these specifications. One of the biggest mysteries was the [emergency] hull release mechanism. It was the first time anything like that was ever conceived."

Paul Fye wrote the Chief of Naval Research:

> We are in receipt of... a *draft* of an audit report by representatives of the Bureau of Ships relative to *ALVIN*....
>
> Unfortunately, the Bureau of Ships representatives who prepared this draft were able to spend only a day and a half in Woods Hole evaluating the project and did not visit any of the subcontractors

Escort swimmers make a final check before *Alvin*, carrying the remote camera package, Jason Jr., in its sampling basket, descends to the *Titanic* in 1986. (Perry Thorsvik, National Geographic Society)

Alvin on an early dive. (WHOI)

Alvin's commissioning in Woods Hole, June 5, 1964. (WHOI)

Alvin returns to its mother ship *Lulu* after a day's work at the bottom of the ocean in 1976. (Emory Kristof, National Geographic Society)

The prow of RMS *Titanic*, dripping in rust, at 12,500 feet, 400 miles southeast of Newfoundland. Only the bronze pedestal remains of the steering mechanism on the bridge; the ship's wheel has been eaten by wood borers. (WHOI)

Artist's conception of *Alvin* approaching the *Titanic*. (Pierre Mion, National Geographic Society; courtesy Michael Winston)

Various items from the *Titanic*: a silver bowl, second-class purser's safe, the iron frame of a deck bench and a lifeboat davit on the promenade deck. (WHOI)

Some 220 different species of animals, all but ten previously unknown, have been found at vents. From off the Galapagos —

Left: Thatches of tubeworms which dominated the Garden of Eden and the Rose Garden. (Robert Hessler)

Below: "Spaghetti," a bizarre conglomeration of worms. (Jim Childress)

Right: At the Mussel Bed, the pink vent fish *Bythites hollisi*, named after *Alvin* pilot Ralph Hollis who captured the first and, so far, the only one. (Fred Grassle)

Below, right: "Dandelions," each a cluster of many tiny pouch-like animals, anchored to rock by dozens of delicate translucent lines. (Fred Grassle)

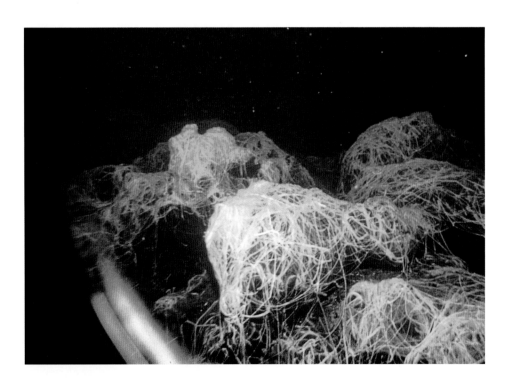

At the Mid-Atlantic Ridge vents —

The "eyeless" shrimp, *Rimicaris exoculata*, swarm around a six-foot-high chimney. (Peter Rona)

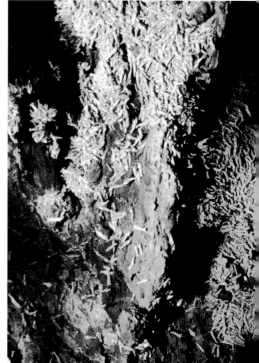

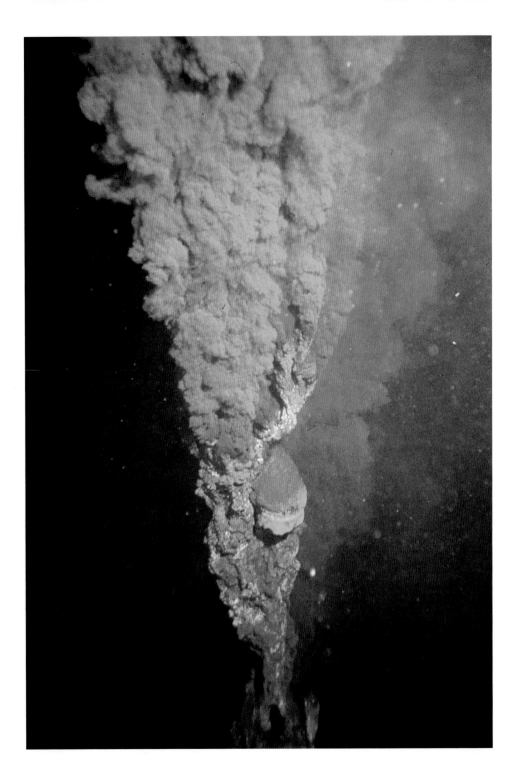

On the East Pacific Rise at 21° north—

Facing: A black smoker gushes polymetallic sulfide mineral deposits and fluid hot enough to melt lead. (Fred Grassle)

Left: The Pompeii worm, *Alvinella pompejana*. (Emory Kristof, National Geographic Society)

At the Juan de Fuca Ridge—

Above: A spider crab feeds on strawlike tubeworms in the Cleft. (Bill Normark)

Right: Proof of the seemingly preposterous-light in the deep sea. With all of *Alvin*'s lights off, an electronic camera using the equivalent of 50,000 ASA film, records light invisible to the human eye at the top of a 662° F black smoker. (John Delaney, Milton Smith, Cindy Lee Van Dover)

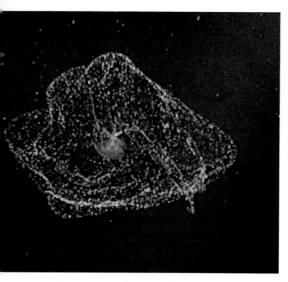
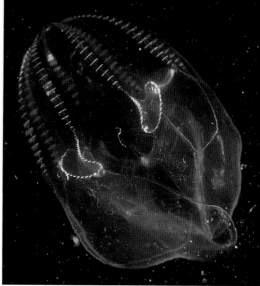

Marine snow: a "house," about six feet across, of mucus, a thumb-sized zooplankton, bacteria, fecal material and other detritus. (Mary Silver; courtesy Monterey Bay Aquarium Research Institute) *Bathocyroe fosteri*, one of several unknown species of ctenophores, gelatinous animals a few inches long, discovered with *Alvin*. This creature was found off New York in 1976 and was named after pilot Dudley Foster. (Larry Madin)

The Mariana vent—

Anemone Heaven. *Alvin*'s temperature probe takes a reading at a chimney covered with crabs, limpets and snails as big as golf balls. (Robert Hessler)

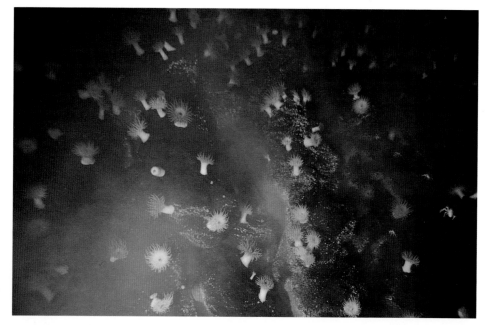

Pen-and-ink drawing of a thriving oasis in the Guaymas basin. Six-foot-tall tubeworms at the base of a towering edifice of sulfide, a black smoker and a mushroom-like chimney about 50 feet tall. "Angel hair" and "styrofoam," bacterial mats, cover the seafloor. (Susan MacDonald; courtesy Fred Grassle)

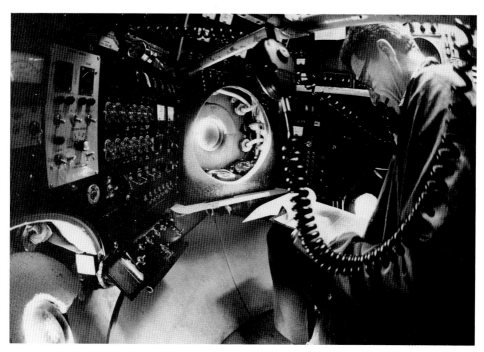

Barely room for three in *Alvin*. Larry Shumaker, at the center viewport, the pilot's position. (WHOI)

Geophysicist Jan Morton, wearing extra layers of clothing, climbs into *Alvin*'s sail aboard the *Atlantis II* for a dive to the frigid deep sea. Note at center the loop of the main line used for launch and retrieval. (Randy Koski)

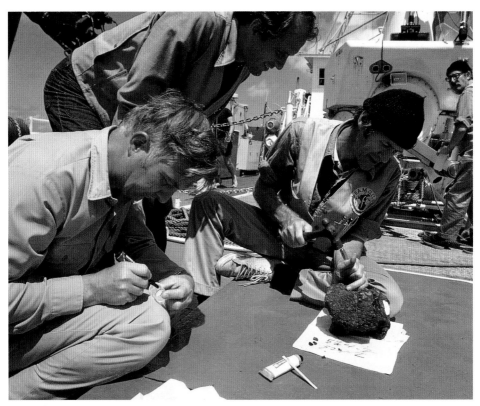

Geologists curate basalt on *Lulu* during the FAMOUS project in 1974. From left, Bill Bryan, Bob Ballard and Jim Moore. (Emory Kristof, National Geographic Society)

The budgerigar "Harris" with friend atop bo's'n Ken Bazner's eyeglasses. (Courtesy Ken Bazner)

Aboard *Lulu*, the early crew wore red berets. Bill Rainnie, third from front with cigarette. (WHOI)

George DeP. "Brody" Broderson, crew chief. (WHOI)

Alvin with the *Atlantis II*. (Rodney Catanach)

Thrusters replace propellers on the modern *Alvin*, here about to be placed on the *Atlantis II*. (Rodney Catanach)

who have been responsible for the design and construction of the submersible. Obviously this is too limited a survey to evaluate properly a project of this complexity. This may in part account for the inaccuracies and misunderstandings contained in this *draft*. The report is so misleading and incorrect that it seems fruitless to discuss it point by point. It in no way represents properly the current status of *ALVIN* and should be withdrawn in its entirety....

ONR made its own case in three single-spaced pages to BuShips, arguing that *Alvin* was unique, too special to fit into any existing category. In fact, ONR said, since *Alvin* would be out of water more than in, it was not a true seagoing craft. How could the navy's rules apply to such an unnaval-like thing? Only its sphere carried people, so perhaps certification should apply only to the passenger hull. The Alvin Group was "small, close-knit, dedicated," and *they* wouldn't jeopardize safety.

With tempers still hot, the Drugstore gathered together *Alvin*'s documentation from Minnesota, Texas and elsewhere around the country and sat down with the BuShips engineers in Washington. Vine knew Dzikowski; he had taught him how to find a thermocline to hide under in World War II. It had saved many lives, many times.

Dzikowski looked across the table and said, "Well, gentlemen, we can do two things. If it works, we can certify it. If it doesn't work, we can fix it and then certify it."

BuShips conceded that *Alvin* was after all well documented and its "conceptual design is adequate"; however, its "detail design of systems and material selection is at best marginal." But the navy agreed to review *Alvin* again the following year when the sub made the first dive to its maximum depth. WHOI didn't argue when BuShips insisted on an extra-deep tethered dive.

Frank Omohundro, one of the BuShips auditors, came to empathize with WHOI. He didn't think *Alvin* was unsafe, just unorthodox. That fall of 1964 Omohundro retired from the navy and applied for a job with the Alvin Group. "I didn't know if they were looking for someone," Omohundro said. "The emphasis of my job interview with Earl Hays was how to certify *Alvin*. Earl said: 'Okay, you're hired. Your first job is to make sure that *Alvin* gets certified.' "

Woods Hole figured *Alvin*'s mother ship would have to resemble a catamaran, with ample space between the long hulls for launch and recovery. They were counting on ONR for the funding, and ONR was counting on its run of good luck. But there wasn't enough money to build a proper mother ship, so they would have to make do with a rather stripped-down version which, they reasoned, would be a working test model. With a bigger budget, expected soon, they could start over and construct the real thing.

ONR and WHOI carefully avoided calling it a ship and kept the whole affair between themselves. A vessel constructed with government money needed congressional approval, and nobody believed Congress would allocate the funds. Also, a ship built with navy money was subject to navy specifications, and enough was enough.

Then WHOI heard from ONR about two big mothballed pontoons, built for minesweeping, that looked like they could serve as hulls for a catamaran. And the Drugstore called Big Dan Clark.

Dan Clark was the Paul Bunyan of Woods Hole. He had his own marine construction business, and most of what he built was massive—which seemed in keeping with his size and reputation. He stood just over six feet four but the folks of Woods Hole village were sure it was much more. Shocks of white hair curled on his forehead, over his shirt collar and down his cheeks to an untamed white beard, frequently singed with ashes from the stogies he smoked.

Clark was sixteen when he first set eyes on the ocean, and he had never been far from it since. He drove pilings, installed docks, retrieved buoys, rode out hurricanes, and for awhile lived aboard an old sixty-foot bugeye schooner in Woods Hole. It made the local paper when it sank in ten feet of water and Clark raised it by packing the cracked wooden seams with seaweed and horse manure. He was the man WHOI called when they needed a barge to launch *Alvin* for its sea trials, and the man they called to fetch the pontoons. And it was he, Big Dan Clark, who from a bunch of junk and surplus parts, made *Alvin*'s mother ship.

The minesweeping pontoons were 96 feet long. Each weighed about 90 tons and the bottoms were inch-thick armor plate. They were clean as a hospital inside, Clark said, and still filled with the machinery for detonating mines.

Two steel arches spanning about 50 feet would connect the pontoons fore and aft. One of the pontoons would hold machinery and the other contain the living quarters. *Alvin* would be parked mid-deck on a cradle lowered by winch for launch and recovery. A ship would tow the catamaran and provide whatever other amenities were needed for life at sea.

That was the idea, anyway. It attracted plenty of unsolicited criticism, and WHOI's engineers didn't disagree with all of it. Their small models of the catamaran crabbed and fishtailed. Engineer Henry Horn wrote:

We have a real problem. The catamaran [looks] like the best approach, not only from a theoretical standpoint but also from a practical one. Not only do these floats represent the bird in the hand but there are no others even in the bush. Our original intent was to design the catamaran ourselves and send out bids for the construction. However, this now looks like more than a two-man job.... As for the state of the art of launching and retrieving delicate little submarines, there is little or no previous experience.

For $25,000 Philip L. Rhodes Naval Architects of New York worked on the blueprints, and with the design in hand, WHOI called for bids. It was not going to be a lucrative project. The deadline precluded the use of a shipyard, and it was midwinter, not a good time to build anything outdoors on Cape Cod.

Dan Clark's bid was scrawled in pencil on a cedar shingle.

DAN W CLARK iNC CAT Bridge Bid
    $42476.oo
    [signed] Dan W. Clark

It was the winning bid.

Clark and his gang, as Dan W. Clark, Inc., was known, began gutting the snow-covered pontoons at the town dock.

Horn's doubts still nagged him. No winch was a match for an angry ocean; a mild swell could bash the submarine against the catamaran-barge. He began to feel better when he spotted an old bomb shelter shell in a shipyard. The cylinder, designed to be crawled under to escape nuclear fallout, was about 12 feet long and wide enough to sit up in without banging your head; it even had a little tunnel extension. It kind of looked like *Alvin* to Horn; he figured they could use it to practice launching and test the line and winch contraption for the 7500-foot dive without passengers.

Dan Clark searched other junk piles and found an old steel boiler to weld to the bomb shelter and fill it out some. At another shipyard Horn saw four huge rusty diesel outboard motors, Navy surplus Murray-Tregurthas more than 16 feet long. They had seen a lot of life and two needed major repairs. But Horn couldn't resist. If the catamaran could move on her own power, launch and retrieval would be safer, and, he reasoned, for short trips it would save them the expense of a tow boat. Woods Hole got all four outboard motors, like the pontoons, for nothing. They paid $400 for the bomb shelter—the price of the steel.

Chain saws burred and wheezed through timber that was

pounded through with large spikes for the deck planking. A makeshift bridge house of welded pipes and plywood was placed at the stern because the rear arch was slightly higher than the forward arch and it was thought that the view of the submarine during launch would be better in back.

Everyone worried about the big screws of the outboards chewing up *Alvin*. It didn't seem wise to put them on the stern, where *Alvin* would be launched. They decided to use only one outboard, which Clark put in a plywood housing on the front in the center of the arch.

Allyn Vine brought his mother visiting from Ohio down to the town dock to see *Alvin*'s tender. They were greeted by Earl Hays who invited her on a tour of the mother ship.

Vine's mother did not like ships or anything else that floated. She was already in her mid-sixties but had managed to avoid stepping on a one of them and wasn't about to now, she said. Hays grabbed her by the elbow and steered her across the deck and back onto the landing. It was a quick tour. Vine's mother was speechless.

"What's your name?" Hays sweetly asked her, trying to make amends.

"Lulu."

"Lulu, what a nice name, Lulu."

It was a nice name; the more Hays thought about it, the better he liked it.

Lulu Vine said nothing about the vessel that was part barge, part cat and part minesweeper. If you didn't know better, you couldn't tell which end was which. *Alvin*'s tender was registered with the state as the biggest outboard motorboat on the books. And the engineers said it was the only front-wheel-drive vessel in the entire worldwide oceanographic fleet. There was no Coast Guard inspection, a requirement for research vessels of 300 gross tons or more. WHOI said the mother ship was 299 tons.

In March 1965 the Deep Submergence Research Vehicle Tender No. 1—otherwise known by Dan Clark and his gang and the Alvin Group as *Lulu*—left Woods Hole under tow, headed for Port Canaveral, Florida, where she would rendezvous with the precious cargo *Alvin*. The sub would make the journey overland by truck. They weren't taking any chances.

Lulu passed her oceangoing test by making it to Florida without incident. With *Alvin* aboard, the crew set off for their test dive site at TOTO—the Tongue of the Ocean, a swath of deep water two miles west of New Providence Island in the Bahamas. They were bound for the good weather to prove that *Alvin* was safe enough to be certified by BuShips and finally to fulfill the mission for which Captain Swede Momsen had justified the sub's existence, to inspect the underwater listening array ARTEMIS off Bermuda.

On arrival in the Bahamas, they tested *Alvin*'s hull release which had been modified to keep from sticking. On the first try, the forebody, carrying the passenger sphere and sail, bobbed to the surface. It would be the last time sphere No. 2 would rise apart from the rest of the submarine.

The dummy *Alvin* bomb shelter was dropped, raised and dropped again until they were satisfied, but not comfortable; nobody felt easy about dangling the real *Alvin* at the end of a 10,000-foot-long nylon line as thick as both Dan Clark's wrists.

The tethered dive to 7500 feet went smoothly, although slowly; it took twelve hours. By the time *Alvin* was safely on deck, high tide had come and gone. The channel into Coral Harbor was eight feet deep, and *Lulu* needed twelve more inches. The crew had no choice but to spend the night with *Alvin* on *Lulu*. Shunning the hammocks in the cramped and suffocating pontoon, they camped outside on deck on air mattresses with life jackets for pillows. Most of the eighteen members of the Alvin Group made the trip. Besides Bill Rainnie, the sub had only one other pilot, Marvin "Mac" McCamis.

McCamis was raised by his father on a farm in Indiana where he developed a special talent for repairing machinery. His blond crew cut and boyish good looks belied his forty years and gave him a collegiate air, but he had never been to college. Two years short of high school graduation, he joined the navy as an electrician. He chose submarines not out of any love for the ocean—he had never seen the ocean—but because the submariners were the elite. He had just retired after twenty years in the navy when he saw this ad:

> UNIQUE OPPORTUNITY. Mile deep oceanographic research submarine requires limited number operating and engineering personnel. Previous submarine experience desirable. Send complete resume to the Woods Hole Oceanographic Institution, Woods Hole, Mass.

He was hired in early 1963.

The man Hays and Rainnie considered their high level brains was William "Skip" Marquet, thirty-five, an instrumentation engineer. At five feet three, Marquet just about reached Rainnie's shoulders. He was working as a manager for Honeywell in France when WHOI heard about him through the grapevine. Marquet felt that with his promotion into management, he had lost a sense of making a personal contribution. A friend suggested he get into oceanography; it was incredibly under-funded and needed all the help it could get. Marquet's friend was Woods Hole's grapevine.

The seagoing crew comprised a core of eight technicians, mostly mechanics and electricians. Job titles didn't matter since everyone did whatever his skills and energy allowed. It was a young group of handymen, with the exception of George de-Pentheny Broderson, forty-eight, a mechanic. They called him Brody.

Born in Connecticut to a Danish father and an American mother, Broderson left home when he was twelve years old. He had learned to fudge his age, landing him, still shy of his teens, a job as cabin boy on a freighter bound for the rest of the world. When he left the ship in New York, he found a way to get paid for doing something he relished, acting. It didn't last long; with the news of war, he enlisted in the air force as a mechanic. After working as a New York taxi mechanic for many years, he settled in Sandwich on Cape Cod and opened his own garage. But he was restless, so he took the job with the Alvin Group. There was something ultimately classy about him. Maybe it was his goatee or the way he wore his straw hat dipped low to the side as he dirtied his hands changing *Alvin*'s oil. At day's end he would change from his filthy shirt and slacks into a dinner jacket and tie and disappear for more genteel company.

The days in the Bahamas were spent in the sun on *Lulu*, taking *Alvin* increasingly deeper and going through the trial and error of getting the submarine into the water. It was not a simple matter of lowering the elevator platform or cradle. Four ropes were tied to *Alvin*, a challenge itself since there were few places to attach a line. When the cradle dropped, the men stood along the pontoons holding the ropes and the pilot, standing in the sail, motored the sub away from *Lulu*. The pilot climbed down into the sphere, and two men in snorkle gear detached the lines so *Alvin* could descend. If there were sharks around, as there often were, the swimmers were picked up in a small rubber boat.

July 20 dawned pristine as usual, and the forecast called for more of the same blue perfection in the deep cul-de-sac of TOTO. Today was the day *Alvin* was going to 6000 feet for the first time and in front of an audience from the navy.

Hays was up at 5:30, and as he stepped between the bodies scattered on *Lulu*'s deck, he thought how like a scene in a gangster movie it seemed. He was worried. On the previous and deepest dive to 3600 feet, the aft propeller mysteriously stopped and nobody knew why. *Alvin* was not unsafe without the prop, just less maneuverable and more like a bathyscaph, he reasoned. But not trusting the official audience to understand that, Hays worked out a code with the pilots so they could converse with impunity about any problems during this critical descent. They had just worked out their secret signals when seven men from ONR and BuShips boarded *Lulu*.

Alvin moved through the pontoons and dropped into the gentle turquoise water. Every 500 feet, about five minutes of descent, Rainnie and McCamis reported in. Their words came over the underwater telephone slightly garbled, as if they were gurgling. The navy visitors craned over the speaker to catch every monotonous word.

"*Lulu*, this is *Alvin*, 3500 feet. All is going well. Except for the stern light, which went out. We are going down, over."

"*Alvin*, this is *Lulu*," Hays said. "Did you say stern light?"

"Stern light" was code for the aft prop. *Alvin* was approaching bottom without it. Three and a half hours after disappearing from the surface, *Alvin* touched down gently. Rainnie contorted his body over the bottom porthole to read the depth gauge. "*Lulu*, this is *Alvin* at 6000 feet," he proclaimed.

Hays gave them ten minutes to check all systems and report back. He noticed low grey clouds on the horizon.

Some 2683 pounds of force pushed against every square inch of the steel ball Rainnie and McCamis crouched in. Every bit of the submarine was moving imperceptibly in response to the immense force. The Plexiglas windows were moving about a thousandth of an inch and the white paint was compressing ever so slightly in the tranquility of the cold black ocean. With only the two side props operating, Rainnie flew *Alvin* to a small knoll teeming with life, but when the fine clouds of sediment settled, the creatures were gone. McCamis switched off the propellers amidships, hoping the animals would return.

Topside the winds whipped up a gale that brought a sudden downpour, and the crew scrambled for their slickers, not for themselves but to cover the electronics on the bridge, the only link to *Alvin*.

"*Alvin*, this is *Lulu*, over," Hays said.

"*Lulu*, this is *Alvin*. It's really dark down here. Over."

"*Alvin*, this is *Lulu*. It's time to start back up, over."

"On our way, out." Rainnie flicked the switch to restart the lift props. Nothing happened. He tried the aft propeller again. Nothing. They had no propulsion whatsoever. He threw the switch to pump the oil in the ballast system. In immediate obedience the pump groaned and *Alvin* began to rise.

Without props the surface recovery would be hit or miss. *Alvin* might crash into the catamaran or ascend beneath it. The pilots did not know about the wild seas that awaited them at the surface.

Lulu bounced and banged hard against the waves. The surface crew couldn't escape their naval visitors to discuss how to recover the helpless sub in the roughest seas they had ever worked in. "Make it look routine," Hays whispered to George Broderson.

Broderson had mimicked everyone, including Hays, who had a habit of carrying a nail with him. You never knew when you might need a nail, Hays had explained. When Broderson saw him twirling his handy nail, he would bellow in his best British accent, "Present nails!" and all would stand at attention holding up a nail. By now, Broderson didn't need any prompting. The crew had gone out of its way to find unusual nails and looked forward to their presentation. If Broderson couldn't make the recovery look routine, nobody could.

Alvin passed through diffused light at 2300 feet, then blue-green twilight. At 50 feet, the pilots blew the freeboard ballast tanks and *Alvin* bobbed to the surface. Suddenly all three props whirred on. The sun burst through the clouds and *Alvin* purred toward its mother ship. The Woods Holers on the catamaran could hardly believe what they were seeing. The pilots heard clapping. There were handshakes all around, and the navy declared DSRV *Alvin* certified.

It was almost embarrassing to McCamis whose tolerance for imperfection was exceedingly thin.

Congratulatory telegrams and letters followed, from WHOI, ONR and BuShips, which decided to make two *Alvin* lookalikes with the spare spheres, one for the Navy and one for WHOI, that is, another sub for Woods Hole. Anxious to call for construction bids, BuShips requested the Alvin Group's comments on the specifications for the lookalikes.

The *Alvin* crew was overdue in Bermuda to inspect the underwater listening station ARTEMIS, but they returned to Port Canaveral and dismantled *Alvin* in a parking lot. Hoover's oil was full of soot, small granules of carbon produced by the electrical relays under pressure. The clinkers gunked up the motors and the Ford automobile relays that switched on and off electric circuits.

When they finally arrived in Bermudan waters in late August, Hurricane Betsy allowed *Alvin* to make only three dives. ARTEMIS would have to wait until next summer when the weather window opened again. It was just as well; *Alvin*'s props continued to stop intermittently below 3000 feet. Tangled messes of cable that littered the ARTEMIS area made it a dangerous place for a submarine unable to maneuver.

But they couldn't go home until they made a perfect deep dive. On September 25, before *Lulu*'s tow ship pointed its bow north, McCamis and Rainnie again descended to 6000 feet. All three propellers worked. They had been down six hours when an ebullient Hays telephoned that it was time to come home. He repeated the call again and again, but still there was no answer, and the exultant mood aboard the catamaran turned grim. At dusk Hays radioed the Coast Guard.

Inside *Alvin* McCamis joked about swimming to Bermuda for a vodka tonic at the Royal Harbor Club bar. They heard everything Hays said. He just couldn't hear them. Their voices on the underwater telephone were deflected by the thermocline, that transitional temperature layer that works magic with sound waves.

The pilots surfaced and waited for dark to release flares which a Coast Guard search plane quickly spotted. *Alvin* was about six miles from *Lulu* and in the other direction from which the mother ship had steamed in search of the sub.

It was long after midnight when *Lulu* tied up at the pier, her passengers exhausted but happy. *Alvin* had made its first trouble-free (almost) dive to 6000 feet; at seven hours underwater, it was also the longest.

Now that *Alvin* had proved itself, Litton could get its money. The balance due the company had been disputed at length. Litton said *Alvin* cost at least $1.5 million, not including the precontract investment.

Woods Hole agreed to about half the $442,752 of "scope increases," born of design changes, bureaucracy, and sundry other difficulties. However, WHOI argued, the sphere was a little too perfect, that is, a little too heavy. *Alvin* weighed two and a half tons more than specified; its ascent rate was 50 feet instead of 200 feet a minute; its speed was two and a half knots, not three and a half; and its endurance was seven and a half hours, not ten. The imperfections were largely contractual and fuel for legal defense. If Froehlich hadn't quit Litton in January 1965, he would have argued that the sub was overweight because of WHOI's additions. In total, $950,529 Navy dollars were given to General Mills/Litton for *Alvin*.

The Alvin Group moved into a borrowed Air Force hangar about twenty miles from Woods Hole and began what they appropriately called tear-down, the annual overhaul and inspection of every inch of the submarine.

Alvin was a surprisingly fragile craft. Vibrations from the overland journey to Florida had cracked a compass case. Paint had chipped off the fiberglass skins during sea trials the previous year. They repainted the outside and painted inside the passenger sphere, pressing into service a terrified canary to make sure there were no lingering toxic fumes. The log read: "The bird lived through the night, although it appears either sick or nervous." To be doubly sure, they locked the bird inside again. "The bird lived through the second toxic check and in fact is in much better condition than he appeared to be after the first check."

A sticking rudder was traced to a faulty valve. The mercury was carefully drained and the trim system's three balls were inspected. A cracked O-ring was replaced.

The propeller mystery was solved. The syntactic foam, which surrounded the aluminum propeller blades, compressed during descents, like everything else on *Alvin*, with increasing external pressure and expanded on ascents as the pressure lessened. But at about 3000 feet, the prop shrouds contracted so much that they touched the blades and stopped them. The crew ground away the tips of the blades.

Alvin would have to wait two decades before brushless motors eliminated the problem of carbon granules. As for the Ford relays, the engineers tried another plan to keep the two metal fingers free of clinkers. Their solution was gravity. By mounting the relays sideways instead of upright, the carbon granules simply fell out. "We had tried everything else that we could think of," Skip Marquet said.

The Alvin Group didn't get overtime for the extra hours and weekends at the drafty hangar. They shared a young esprit de corps. The families and friends who greeted them at the WHOI dock that fall saw them wearing red berets. Val Wilson, who was hired as *Lulu*'s skipper but decided instead to be a pilot-in-training, bought one in the Bahamas, and soon they all had one. They wore them like a uniform as they worked over the guts of the sub. Those who stayed loved the frontier.

It was already dark when Hays left the hangar one bitter cold Saturday in January, weary with the long hours, but satisfied with their progress. They were only two weeks behind schedule. The plan was to leave for TOTO February 1 to inspect ARTEMIS and begin, finally, their first diving season for scientists.

Hays' wife Charlotte waited dinner for him. Her husband's research in physical oceanography had already taken him on long cruises away from her, and now it was the submarine. Hays poured himself a mug of warmed cider laced with rum and soon felt himself begin to thaw.

The phone rang. "It's probably Bill wanting to work tomorrow," he said, letting his wife know that he would not interrupt his Sunday.

But it was not Rainnie. It was the new ONR project manager who had replaced Swede Momsen.

"Earl, the air force and the navy have a problem and we need *Alvin*," he said slowly and distinctly. "I can't discuss it over the telephone."

Hays recalled a brief item in the local newspaper about two U.S. Air Force planes colliding in midair while refueling over Spain.

"Is this about a trip to Spain?"

"Right."

"You know, we're just finishing up the overhaul and we haven't even had a test dive."

"We'll just have to pray that *Alvin* works."

9

January 17 was the feast day of Anton the Hermit, patron saint of Palomares, a poor Andalusian village that for eight years hadn't been blessed even with rain. On that feast day morning in 1966 the skies thundered with the fiery wreckage of a U.S. B-52 bomber, its KC-135 tanker and four hydrogen bombs, each many hundreds of times more powerful than Little Boy and Fat Man dropped on Hiroshima and Nagasaki.

An old man searched the sky as one of the bombs plummeted and cracked a wall of his house and shattered the windows. Chunks of metal rained down on the road in front of a schoolhouse, and dozens of boys and their teacher ran outside.

American troops rushed to southern Spain to put out the fires and to pick up the pieces and the bodies—seven of the eleven fliers perished; the wristwatches on two of them were stopped at 10:16.

The H-bombs had been "unarmed," that is, uncocked and therefore incapable of producing a radioactive explosion. However, two of the three bombs which were found on land had cracked their casings on impact, and the radioactive material had leaked out. Plutonium was spreading over the cacti and tomato fields of Palomares. Some 600 acres were closed off as American soldiers began to dig up thousands of tons of contaminated dirt.

Spain would not publicly acknowledge the accident. The third day after the midair collision, the United States admitted that the B-52 carried "unarmed nuclear armament," but did not say—and would not say for forty-four days—that a fourth of the armament was missing and about to become the object of the biggest, longest and costliest search in U.S. military history.

At the Cape Cod hangar the Woods Hole team prepared for the trip to Europe on Air Force cargo planes. They had lots of unanswered questions. How long would they be away? Would *Alvin* work? Was the bomb radioactive? Even Earl Hays' friends at Los Alamos laboratory, where he had done classified research, would not discuss it with him. Before leaving, he grabbed a radiation counter from one of WHOI's labs.

At Rota on Spain's Atlantic coast, *Alvin* and crew were met by a navy landing ship dock (LSD), whose skipper had orders to take them to Palomares immediately. But *Alvin* would not be rushed. The vibrations from the trans-Atlantic journey had loosened the covers on all three battery boxes. Not knowing that, the Woods Hole team put the sub in the water for a test dive and the batteries flooded.

Within twenty-four hours, a repaired *Alvin* went without a hitch to 40 feet in Rota harbor—hardly a thorough test dive, but time was up—and motored through the LSD's gates into a large flooded well, lining up beside a 51-foot long, bright red submarine with four small windows in its nose. It was the *Aluminaut*. By morning the world's two deep-diving submarines and their crews joined the more than 30 ships and 3000 men of Task Force 65, and an uninvited Russian trawler that treaded nearby.

It was February 11.

The United States government used every possible piece of equipment available and wished for more but nothing better existed that was not in the southwest Mediterranean Sea off Palomares trying to find Nuke 4.

Fishermen said they saw "half a man" strapped to a grey parachute drop into the water about five miles southeast of Palomares. It had to be Nuke 4 because the bombs' parachutes were grey, not white as they were for the fliers. The fishermen boarded a navy ship and "fixed" the spot, using landmarks and their experienced sense of where things ought to be after they turned their back and looked again at miles of monotonous flapping water.

Task Force 65 placed that fix in a circle about two miles across and called it ALFA I. The navy's many advisers who analyzed currents and winds to calculate possible trajectories thought Nuke 4 could also be in ALFA II, BRAVO or CHARLIE, all told about 135 square miles of possibility. Rear Admiral William Guest, who directed the task force, likened their mission to using a penlight to look for a .22 caliber bullet in a muddy, water-filled Grand Canyon.

Guest held daily briefings aboard the USS *Boston*, where the men from Woods Hole saw sailors stand at rigid attention while speaking to him on the telephone. The fishermen's fix seemed a little too magical to Guest, and the search a little too hopeless. He was also running a fever, diagnosed as pneumonia by a corpsman who kept taking his temperature.

The underwater terrain and visibility were terrible. On a good day the pilots could see about twenty feet. The bottom mud was gooey and yet easily stirred up into blinding sediment storms.

Charts of the seafloor were usually about a half mile off. On one dive to where the bottom was labeled as being much deeper,

Aluminaut violently smacked into the seafloor and slid out of control, picking up several thousand pounds of mud in her keel. Where the terrain was rocky, the Reynolds submarine was too unmaneuverable; it had to stay in the flattest areas.

Equipment for navigating was still grossly inaccurate, except for the navy ship *Mizar*'s new camera-sonar system. But even it was still not free of bugs. Navigation under water was mostly by instinct. *Alvin*'s pilot Mac McCamis likened it to walking around your own back yard in the dark—you eventually got familiar with it.

The ships could track the subs to within about 400 yards. The pilots counted aloud on their underwater telephones, clocking the seconds it took their words to travel underwater to determine their distance from the surface control ship. It was crude but the most reliable way to get their range.

Fishing boats crossed into the restricted search area and one rammed into the USS *Pinnacle*'s towed sonar. The towed sonars frequently smashed into the bottom and three were lost entirely.

With the heavy communications traffic, many course orders were garbled and misunderstood, especially messages involving three particular naval officers—Mooney, Moody and Loony. So much confusion resulted that no one was permitted to use the names on radio unless they were accompanied by a function— such as, Mooney Diving or Moody Public Affairs.

The danger of radioactive contamination seemed the least of their concerns. The navy took daily readings of radioactivity and reported nothing unusual. Hays checked *Alvin* after every dive but his counter rose above normal only when it was aimed at a wristwatch with luminous dials. The government's experts said the 30-knot winds that blew on January 17 had pushed the plutonium cloud out to sea, dissipating the radioactive grains.

Weather was the worst problem. Early in the search it began to rain as if to make up for the eight-year drought. High winds twice knocked down the navigation shore antennas the task force had erected. A 70-knot mistral forced *Alvin* to remain moored with two pilots for twenty-one hours in the heaving, nauseating surface waves.

There had been several close calls while retrieving *Alvin* in bad weather, the worst on February 23. Seas were running too high for *Alvin* to maneuver into the LSD without banging the sides of the well, so sub and passengers were lifted by crane onto the ship. Several people, including Hays, moved just in time to escape serious injury. Hays felt lucky that everyone was still alive and the submarine was intact. He hadn't wanted to launch in the first place. If he had followed his own good sense, this wouldn't have happened; and he hated to think what would have happened if he hadn't cut short the dive. He sent a message to the flagship, saying he would not allow any more transfers of his men or submarine in rough weather. The response was unsigned:

> I want a full written explanation as to why Alvin's dive on 23 February was only 2 hours and 55 minutes. Certainly sea conditions did not provide another excuse for such a short dive. We have very sizeable resources tied up.... Our progress has been almost nil with Alvin and Aluminaut. When can I expect daily dives, weather permitting, of reasonable operational duration!

Things had indeed gotten off to a slow start. The *Mizar* with her critical tracking gear had only arrived February 19, the new charts had just been completed, and a storm had knocked out the new navigation system.

The eleven men of the Woods Hole team were giving it all they had, working twelve and fifteen hours a day, every day. George Broderson found Skip Marquet asleep standing up on the deck of the LSD, his arms around his ever-present clipboard. Broderson gently led the young engineer to his bunk. Exhaustion overtook many more.

Alvin had never worked so hard or so well. Rainnie wrote in his log for the week ending February 28:

> ALVIN now commenced a series of five consecutive dives that breaks all records, even in shallow water and exceeded even our fondest hopes for reliable performance.

Rear Admiral Guest wasn't used to having his orders questioned, least of all disobeyed. He had already yelled at *Aluminaut*'s chief pilot Art Markel for straying out of his assigned square to follow a possible lead. At a flagship briefing following the February 23 message exchange with Hays, Guest erupted again, this time at the WHOI physicist.

Alvin would resume diving in the morning, Guest said. That was all.

Weather permitting, Hays added. He would not endanger the lives of his men.

When I tell you to dive, Guest exploded, *you dive!*

The navy would not give the sub pilots any details about Nuke 4 because they lacked secret clearance. After another frustrated briefing aboard the flagship, an Air Force sergeant took McCamis and Val Wilson aside. "You fellas want to see what that bomb looks like?" he said. "Come on." They followed him to another cabin and were shown a photograph of the bomb. The sergeant didn't say who he was and they didn't see him again.

As they had expected, Nuke 4 looked like a large missile. The bomb was unarmed, the pilots were told, but even if it could not explode with radioactive force, Nuke 4 held lots of TNT and could explode in a more conventional manner. If its casing had cracked, it could be leaking plutonium. The pilots were advised repeatedly not to touch Nuke 4—that is, if they ever found it.

They didn't want to touch it or the parachute which presumably was still attached. If the chute engulfed *Alvin* from above, the emergency hull release would be useless.

The pilots crept through the murky depths, straining to see the grey netting big enough to wrap up two *Alvin*s, and the big can that might obliterate them with TNT or shorten their lives with radioactivity. And they shivered in their own sweat, knowing that they could search every inch of ALFA I for ten years and never see their prey, and it could be right there, near enough to touch but not see in the confounded mud-filled night.

Sometimes *Alvin* suddenly sank in the mush beneath, covering its lights. Rainnie's big body ached from the tension and the contortions of being in a seven-foot sphere. In their long years of riding submarines, neither man had ever smelled his own death so near as there in the Mediterranean.

The steering rod was too delicate and sensitive for Rainnie's large hands and he knew it. His wedding ring was size twelve and a half. And on one of the dives, instead of holding the joystick with a finger and thumb, his whole fist was wrapped around it, and his fist did not obey his brain and what his eyes saw; he drove *Alvin* into the side of a cliff. The impact, which didn't seriously damage the sub, brought a torrent of sediment, and in the awful minutes it took the mud to settle, the terrified pilots thought they were being buried alive. Rainnie couldn't stop his hands from trembling. He handed the joystick to McCamis.

At the end of *Alvin*'s tenth Mediterranean dive on March 1, the submarine crossed a track and pieces of eggshell at 2600 feet.

"Wait a minute, I see something," Rainnie said.

"What?" Wilson asked.

"I'm not sure, a little to the left, that's it, no, dammit, you went over it, to the *right*."

"What?"

"To the *right*, dammit! That's it, right on target."

"What is it?"

It was nothing, Rainnie said, just pieces of eggshell and another track left by a trawler.

But it wasn't just another track. They didn't know until they examined *Alvin*'s photographs late that night that this track was wider and deeper than the others.

It took seven more dives to relocate the track sprinkled with eggshell but when they at last spotted it, they couldn't follow it because *Alvin*'s power was nearly exhausted. But they knew they were hot. McCamis was positive; he had hunted in Indiana in the winter, he said, and he knew how to read tracks in the snow, and this ooze at the bottom of the Mediterranean was no different.

On the next dive he and Rainnie looked for almost seven hours, but they couldn't find the track or any eggshells. They cursed their crude navigation and the flagship which assigned them to another square of seafloor away from the clue, the first and only clue.

Despite the official silence from the United States and Spain, *La Bomba* was discussed openly on the beach, which was swarming with American soldiers in white coveralls and face masks. Rumors flourished. Out-of-town markets refused to buy Palomares tomatoes because they were *radioactivos*, and at least one villager said he was thrown out of a cafe in a neighboring town for being *radioactivo*.

Reporters set up radios ashore to eavesdrop on the busy communications of the classified sea search, and the navy was sure that the Russian trawler was doing the same. When the flagship realized it was being spied on, voice communications were kept to a minimum and the submariners spoke mostly in code, using for certain information colors and Peanuts comic strip characters. The monitoring press and Soviet trawler now heard: "Snoopy purple." "Charlie black." "Linus red."

The Woods Hole Oceanographic Institution was as anxious as the rest of the world for news of the task force. In early March, Hays called to request parts for the sub and report that Broderson had developed a hernia and might need surgery. All was going well, except that they hadn't found what they were looking for, Hays said cryptically.

Was everyone behaving?

Yes, they had to; there was no liberty.

Where were they?

Hays couldn't say; it was classified. The communications patch was made via a radio station in Philadelphia.

But of course Paul Fye and his aides and everyone else knew, they were on *Alvin*'s LSD *Fort Snelling* somewhere in the Mediterranean looking for the unmentionable.

The flagship prepared for what it considered inevitable and began to plan how to tell the world that Nuke 4 would never be found. They would talk to the press and the United States Ambassador to Spain would take a quick swim in the cold Mediterranean to prove there was no danger from radioactivity. But for the time being, the submersibles would keep diving.

On March 9, Fye received a letter from John Bruce, a WHOI physical oceanographer helping the flagship analyze currents.

Dear Paul,

...nothing spectacular has happened.... The Navy has divided an area (approx 12 mi × 10 mi) from the shore out in 1000 yd squares and assigned daily search jobs.... They get daily reports...put [them] on a big board in the operations room aboard Boston (as one might plot battle maneuvers) and have a statistician here give each a "grade" (probability of having found the bomb if it had been there, etc.).

How the "grade" is determined is somewhat involved and a bit of a mystery to me since the [sonar] sweep paths are functions of so many variables not well known (roughness of bottom, clarity of water, softness of bottom, how well sonar is working, etc.) but they are more or less lumped together and a probability factor then marked in each square....

The Navy people here seem quite impressed by ALVIN's and her crew's performance. She seems better suited for this job than ALUMINAUT since she is more maneuverable in the tight spaces and can adjust her trim easier....

The press arrived yesterday for the ambassador's swim off the Palomares beach and today came aboard the ships for a look around and a Navy lecture.

Sincerely,
John

Bruce believed the fishermen, as did *Alvin*'s pilots, although the navy's mathematicians calculated that there was only a 66 percent chance that the H-bomb fell where the fishermen said it fell. But now at least one officer on the flagship had come around to agreeing with them. Lieutenant Commander Brad Mooney changed his mind about the fishermen's fix after seeing photographs of the track strewn with eggshell fragments in ALFA I. Mooney spent days trying to convince Guest to let *Alvin* dive again in that area. Guest, exasperated, finally agreed to "just one more dive."

It was *Alvin*'s nineteenth dive in the Mediterranean. It was March 15.

Rainnie and Hays were lunching on the minesweeper *Ability*, whose cook had earned a reputation for being the best of the entire fleet, when a sailor interrupted them.

"Sir, *Alvin* wants to talk to you."

Alvin was in deep water with McCamis, light finger and thumb against the joystick, gliding over bits of eggshell. Wilson and one of *Alvin*'s electricians, Art Bartlett, were on their knees straining to see and occasionally bumping one another.

"This looks damned familiar, man," Wilson said.

"What the hell is that?"

"You've got it right in front of you now. That's the track, that's it, that's the son of a bitch, I'm sure it is, it's the same one. Don't touch down and get it stirred up so we can't see."

"Yeah, that's it. OK, I'm going in. Better pitch 'em real fast," McCamis said to Wilson who was taking pictures.

"I'm clickin'. I don't know what it is, Mac, but there's two tracks that come down here and converge."

"OK, OK, snap pictures."

"If we hit the son of a bitch, we'll be in great shape. I want to make sure this is it. It sure looks like it."

"That's it, that's it all right."

"Echo, this is *Alvin*. We have found the track."

"I told you I saw the son of a bitch as soon as I saw the bottom."

"That's right."

"Goddamn, it's steep... they told me it was a 10 degree slope!"

Alvin was creeping down what McCamis judged to be a 70 degree slope, almost vertical. The big aft propeller kept hitting the side of the mountain as they edged down, and the slightest touch stirred up the sediment and forced them to stop until the water cleared. Topside was anxious to talk.

"You just tell him to wait. Tell him we're taking soundings. Wait!"

"Echo, we're taking soundings. Wait."

"We're going down, we're going down fast...."

"OK, I'm going to dive down now. Hold on to your hat."

Clouds of sediment swirled.

"Goddamn, I've lost the stinking bottom. Christ. What happened to the bottom?

"What's the depth?" a weary voice asked.

"Two-four-five-zero."

It was Art Bartlett's first dive. He had spent a good deal of time trying to get comfortable. For the descent, he sat with his back up against the front of the sub, facing the stern where there was no porthole. It was fine for the ride down; to conserve battery power, they kept the lights off, and after the first 500 feet, Bartlett couldn't see anything anyway; it was like sinking through a barrel of tar, he said. But once they touched bottom and turned on the lights, he had to get another spot with a view. He perched on a boat cushion by the bottom viewport, which angled forward slightly. McCamis sat with his feet against the cold hull before the center porthole. Wilson was at a side window. For the next twenty minutes, they looked for the track with eggshells.

"I see something...."

"The track?"

"Eyup, that's the baby...."

"Ouch!"

"I can see the track! I can still see the track. It's swinging to the left. It's swinging south now. The track is...."

"I can't come too far, I'm running right into the fucking slope...."

The slope was too steep to follow and at the same time keep the track in sight, so McCamis drove backward, wagging slowly from side to side, down into the dark ravine. He was driving blind, fully aware of the risk of backing into the parachute. He kept asking Wilson and Bartlett what they saw. But his passengers had no view astern either; they were watching the track.

"You're coming right into a deep hole...."

"Yeah, I know...got to do something about this rudder.... How'm I doing now?"

"It's on the right hand side now...no, no, no, you've got to back straight up to get on it...back up like a son of a bitch, right rudder down some.... OK, now drive down...."

"It's on that side now."

"I can't be sure."

"I tell you I can see it."

"Coming down...coming down."

"About two inches you're going to hit...."

"That looks like a parachute!" Bartlett shouted. "A chute that's partly billowing."

"Could be!"

"Open up with the pictures...."

The electrician scrunched his nose against the bottom porthole. "It's right underneath me."

"What is it?" Wilson knelt near him but the window was too small for two pairs of eyes.

"You know what it looks like, it looks like...get that fucking squid out of the way. You're spitting all over me! I can't see a thing.... That's *it!*"

"That's it! I've seen a lot of parachutes and this is a big son of a bitch."

"What a big bastard!"

"That's where the stinking bomb went! Down this gully."

"No, it's going to be under the chute, I think," McCamis said. "Let's go up and take a look."

"I bet it's gone down the gully."

"No. Let's look under the chute first."

"Something sure as hell has fallen down into this gully."

McCamis edged closer, mindful, as were his passengers, that the current could push them into the parachute.

"It's an awful big chute, isn't it."

"It sure is."

"I wish I could reach out this window and pick it up."

Topside signaled again.

"This is *Alvin*, don't bother us now, Bill, we'll call you back."

"Take a good look at this one, over the edge. This is where I think the bomb is right here ahead of you."

"Can you see it?"

"There's something under there."

"See this wreckage right here?"

"Right here?"

"You know what that is? That's a fin! Mac, that's what a fin looks like on a bomb." Wilson grabbed the phone. "Echo, this is *Alvin*. Bill, get as good a position on us as you possibly can. I think we got a big rusty nail down here," he said, confusing Hays' code, bent nail, and forgetting altogether the flagship's code, kitchen sink. But protocol didn't matter after he excitedly delivered the rest of the message. "We found a parachute and we believe we have a fin of the bomb in sight. It's underneath the parachute."

"Ro*ger!*"

Mooney ordered *Aluminaut* to relieve the midget submarine at 2550 feet and bring to the scene a pinger which emitted constant pulses of sound that *Mizar* could track to within about 130 feet. The advantage the Reynolds submarine had over *Alvin* was endurance. It could remain submerged for thirty-two hours, while *Alvin* had only about eight hours of power.

To *Alvin*, Mooney said: "Don't move."

The three men sat quietly, listening to the whirring blower that was pushing the air through the canister of lithium hydroxide to clean up the carbon dioxide in the passenger sphere. Suddenly they realized that the current had moved them. They were disoriented and the parachute was out of sight. They were too depressed to answer Mooney's plaintive: "Why did you move?"

"Jesus Christ. . . ."

"Where in hell is it?"

"Ouch!"

"Oh, pardon me."

"I *saw* it."

"Yeah but it's just a chute."

"I haven't the slightest idea now where the son of a bitch is."

"We're too deep...."

McCamis crept down, down. "I've got the bastard again, I think... is that it right in front of us? Shoot it.... I'm going to catch that son of a bitch.... Take pictures hard as you can, take pictures."

"I'm moving lots of film through that camera, Mac."

"I see some blue...."

"I see something blue too...."

"It's a blue and gold insignia of some kind...."

"This might be the very fucking nose of the thing. *You're going to set down on it in just a second....*"

They had talked about what to do if the sub got snagged on the parachute straps. Short of releasing the passenger sphere, McCamis said they would dump the batteries and drag up the bomb with them. That idea was based on his assumption that Nuke 4 weighed much less than it did. He knew it might not work, maybe, it was complicated. His shirt was soaked through; he could feel the perspiration on the back of his neck. He couldn't back away yet, not yet.

"Look down there and see if you can read that insignia."

"I'd rather go up and see the other end of it."

"God, this muddling rudder.... That has got to be the son of a bitch," McCamis said, looking at the sonar screen. "We're losing it. *Oh damn it, we're going to murder it!*"

The submarine brushed against the parachute and quickly backed off. "There's nothing I can do with this fucking rudder," McCamis said, as if by way of apology.

"OK, park it, Chrrrist."

"Ouch!"

"Glad we got out of there!"

"I can see the bomb nose. That's it right at the end, yes, that's the bomb, no doubt about it...."

They saw a portion of parachute which seemed to be wrapped around a large object hidden entirely except for a section of metal with holes in it. It looked to them like a bomb rack, what Wilson had thought was a fin. Parachute straps undulated like cobras.

"I've got to get out of here before I tangle up in that stuff. Echo, this is *Alvin*, how do you read me?" McCamis said.

"I read you loud but not clear."

"I think we have enough identification. We'd like to skip clear of this area. There's several straps hanging down loose. There isn't any doubt in our minds about what we see. It's wrapped in the chute but part of it shows. The thing is still lodged on a very steep slope...."

"*Alvin,* this is an A-one job, outstanding."

"Thank you."

McCamis wedged *Alvin* sideways into a crevice just below the bomb so the submarine would not drift with the current, and there they waited the long hours for *Aluminaut,* most of the time with the lights off to conserve power.

The three men were smokers and their craving got the better of them, despite their knowledge of the rules—smoking in the sub was prohibited—and the possible consequences—asphyxiating or starting a fire and incinerating everyone and everything inside the sphere. There was plenty of time to rationalize and figure out how to get away with it. The oxygen and carbon dioxide monitors were not entirely reliable anyway and sampled the air only intermittently; the pilots had come to gauge the air quality by the frequency of their headaches. When they had a headache, they turned up the oxygen.

Before lighting up, they fully opened the oxygen valve for a long burst, shut it off, and set the blower on high. A cigarette was passed like an illicit joint of dope from one man to the next and then it was carefully extinguished. Someone had an envelope; the butt went in there and into one of their pockets.

At 6:44 P.M.—almost seven hours after *Alvin* left the surface—faint light glimmered in the eternal night. *Aluminaut*'s lights flooded the craggy moonscape, and for the first time, the men in *Alvin* had a panoramic view. Nuke 4, much longer than half a man, was partially sunk in the soft mud; its parachute stretched tightly over the down slope end. Beyond was the blackness of a cliff drop-off. The antagonist lay bested, stripped of its darkness by the lights of a long red monster from another planet. It was like a dramatic final act. But this was not make-believe. The Reynolds sub hovered overhead in the world's first rendezvous with another vessel in inner space.

"OK, boys," *Aluminaut* said, "turn on your lights."

The six-man Reynolds crew had wanted this victory and their boss, J. Louis Reynolds, wanted it even more. Of course, they were glad the bomb was found, but still the mood was bittersweet.

The subs, shimmering with phosphorescence, moved in a slow, awkward pas de deux.

"OK, that's far enough." McCamis was out of his trance.

Guest's voice reminded them not to touch that bomb. *Aluminaut*'s pilot, Art Markel, still smarting from the Admiral's rebuke for disobeying orders, silently walked his tall 200-pound body to the stern to urinate. As he did, *Aluminaut*'s rear end dipped with his weight, to the concerned shouts of the three men underneath who were squinting out the two-inch window in *Alvin*'s hatch.

"Hey! What the hell do you think you're doin'!" McCamis yelled.

Markel replied calmly that a whole five feet separated them. It did not appear so from *Alvin*. All McCamis could see was *Aluminaut*'s big belly. He pushed the joystick and positioned *Alvin* above the Reynolds craft.

The relief and elation that followed *Alvin*'s longest dive of ten hours and twenty-three minutes deteriorated quickly during the debriefing. Rear Admiral Guest wasn't convinced. The photographs showed a parachute. It was tapered at the one end showing in the pictures, yes; but still it was all parachute. The camera did not see all the pilots saw, and Guest did not believe the pilots.

The session ended in argument, with McCamis insisting that it was the bomb, and Guest replying, "But how do you know it's not a parachute full of mud?"

"You can't photograph something as well as you can see it," McCamis said. "Plus, what else is going to be down there with a parachute and a bomb rack hanging on to it?"

Guest ordered routine search operations to continue.

But news like that would not be quashed. It took only two days to make the *Washington Post*.

U.S. Midget Sub Finds Missing H-Bomb Off Spain

REAR GUARD SENDS CONGRATULATIONS TO ALVIN
GROUP. WE ARE PROUD OF YOU.

> FYE
> WHOI

PLEASE CONVEY MY CONGRATULATIONS TO "MY BOYS"
ON ALVIN SO PROUD I COULD BUST

> MAY [Reese]
> KETCHIKAN ALASKA

Alvin?

Alvin, the *New York Times* explained, was "a curiously shaped midget submarine, [that] somewhat resembles a chewed-off cigar with a helmet."

The following day *Alvin* took down a nylon line, one end tied to a big cleat on *Mizar*, and the other end affixed to a fluked stake, and stuck it in the mud near the bomb. But *Mizar* could not keep still in the rolling waves and the stake pulled free.

Carrying pingers and two new lines with different grapnels,

Alvin tried again. *Mizar*'s winch turned and in moments the line snapped. The same day, the USS *Boyce* departed with 4500 drums of contaminated Palomares dirt to be buried beneath the topsoil of South Carolina.

It was March 24.

"Tonight," a *New York Times* reporter wrote, "shrouded in the grayish parachute that clings to it tightly as a wet dress clings to a woman, the bomb still lay on the side of a steep slope...as submariners gently tried once more to clamp a line around it."

But in truth nobody knew where the H-bomb was. When *Alvin* descended again, Nuke 4 was nowhere to be found. The subs were assigned to search upslope where the flagship team, based on the presence of a new uphill track, thought the bomb had gone. The pilots obeyed but unhappily. How could the bomb fall up? For five days they looked uphill and grumbled about what nonsense it was. Finally on April 2 Guest relented and allowed *Alvin* to go down the slope. At 2800 feet, McCamis saw it. The bomb had rolled about 120 yards from its original resting place.

"There it is," McCamis said quietly.

"What is it? What do you see?" Rainnie asked.

"There it is right there."

"Well, what? What's there?"

"*It's* there." McCamis moved so Rainnie could look out the center porthole. "This is *Alvin*," he said calmly, and pronounced the new secret code. "Benthosaurus."

Now, the goal was to pluck the chute off the bomb and dig the grapnels into the parachute straps. The idea was to get close enough to get at the chute, but not too close. It was complicated, extremely dangerous and, McCamis finally decided, impossible. Hovering above Nuke 4, *Alvin*'s pincers grasped a piece of parachute, but when the arm pulled, the whole submarine moved. It was like being in the zero gravity of outer space. So, McCamis set down *Alvin* right on top of the bomb. (You *what*? Rainnie later asked. I *had* to, McCamis replied.) And little by little, McCamis and Wilson painstakingly plucked off the entire parachute, laying it in long folds on the slope. Then they dug in the fangs of a grapnel. The line to the other grapnel was hopelessly tangled.

This time the navy sent for CURV, Cable Controlled Underwater Research Vehicle, a remote sled designed to recover spent torpedoes. CURV had TV cameras, lights and a big claw, and CURV carried no passengers, which was a relief to everyone.

CURV's crude claw dug a grapnel into the parachute, and *Alvin* dived to confirm it. Topside gave the line a stern yank, hoping to expose the hidden bomb for CURV's cameras. The yank stirred up a cloud of sediment that engulfed *Alvin*, and all Wilson and

McCamis saw was parachute billowing like a circus tent. Wilson alerted surface control who heard the fear in his voice and thought the worst. But before he shouted the last of his message, *Alvin* was clear of the parachute. McCamis had moved *Alvin* quickly enough and in the right direction.

CURV was lowered again to dig in the second grapnel. But when *Alvin* descended to investigate, Nuke 4 had disappeared again. The following day *Alvin* dived and found the bomb—it had rolled down another 300 feet to the edge of a cliff.

It was after midnight when CURV was repositioned and began to dig in the third grapnel. Mooney told Guest they were ready to take the next step and haul up the bomb. They did not have much choice; the life of CURV's motors was about up.

"Cut it loose," Guest said.

"Sir?"

"Cut it loose."

Nobody else within earshot thought he had heard Guest right either. Surely, his order was not to sever the lines attached to the object that had been the excruciating focus of their lives for the past 75 days.

It was. Guest was unwilling to risk another lift. The chance of the lines breaking and Nuke 4 rolling away again was too much for him. This way, he said, he at least knew where the bomb was. *Cut it loose!*

Mooney followed him into his cabin. "Sir, this is dumb as hell."

Rear admiral and lieutenant commander argued until 4:00 A.M. Just before dawn, Guest gave in. They would try Mooney's new plan.

Mooney ordered surface control to drive CURV, with every bit of its little remaining strength, into the parachute. The six-foot-long tethered frame whirred to a final burst of life, entangling itself in parachute straps. Its small motors sucked in parachute and then it stopped for good. If Nuke 4 got away again, it would take CURV and its navigational pingers with it.

At 7 A.M. April 7, *Alvin* hovered at a safe depth of 1925 feet and the topside winches started to turn. In an hour the whole sassy package was at the surface—CURV, the parachute and Nuke 4, only slightly dented from its 30,000-foot plummet through the sky.

It was over.

The search cost the United States some $12 million. For the services of *Aluminaut*, the government was charged $304,000.

Alvin was a hero. Woods Hole played down the dangers and denied any close calls. Rainnie did not want to attract any unnecessary attention from the navy. He told inquiring reporters that their mission had been a lot safer than flying or driving down the mid–Cape Cod highway on a summer Friday.

The Alvin Group's new secretary Charlotte Muzzey posted the clippings and the letters of good wishes behind her desk at the Drugstore. Secretary of the Navy Paul Nitze telephoned and sent a telegram.

THE RESULT OF YOUR EFFORTS TO RECOVER THE MISSING UNARMED NUCLEAR WEAPON IS A SOURCE OF PRIDE TO ALL OF US IN THE NAVAL SERVICE. WITH COMPETENT PROFESSIONALISM THE PERSONNEL OF YOUR TASK FORCE CONDUCTED AN INTENSIVE SEARCH UNPARALLELED IN HISTORY.... WELL DONE.

Rear Admiral Guest's telegram arrived a week later.

... IT HAS BEEN A RARE PRIVILEGE FOR ME TO HAVE SERVED WITH MEN, MILITARY AND CIVILIAN, WITH SUCH SINGLNESS OF PURPOSE AND SUCH SUPERB MOTIVATION TOWARD ENHANCEMENT OF THE INTERESTS AND PRESTIGE OF OUR COUNTRY....

Dear Earl
The ... Houston Chamber of Commerce has honored me by electing me the Chairman of the newly created Oceanographic Section. How about that for an iron bender?... They want a story about ALVIN's spectacular achievements in helping to locate that pesky bomb....

Larry Megow

Mr. Speaker
... I believe it is important to again call attention to this nation's woeful lack of preparedness.... It comes as a surprise to many of us that in this day of manned space exploration, the U.S. does not have a capability of exploring the ocean's bottom; that while we are spending billions each year to send men to the moon, we have developed only a handful of research vehicles for penetrating the ocean and even these are in the model T stage of development....

Rep. Paul Rogers, Florida

Everyone was calling, writing or wiring, everyone but the President of the United States. Surely, Charlotte Muzzey thought, it wasn't too much to expect a simple, thoughtful thank-you. But no call or telegram came from Lyndon Baines Johnson.

April 14, 1966

Dear Mr. President

Perhaps you might consider me prejudiced, being the secretary of the Deep Submergence Research Vehicle Group of the Woods Hole Oceanographic Institution, but the culmination of the operations at Palomares, Spain to recover the missing nuclear bomb was due in great part to the little submersible ALVIN. The men who operate this small craft are dedicated and hard-working and when called upon to take part in this Task Force 65S complied with the request.

There is always a great deal of honor given to those researching outer space and I feel credit should be given to these men who helped our nation in 'inner space'. Too many people forget the benefits from the seas.

I think it would have been nice if the White House had recognized these men with a congratulatory telegram.

Very truly yours,
(Mrs.) Charlotte A. Muzzey.

Two weeks later a letter arrived from an assistant secretary of the navy.

Dear Mrs. Muzzey

The President has asked me to reply to your letter....

I appreciate your expression of interest in insuring that proper recognition is extended to the crew of the ALVIN for their contributions to the successful recovery of the nuclear device submerged off the coast of Spain. You may be interested to learn that the authorities in the Department of the Navy are currently reviewing the contribution made by all personnel involved in the recovery operation, and it is expected that appropriate recognition for individual efforts will be forthcoming in the near future.

Thank you for bringing the matter to the President's attention.

Sincerely,
Charles F. Baird.

10

As deck senior, bo's'un, honcho, crew chief, whatever... I attempt to keep the crew happy. So long as they get a kind word, they put out their best effort. I take the check-off list and I sign it. That means I have to trust each individual plus the fact that I double-check it anyway. Makes it a little bit more safe.... It's not exactly dangerous, but it can become hairy. It takes a combined effort to launch and recover the submarine. And the sun beats down, the hours go on.

You prepare the sub for a dive and you know when it's all through, you're going to have to postdive the sub because it cannot stand not being taken care of.... A prototype. Even the cat is a prototype. In fact we're all prototypes. I mean, who could actually have a background in this field?

Alvin sort of devours us. When we come into a port we look like a cult. Here we are, all around this little white character sitting in the middle of the deck, like we're paying obeisance to it, a rather big fat nasty child. It's perverse.

As soon as you've got everything cornered, it finds something of its own. Little fat devil. I swear sometimes I think it's trying to defeat us. It sits there in all its white elegance trying to think of something new, a little oily leak, a little air leak, something, just when you think you've got it licked.

When people see it for the first time, they're sort of let down. See, they have this feeling it should be a long black sleek thing. Instead they see what looks like a big white toilet and they say is this the submarine?

Yeah. Yeah, the thing has a grip on us. You don't want to yell quits because you know that you can lick whatever it is, and in doing so, you put yourself in a special class, you know you've done something important.

Basically we're an adventure outfit. Takes some of the sting out of the long hours.

George Broderson

Broderson didn't want to go to the military hospital in Torrejon but there it was, the offending left side of his belly that bulged as he was tending to *Alvin*, and the navy physician predicting strangulation and blood poisoning. It was bad enough not being there when they found the bomb. He would just as soon have had the surgery done on the *Fort Snelling*. They had male nurses at that military hospital. Of course he didn't need to recuperate, but he went home to Cape Cod and obeyed. He began to feel much better about his unfortunate twist of fate when WHOI asked him to return to the Mediterranean to accompany *Alvin* home on its trans-Atlantic crossing; the rest of the group was home or en route by air.

Before the hernia, Broderson had worked as hard as any of them: prying off the taffylike mud from the submarine and washing it down with fresh water after each dive; recharging batteries; sponging out the water that collected from their sweat and condensation in the bottom of the passenger sphere—after some dives they could fill a bucket; putting in a fresh carbon dioxide scrubber every morning. He was a good mechanic and he was good with people.

Some men in naval uniform boarded *Alvin*'s LSD in the Mediterranean and demanded: "Who's in charge here?" Hays, who was chewing tobacco as he spliced line, sent a stream of spittle toward Broderson and replied: "He is."

Not one to miss a cue, Broderson delivered his line as if he had rehearsed the scene. "Sirs, what can I do for you?"

The *Aluminaut* crew came by, too. As Broderson remembered it: "Those *Aluminaut* guys were rather obnoxious until they needed my spare compressor." The Woods Hole team felt a smug sense of satisfaction that the Reynolds crew needed something from *Alvin*; they couldn't have imagined that those roles soon would be reversed and they would desperately need *Aluminaut*.

Following the mission in Spain, Broderson was made crew chief. His purview would extend beyond, to peacemaker, safety officer, teacher and all-around man in charge of *his* submarine.

That itchy woolen red beret precisely angled over his balding head, Broderson and DSRV *Alvin* were met in May 1966 at Boston harbor by a spiffied-up *Lulu*, which would be home for Broderson and *Alvin*, the mother ship's longest staying tenants, for the next eighteen years.

Lulu now had a permanent bridge house, still at the stern, and over it were a tower and antennas. The starboard hull had a new air conditioner and more bunks with small lockers beside each. A 40-foot house trailer on the foredeck held the galley, mess and a head. The aft arch stored fuel, while the forward arch carried fresh water for the sub's daily wash down. But most surprising, *Lulu* was alone; she came on her own power. Two more of the junkyard outboards had been added to the stern of each pontoon, doubling the tender's calm water speed of 3 knots.

Paul Fye felt a foggy déjà vu at hearing the catamaran called *Lulu*. He wrote Hays and Rainnie:

> It is customary to submit suggested names for the research ships to the Executive Committee [of the WHOI Trustees] for approval. My own reaction to *Lulu* is that it does not connote proper scientific "flavor."
>
> For your consideration I would suggest Benthosaurus or Nematognathi (catfish).

But they were serious about *Lulu*. It was Al Vine's mother's name. He didn't want to hurt her feelings, did he?

Vine confirmed that Lulu was indeed his mother's name. *Lulu* seemed reasonable to him, he said. There were two pontoons. Lu Lu.

Fye did not feel his surrender nigh when he left the giggling scientist. WHOI's port office assiduously avoided using the name *Lulu*, preferring instead *Deep Submergence Research Vehicle Tender No. 1*, or simply "the catamaran."

The Alvin Group added little *Alvin*-shaped aluminum pins to their red berets, and following a quick change of batteries and other freshening of the sub, they were ready to get on with what remained of the abbreviated 1966 diving season.

Lulu with *Alvin* on center deck left Woods Hole for Bermuda in late June. The tow ship *Narragansett* cruised at her side. Midway, *Lulu* took on fuel from the bigger ship but made it untethered the whole stormy way. The mother ship did not have to exert herself so. WHOI had to pay $3350 a week for the tow ship whether it towed or not. But the Alvin Group wanted to prove that their *Lulu* could make it on her own power—piddling though it was; with a top calm water speed of 6 knots, it took six days to reach Bermuda.

Most of *Alvin*'s commitments were for the navy, beginning with an inspection of the listening array ARTEMIS, Swede Momsen's five-year-old justification.

Before leaving Bermudan waters, *Alvin* made the first deep science dive with Robert Hessler, a thirty-three-year-old WHOI biologist. Off and on for two years, Hessler and his colleagues had lowered dredges from ships off Bermuda to capture the creatures of the deep sea. While eight dredges came back full, many more were empty or full of rocks, and some dredges never came back.

The experience was so frustrating that the scientists chose a new site, a soft bottom off Cape Cod.

On his dive to 5850 feet on July 17, Hessler realized that they had been lucky to get even eight dredges full. "I was awed by the tremendous vertical precipices, and I finally understood why we had so much difficulty ever taking any samples from that area," he said. "That dive really taught me something. From then on whenever I lowered a dredge into the ocean, I could close my eyes and picture what the bottom of the deep sea looked like."

Off the Bahamas, navy divers inspected another underwater array, and in the Tongue of the Ocean, officers from the Naval Oceanographic Office made fifteen dives to evaluate *Alvin*'s "usefulness." They tried to drive *Alvin* in a straight line and in circles, using as a navigational guide a pinger stuck in the bottom mud. *Alvin* stayed on course to within about five feet which was considered good, and was able to find and retrieve various items dropped from *Lulu*. The Naval Oceanographic Office was impressed. "There is no better way," the officers concluded, "to assess the bottom environment than that of direct viewing."

Alvin's crew did some experimenting, too; they attached to both sides of the sub small chunks of lead releasable from inside the sphere. By descending with enough weights, *Alvin* could sink without having to use battery power to pump ballast; to ascend, the pilot simply dropped the weights.

In late August, two more civilian scientists dived; the first was Allyn Vine who spoke into a tape recorder on his plunge to TOTO:

>We are on the bottom at 5850 feet... moving dead ahead to get into clear water.... Every now and then a big glob of this mud will come off oh! just saw a beautiful string of it going up, gives an idea of the circulation... clouds... like you're in an airplane flying around....
>
>Now here's about a five to six inch type that could sure pass for a small black bass. Tiger-striped daddy long legs about a foot long.... This is one of those jug-like sponges... like an extremely large football with a friar's beard....

John Schlee, a thirty-eight-year old geologist from the U.S. Geological Survey in Woods Hole, was next to climb the three handles on *Alvin*'s sail and swing his legs over. He swallowed hard when the hatch closed and assumed a fetal position by a side porthole. He was unnerved when the submarine began to jerk back and forth.

Pilot Val Wilson explained that the freeboard ballast valves, which had to open to let in heavy water, were stuck, and the two swimmers who escorted *Alvin* off the catamaran were rocking the sub to jog them free, that's all.

At least it worked, and soon *Alvin* disappeared into the darkness with two men and a wire milk crate carrying Schlee's tools: a pick, pry bar and quivers that held hollow tubes to take cores of mud. Schlee hardly ever used a pick, which was the traditional tool of the land geologist; he preferred a sledgehammer for breaking up the rocks dredged from the sea. But he understood as few of his colleagues did that the submarine was going to revolutionize the way oceanographers sampled the ocean. His unused pick secured to the milk crate under *Alvin*'s center porthole was a poignant sign of imminent change.

As they dropped at about a foot a second, Schlee felt no sensation of motion. He was aware only of the fading light and the steady climb of the depth gauge needle as they dropped gently through dustlike particles. It was an odd sensation because the tiny particles appeared to be falling up. The temperature in the sphere dropped from 78° at the surface to about 50° F and moisture formed. It was a damp cold, and Schlee was glad he brought a sweater.

An hour after leaving the surface, *Alvin*'s floodlights illuminated the bottom of TOTO 5550 feet down. TOTO was a huge canyon, really more of a valley that cut a U-shaped swath 40 miles wide and more than twice as long. Geologists didn't know how TOTO and the more conventional underwater canyons had formed. Some argued they were hewn by ancient rivers when the seafloor was above water. Others thought canyons were more recently formed by underwater avalanches. TOTO was thought to have been hewn during an Ice Age some two million years ago.

Schlee wrote of his dive:

> With all outside lights on, we both watched through the watery green haze for the first indications of bottom.... About 15 to 20 feet from the bottom, the water cleared and there before us was a pale olive-grey carpet of carbonate sediment....
>
> About 30 or 40 feet away from our initial landing site, the slope steepened perceptibly, and rising upward was a rocky precipice at least 50 feet high (we could not see the top). The cliff had bold, blocky and angular aspects similar to a quarry face or rugged mountainous terrain.... the whole scene could best be pictured by imagining a series of steep rocky alpine cliffs dusted by a freshly fallen powder snow and bathed in green moonlight....

Schlee was amazed that the sediment stuck to the nearly vertical slopes. What kept it from sliding off?

Wilson handed him the control box of twelve toggle switches

to operate *Alvin*'s arm: GRIP. WRIST ROTATE. ELBOW PIVOT. SHOULDER ROTATE.... It was a real trial. Schlee worked at it for half an hour. When he twisted the robotic hand, the whole submarine twisted with it. No one could do any better as long as *Alvin* hovered neutrally buoyant without a second arm to brace itself.

In an hour Wilson found a ledge and wedged *Alvin*'s basket of tools under it, and, using the pincers, dislodged a piece of limestone.

From his two dives, Schlee brought back three sediment cores and three pieces of rock, one of them dated at twelve million years, proving that the huge submerged valley was much older than previously thought.

On one dive in southern waters Mac McCamis asked a passenger, "How would you like to go deeper than anyone has ever been in a submarine?" The guest eagerly replied in the affirmative, and sworn to secrecy, dropped with McCamis to 7000 feet.

The former Indiana farmboy had a reputation for being a daredevil. Broderson, who called him Mad Mac, had watched him yank wire stitches from an old operation out of his stomach. McCamis had also ripped off a dog's nose with one hand. The dog was attacking his child, but if it had been a tiger, Broderson was sure, McCamis would have done the same. He also watched the pilot work at night in *Alvin* on *Lulu*, practicing the controls in the dark.

McCamis had never known anyone like Broderson. The few times they condescended to go out at night together in the Bahamas, the crew chief embarrassed McCamis. The pilot knew Broderson wanted to be an actor, and it was too bad he missed his calling because Broderson was damn good. McCamis didn't mind when Broderson put on his limey accent or pretended to be the hotel manager. What really bothered him was the death scene in which Broderson would strut into a crowded hotel lobby, collapse and pretend to die. And he did it well. Women would scream and someone would rush for a doctor. "Damn it, Brody," McCamis would whisper into the crew chief's ear, "what the hell do you think you're doin'?" And he came back to life just as well as he died.

The bond of reluctant admiration each man felt for the other was cinched by their mutual love, which bordered on obsession, of *Alvin*.

Alvin's supporters still consisted primarily of its crew. No one was lined up at Bill Rainnie's door to get on the schedule for next year's diving season. The Drugstore had sent memos announcing diving opportunities to WHOI scientists and the U.S. Geological

Survey in Woods Hole. Most of them needed to be reminded that there was a deep-diving research submarine right outside their laboratory windows and they could use it at no cost whatsoever. The Alvin Group would even provide them with the film for underwater photographs.

Only five of the eleven scientists on WHOI's 1960 proposal to Swede Momsen requesting funding for a research sub ever dived in *Alvin*. Three of the eleven said they hadn't given Vine permission to put their names on that proposal in the first place.

But the two-year-old submarine had starred in two big productions—the H-bomb search that was played out on a world stage, and a TV documentary, "Flying at the Bottom of the Sea," filmed by National Educational Television in the Bahamas. And the Drugstore, especially Charlotte Muzzey, felt vindicated that November 1966 when Bob Frosch, the new Assistant Secretary of the Navy for Research and Development, awarded Navy Meritorious Public Service Citations to Hays, Rainnie, McCamis, Wilson and Marquet for their participation in, as the award read: " . . . the search for and recovery of wreckage including a nuclear weapon. . . . "

Frosch was one of Fye's three peers back in 1962 when he wobbled on the edge of decision. "We know more about the moon's backside than the ocean's bottom," Frosch had said. "Of course it's a risk, but someone's got to take it."

11

There's several rattail fish that hang out in the deep.... I seen a jellyfish out there one day and it looked like an octopus. You get tremendous distortion.... Then I seen this monster or somethin'... I turned around sharply and it's gone. Kind of shook me up. Could it be possible? This was a living creature. I seen at least 40 or 50 foot of it. I swear to God.

Like the other day, a ten-foot worm. The first few feet, it looked like a kind of caterpillar with fuzz on it and the next few feet, it looked like a bunch of Ping-Pong balls glued together. And this thing was easing out, contracting. Some kind of worm, a real weird worm.

There was a five-foot fish at 6000 feet, it looked like a wahoo, eyes white like a blind horse. It swam around *Alvin* several times. Turns out it has teeth, like a barracuda.... This biologist, he told me all about prehistoric animals in the sea, and the amazing thing is I ask him what type of fish this is and he's as much in the dark as I am.

I seen no creature that would attack *Alvin*, other than squid, which I think are more excited by the light.... The large fish stay away.... Turns out these large fish don't like to be photographed. When you turn the lights on them they scamper away.

The scientists, they don't say a word. I've noticed, seems like all oceanographers, it takes them so long to find out what they want to know, maybe we can speed this up a little.

Mac McCamis

Life in the deep ocean. It was almost a contradiction. By now, scientists knew there was life down there but not much, and what existed succumbed to the Procrustean rule of survival of the fittest. The pressure, at many thousands of pounds per square inch, was fantastic; the temperature hovered just above freezing; and no sunlight ever penetrated to allow the nutrient-producing process of photosynthesis. And there was hardly any current in this vast graveyard for the dead and decayed remains of plants and animals that wafted down to the seafloor in an eternal shower of fine particles that Japanese scientists in the 1950s had dubbed "marine snow."

Scientists who were trying to learn about the deep sea were forced to work without benefit of the insight gleaned from simply being able to see something with their own eyes. Their knowledge was based on photographs and sonar and what fell into their traps, all blindly lowered into the sea. The pictures depicted a paucity of life, mostly diminutive. It was a rare photo that showed an animal as large as a shoe. Most creatures were smaller than bees, much too small to be identified from the photographs.

Not all oceanographers laughed at McCamis for what he said he saw, especially not those who dived in *Alvin*.

In late spring of 1967 *Alvin* began its first full season of diving for science. From the Bahamas, *Lulu* was scheduled to work her way north, picking up scientists along the way.

The trouble started as soon as *Alvin* and *Lulu* arrived in the Bahamas. The new hydraulic system that lifted the cradle did not work. The mother ship's starboard engine broke, and the new radar tower trembled so much that it seemed about to topple. The first mate quit and Rainnie fired the recently hired cook for "lack of experience." He hired a new first mate who quit three days later. By then only one air conditioner on *Lulu* was working. The crews' shoes were covered in green mold in the morning. Lacking the needed parts and facilities, they went to Miami.

ALVIN, THE SNOOPY SUBMARINE, ARRIVES WITH ITS UGLY MOTHER

Alvin is a cutie... *Lulu* is something else... a sea-going nightmare

Snug in its center deck cradle, *Alvin* got all the good press from the *Miami Herald*. The submarine looked slightly different. An angular brow of the buoyant syntactic foam had been added above the center porthole, increasing by about 600 pounds the carry-on weight allowance. Now, *Alvin* could dive for the first time with three people and the mechanical arm. The brow was also a handy and protected place to sink in lights and cameras. Hooked to a tiny monitor in the passenger sphere were two TV cameras, one looking forward and the other aft. The pilots could use the joystick or the new control box of toggle switches.

Teardown the winter of 1966–67 had been almost as invigorating as diving. Everyone, it seemed, had his own wish list, limited by only time and money. "In those years we were living and breathing *Alvin*," Skip Marquet said. "There was a surplus of things we wanted to make better. Since the textbook hadn't been written yet, it was a period of very high learning experiences. Whenever we went out, we learned things we didn't know the day before. It's what made it so exciting."

The aluminum ballast spheres had been replaced with fifteen titanium balls, each two feet across. Unlike most other metal, titanium does not corrode. Six of the spheres were for the oil ballast system, and the other nine were encased in syntactic foam.

That was all those nine balls did, just sit there and be buoyant. It occurred to the engineers that they really didn't need the two aluminum scuba tanks for the freeboard ballast system. They piped lines to two of the idle titanium spheres to store the air.

The big, heavy valve in the freeboard ballast system was gone. How McCamis hated that valve and his inability to keep it from sticking. He had brooded about it alone over a vodka collins at a fancy club in the Bahamas when a stranger approached him. "You look pretty depressed, fella," the man said. "I am," McCamis replied, and told him about *Alvin*'s sticky valve. The stranger said he helped develop a new kind of valve for the Navy's Terrier missile and he bet that was exactly what *Alvin* needed. With the stranger's help, *Alvin* got a Terrier missile valve for the freeboard ballast system. McCamis couldn't wait to try it out. *Alvin* was ready for action. It was the mother ship holding things up.

It wasn't until the beginning of July that *Lulu* was ready to pick up the scientists awaiting her in West Palm Beach and head for the first dive site—a plateau on the continental shelf, the wide ledge of seafloor along the coast. The geologists hoped to collect bits of rock and muddy earth and gain from these some insight into the origin of the shelf, and of course they wanted to do something they had never done before: see it.

The pilots took each scientist into *Alvin* before every dive to explain the emergency features—the switches to disconnect the arm and the sample basket, the fire extinguisher, scuba, flashlights, life vests. The T-wrench to release the passenger hull was in a box the pilot sat on at the center porthole. The rusty saw Rainnie carried in case the hull release shaft became stuck had been replaced with a drill. In an emergency they could use it to make a small hole to break the vacuum and open the hatch. Or so they reasoned.

Those wearing glasses were cautioned to pad the frames with tape so they wouldn't scratch a porthole. All were advised to go easy on the coffee and orange juice at breakfast; everyone was responsible for emptying his own Human Element Range Extender or HERE flask. The pilots had learned to shun all liquids before diving. Bag lunches were packed for each dive.

The idea was to put the passengers at ease. Almost everyone felt anxiety at first. As McCamis put it: "There was a lot of swallowing." One scientist wrote the name of his next of kin on the large blackboard with the day's dive assignments.

The pilots came to tailor their briefings. If they knew the passenger was a swallower, they would omit their spiel about what to do if the pilot had an unlikely heart attack. And McCamis would skip his tease of powering over a cliff drop-off; most people cringed, not realizing that they could not fall off a cliff. They were *flying*. The most dangerous part of the dive, the pilots said, was launch and retrieval; and in the deep sea, the biggest dangers were put there by humankind—mostly cables and thousands of tons of unexploded mines and other weapons.

Anxieties had a way of disappearing after the few moments of sinking through the turbulent surface to the tranquility beneath the top waves. And the sights.

The geologists diving to the continental shelf quickly became engrossed in the oceanic milieu that surrounded them. They saw that the supposed low bank of coral was really a series of huge troughs and ridges that rose higher than 150 feet. Although they dived no deeper than 2300 feet, they were amazed at the abundance of life and the swift current.

Frank Manheim, a thirty-six-year-old U.S. Geological Survey geochemist, was especially intrigued with the marine snow. Little was known about its distribution or ecological importance. For several years he had been filtering seawater and weighing the dried filters to determine how much marine snow there was. Under the microscope, he had found a lot of soot, fly ash, toxic pollutants and a peculiar processed cellulose, eventually identified as toilet paper.

On his dive, Manheim excitedly sketched the odd shapes raining by his window—large ragged particles mixed with tiny planktonic organisms, but mostly "unrecognizable aggregates, flocs or blobs," he wrote.

When he filtered the water collected from *Alvin* and weighed the filters, he realized that this method of quantifying marine snow was not accurate. He had seen a heavy snowfall but the weight of his filters indicated otherwise. In the mud cores he brought up, there was little organic material. This seemed to indicate that the animals were extremely adept at food gathering; they were consuming the fine rain of organic matter as soon as it hit bottom.

Lulu pushed up the coast for the next dive about 100 miles off Savannah, Georgia. It was WHOI geologist Rudy Zarudski's turn.

"Man your diving stations," Rainnie announced from the bridge at the stern. Everyone in the crew, even the cook, was needed for launch and retrieval. With Rainnie's call, crew chief

George Broderson looked to make sure the six line handlers were in position on the pontoons and signaled the pilot. "Have a good dive, Mad Mac."

McCamis stood in the sail as *Alvin* dropped on its cradle into the ocean and moved away from *Lulu*. Two swimmers detached the lines, made sure the hatch wasn't leaking and gave a thumbs up underwater to McCamis at the center porthole.

Zarudski looked at his watch. It was 1:40 P.M.

Rapid temperature fluctuations [he wrote].
 Fine "snow." Microscopic creatures darting about.... One sardine-like fish; small jellyfish.... Fluorescent ribbon passed the window; 8 inches long.... 1,985 ft. Landed on bottom....

It was 2:22 P.M. According to the meter in *Alvin*'s sail, the current was moving at a half knot, and from the orientation of the ripples on the seafloor, Zarudski judged it to be from the southeast.

Moving toward a lone branch of coral, 18 inches high. Photograph coral.

Then Zarudski heard a loud rasping, and thinking it was their skids scraping over rock, he looked through the bottom porthole. But the submarine was stationary.

"We've been hit by a fish!" Wilson shouted from the starboard porthole.

The fish was bigger than any of them and somehow captive, for it was thrashing violently. A piece of its flesh had ripped off its back.

It was a swordfish, bleeding profusely, and its entire sword was wedged into the gap between the sphere and *Alvin*'s fiberglass skin. Suddenly the leak detector light blinked on, indicating seawater in a line or battery. Electrical wires snaked around the passenger sphere; it was possible that the sword had cut one, although all systems seemed to be operating normally. The men in the submarine wanted to continue the dive.

Rules and policy were evolving with experience, and there was already a rule to govern this situation. If the surface controller, *Lulu*'s skipper or anyone in *Alvin* thought a dive should be aborted, it was. Disagreement was permitted but only after the sub was at the surface. Rainnie, the surface controller, ordered *Alvin* back home.

Throughout the ascent, the swordfish struggled and in one final convulsive wrench at the surface, ripped itself free of its own sword. The escort swimmers lashed it to *Alvin* and the fish arrived on *Lulu* in two pieces.

No ichthyologist or any other kind of biologist was on board to tell them about *Xiphidae gladius* or to appreciate its beauty even in death. The creature, which was eight feet long and weighed 200 pounds, was in a sorry state. At the enormous pressure change from 2000 feet to sea level, the swordfish erupted inside, spewing its innards onto the deck.

It took two hours to remove the foot and a half long sword which had missed a sheath of electric cables by a thumbnail. The scraping Zarudski heard was the sword against the steel passenger sphere. The leak alert light, it turned out, was unrelated to the swordfish attack; it indicated a slight leak in a battery box, not serious enough to abort a dive. If the sword had cut the electrical lines, the passengers would not have been in danger. What worried everyone was the porthole. The sword missed it by a few inches. Could a swordfish penetrate Plexiglas? A leak in the sphere under deep-sea pressure would be catastrophic. Would swordfish be a routine hazard? They didn't know.

There was plenty of swordfish to go around for supper which was easily the best meal ever served aboard *Lulu*. WHOI's front office was piqued to learn that its own oceanographers had eaten the "specimen."

In the following dives, off Virginia, other creatures of the ocean reacted to *Alvin* with the same offense as the swordfish. Squid squirted yellow ink and crabs faced *Alvin* with claws raised in battle stance. But a fat clam ran for its life on one stumpy foot. It startled WHOI geologist, K. O. Emery, and Bob Edwards, a biologist from the Bureau of Commercial Fisheries in Woods Hole. They had dropped to about 200 feet where the seafloor, highlighted by bluish sunlight, reminded them of a Salvador Dali painting. "There is nothing silver about silver hake in their natural environment," they reported. Silver hake are brown.

Then they saw a "sea monster," as they called it. McCamis was the first to spot it, undulating like a translucent python, its long body seemingly unending. He grabbed it with the mechanical arm and, loath to let go a sea monster dragged it along for more than an hour until Emery wanted to pick up rocks.

The sea monster, later identified from photographs, was a 25-foot-long chain of gelatinous creatures called salps, each not much bigger than a hand. Salps are a common marine creature, although nobody had ever heard of so many linking up. These were *Salpa vagina*, named for their shape.

Of all the research underway at WHOI, the physicists' work was the most theoretical, and most physicists could not imagine how *Alvin* could help them. But a few of them—Claes Rooth, his student John Van Leer, and a visiting scientist, John Stewart Turner—tried. These physical oceanographers specialized in fluid dynamics; they spent most of their time in the lab experimenting with the movement of fluid to understand the fundamental behavior of ocean currents. They knew that the ocean was structured like a giant layer cake. What differentiated the masses of water was different temperature and salinity. How much the layers mixed with one another was far less understood.

For his dive Van Leer built the "six shooter," which dropped pellets of dye from a revolving barrel, like the pistol. From *Alvin* he watched chartreuse fingers spread through several different layers of water, confirming what until then had been inferred—that mixing occurred on very fine scales. The experiment became the basis of Van Leer's Ph.D. thesis.

Bob Hessler hadn't planned on getting into deep water, and neither had Howard Sanders, the distinguished biologist who hired him in 1960. When Sanders began his career, he poked around for life in shallow water. In 1955 he found a tiny shrimplike creature no one had ever seen before; he called it Cephalocarida, the most primitive of all living crustaceans, what one could expect to have seen 600 million years ago. It was a missing evolutionary link. Sanders wanted Hessler for his expertise in ancient marine life.

But Hessler did little paleobiology at WHOI. He and Sanders followed their curiosity into deep water and the fledgling field of benthic or deep-sea biology. For seven years they lowered dredges and dragged sleds through the mud and over the rocks in the deep ocean, and spoonful by spoonful, counted and examined several thousand animals, most of them microscopic. By 1967 they had reached a startling conclusion: life in the deep sea was remarkably diverse. Life might not be abundant, but the deep held many more different kinds of animals than some shallow-water areas.

In late September Sanders and Hessler and two other WHOI biologists, George Hampson and Rudy Scheltema, packed their heavy sweaters and long underwear for a week-long expedition to their soft bottom sampling site about a hundred miles off Cape Cod. Their plan was to take photographs from *Alvin*, collect animals and above all, they were going to look—to finally see what they had studied for seven years in the vicarious fashion of blind sampling and educated guesswork.

The young Hessler and the senior scientist Sanders wrapped themselves in blankets and crouched at the side portholes, oblivious to the pain in their cramped legs.

"It was like sinking into a womb," Sanders remembered.

"There are so many things to see, all these things, new phenomena, and that is all that matters; you finally have no fear. I felt like I was in the middle of the Milky Way. We had no idea what the life was. The exciting thing was when you would go down in the submarine, you would be where no one had been."

They were ripe for surprise. There was much more life than they expected. Deep-water fish, unlike their shallow-water brethren, were lethargic; they moved a short distance and suddenly stopped. Some of the animals were not actually on the bottom but hovered several inches above it; others were almost entirely buried in the mud. Still others that filled them with wonder were common; they were seeing them for the first time in action.

"There's an abundance of fish," Hessler said into a tape recorder at 4202 feet. "No matter where you look, you'll see fish. ... We have a hole over here. It's actually ... four holes in a circle. ... The impression ... is that in the center of this complex is where one animal is living perhaps, and that it's capable of coming out of any one of these holes. Again, I'm impressed by the numerous fish we see here. We're going through some slime right now."

"Oh, my god!"

"Slime is the best way to describe it ... suspended in the water here ... about eight inches long. ... The bottom is just pockmarked with small holes. ... There's really all kinds of biological activity. Every square inch shows indications of it."

Alvin hovered over a patch of sea urchins which made up a carpet of black-eyed Susans, petals waving gently.

"The so-called red sea urchin appears to be quite white...."

The urchins were also covered with strange puffed-out polyps. The scientists reasoned that their nets must tear off the polyps. Sanders likened the creature to a bunch of golf balls with bayonets sticking out of them. What would eat that? he wondered, and in a second he had an answer. A big fish swallowed one whole. Sanders watched an octopus disappear into the mouth of another fish. Two hours later he saw what appeared to be the same fish hovering peacefully near the same species of octopus. They were both ignoring each other. How often did they eat? Sanders wondered. Once a day? Once a month? His questions were fundamental and still unanswered.

"Oh!" Hessler said. "A beautiful deep-sea arrow worm with large round gonads just passed by the porthole." The arrow worm was transparent.

"Right now Mac is trying to catch a large [crab] that's in the typical crab defense posture, all hunched up with its body tilted upward, with the claws open, ready to attack anything that comes down on it from above. Every time Mac tries to grab it with the mechanical arm..."

"You son of a bitch."

"... it side-steps. It side-stepped Mac again. It's quite capable of eluding the mechanical arm so far."

Alvin went deeper and deeper—to 6112 feet, 6200 feet and 6242 feet, three times beyond its limit. The biologists didn't mean to break any rules; all they could think about was what they saw and if it went deeper, they wanted to follow it. The pilots, who knew *Alvin*'s sphere had a crush depth of 16,100 feet, were glad to oblige.

Time seemed to stand still above the water, too. During George Hampson's tape-recorded debriefing in *Lulu*'s tiny mess, he talked excitedly about a creature that resembled a brittle star, a kind of skinny starfish, that moved its long legs like oars, as if it were rowing in the water.

"Were the arms very flexible?" Hessler asked.

"And the arms are thin and long?" Sanders.

"Yeah.... But the strange thing was this animal *wasn't living on the bottom, it was living on things on the bottom!* I also saw small white sea urchins."

"Were they climbers?"

"No. They colonize in clumps. I saw twenty, twenty-five at a time."

"What do you mean by clumps?"

"Well they weren't on top of each other, they were about two inches apart. It would be a very long time before you came upon another of these clumps. Why? I can only assume they were feeding on something but I couldn't see what."

"Unless they were aggregated for some other purpose."

"Right."

"Right. Like to have a meeting."

Hessler spoke through the laughter. "Now, that was the most common?"

"Right, the white urchin.... And the bottom was very, very flat.... I mean as flat as you could ever imagine.... The coelenterates were so spaced, you go boom boom, take a stick and just hit one...."

"Wait, *what* coelentrates?" Hessler asked of the phylum that included jellyfish, hydra, coral and anemones.

"We haven't gotten any of those in our samples," Scheltema said.

"What color was it, white?"

"No, dark purplish-red."

"When the tentacles were extended, were they long and slender or short and stubby?"

"Neither of those. I would say they were in between."

Hessler reached into the bowl of mixed nuts and poked around, as everyone else had, for something other than a Brazil nut.

"The sea spider," Hampson continued, "that creature walks on the tippy toes of its legs. I mean it's just sort of neutrally buoyant. It seems the abdomen goes right down and touches the bottom as it moves. I saw this one individual just, just bouncing, sort of just floating along, just beautiful."

Hampson could have said that he had seen a dog floating in the air. How could a sea spider—what Vine on his deep dive called a daddy long legs—be buoyant enough to hover? Most of its body was made up of long sticklike legs.

"Did these sea spiders seem to be associated with any feature of the bottom or . . . with bits of animal life?" Sanders asked.

"No. They were just walking along. . . . They're so delicate, the way they walked. . . . It's as if they . . . well, I don't know."

"That's the spirit," Scheltema deadpanned.

"Every once in a while I saw what seemed to be a very white, a porcelain white creature scooting around along the bottom. This was very unique. They moved so damn fast I couldn't see what it was. You get a very good opportunity to see these creatures. I would even bring down a pair of binoculars. Beautiful opportunity. . . ."

"What?"

Hampson was too tired to think of anything else, except to mention the sneaker and the orange peels he had seen.

"We didn't see any human artifacts at all in our dive," Hessler said.

"Except those in the submarine," Sanders added.

"Right. And we didn't have turkey sandwiches either."

"What the heck did you have?" Scheltema asked.

"Underwood canned paste. Also Tootsie pops and atomic fireballs."

They talked each night until they were too exhausted or punchy to continue, hoping the weather would hold out for another dive in the morning.

Good weather was the prayer of all sea-going scientists. But in the North Atlantic, it was naive not to anticipate at least some interruption. Fast-changing weather worried Rainnie most. *Alvin* was powerless to fight all but the calmest seas. He had called *Alvin* home many times when the waves grew. The trick was knowing when to abort a dive; it could take two hours to surface. He always played it safe. If there were only whitecaps but 20-knot winds, he would end the dive.

On the benthic biologists' fifth dive, *Alvin* surfaced to ten-foot swells and 35-knot winds. The sub wallowed in the trough of a wave and rammed against *Lulu*. The impact wrenched off the mechanical arm and the sampling basket, which disappeared into the stormy ocean.

It wasn't until two days later that the seas subsided enough to allow another dive. But there was no sign of the lost items. The crew had no choice but to return to Woods Hole to pick up the scientists next on the dive schedule.

As *Lulu* steamed home, its echo sounder, which recorded images on the seafloor, drew a peculiar picture. The machine commonly drew horizontal and angular stripes to mark the echo of the sound pulses striking the seafloor. The echoes from schools of fish with swim bladders or other pockets of air were shown as sharp V's. But these lines had no sharp angles; the recorder drew arches. When the biologists reached Woods Hole, they told Dick Backus that Alexander's Acres was nearby.

For a dozen years Backus, who had combined his specialty of ichthyology with acoustics, had been trying to unravel the mystery of what everyone called "Alexander's Acres." Sid Alexander was a Coast Guardsman who years before happened to wander into some oceanographers' laboratory at sea. Backus was there, Earl Hays and others. When Alexander saw the echo sounder draw a field of mounds, he asked what it was. The scientists didn't know; in fact, they hadn't any idea. Alexander said, in the vernacular, that it looked to him like acres and acres of breasts and how he would love to walk all over them barefoot, and so on.

On October 2, Backus had his fingers crossed when he and eight others boarded *Lulu*. They planned to net marine life and see for the first time the area they had been sampling from ships for so long. But if Alexander's Acres was there, there was no question about diving on it immediately.

At the dive site the echo sounder drew the strange parabolas. Backus and Jim Craddock, another ichthyologist, climbed into *Alvin* and Broderson waved his sendoff. "Have a good dive, Mad Mac."

Since there was no arm or basket for samples, an open net was attached under *Alvin*'s chin to capture whatever it was that bent sound waves into parabolas. A big shark? A school of jellyfish? Unknown creatures? Whatever it was, they were going to have to sneak up on it. The lights were out. They watched the pulsing green blotches on *Alvin*'s sonar screen. McCamis steered toward a target that showed as a good-sized blob on the scope.

"We are closing [in on] a lovely sonar target now with our lights off," Backus said. "About seventy yards now... twenty-five yards... lights on... I don..."

Backus didn't finish. He and Craddock were mute with awe. McCamis had the presence of mind, for which his passengers were grateful, to take pictures and position the net by dropping *Alvin*'s nose.

They were surrounded by thousands and thousands of three-inch-long lantern fish, a common creature of the Myctophid family. The fish had swim bladders and large eyes encircled with a reflective material that resembled aluminum foil; smaller circles of the reflective material also lined their undersides. These were actual bioluminescent organs called photophores.

The lantern fish had their lights off now. They hung upside down and right side up and at all kinds of angles. They weren't feeding or mating or fighting or doing anything the biologists could determine; they were just there. They were in Alexander's Acres; they *were* Alexander's Acres.

"*Oh!*" Backus gasped. "A tremendous cloud of Myctophids ...tremendous cloud! Beautiful, beautiful fishes, silvery, head down, head up, swimming up, swimming down, lovely fishes. They are very close to, very close. They're getting thicker and thicker. We are sitting motionless, right in the middle of a swarm of these fishes. You couldn't call it a school because they are all moving in different directions, but it's a fantastic aggregation, fan*tas*tic numbers...."

Since *Alvin* was diving in midwater, that is, not deeper than 3000 feet, it made an unprecedented two trips daily, giving everyone at least one chance to dive. Only Asa Wing, Backus' longtime electronics technician, would not get into the sub; he didn't trust it. *Alvin*'s new pilot, Ed Bland, a retired navy submarine radioman who had been in training all year, got to take the sub down on his own.

The biologists had thought that photophores, which were usually on the underside of fishes, were meant to shine down and blind predators. But the photophores of these lantern fish were aimed in every conceivable direction. Rarely had their nets captured a lantern fish, and yet *Alvin*'s net brought up 774 of them. How could that be? What else had they missed? The same species of lantern fish, *Ceratoscopelus maderensis*, existed throughout the world ocean. Why had they seen Alexander's Acres only off the northeastern United States? And why did these fish get together?

Backus asked for two weeks of diving the following year.

It was mid-October and the weather window for diving in the North Atlantic was all but closed. *Alvin* had only one more cruise of the year, the last chance of returning to try to find its arm. Nobody had much hope of finding it; besides, it was probably mangled after the 4365-foot drop. But they had to try. There were no funds for a new one.

Two WHOI geologists, K. O. Emery and David Ross, were signed up for this expedition. Ross lost all his samples while surfacing from a dive in August; Emery's second expedition of the year was canceled because of mechanical problems with the submarine. They didn't mind that they would be looking for the arm as well as doing geology.

Two days passed with good weather but no sight of the arm. On the third dive after a half hour on the bottom, *Alvin*'s sonar picked it up. The upper arm was almost completely buried in the soft bottom ooze; the forearm stuck out in an absurd pose, pincers pointed upward, clutching a twisted piece of the sampling basket. McCamis pumped mercury forward, and with *Alvin*'s nose down lifted the arm with a hooked piece of piping under the center porthole. On the next dive, the sub returned to retrieve its milk crate.

"Dumb luck," Skip Marquet said. And it was. There was luck all around. The arm had fallen in an undiscovered canyon, and the geologists hadn't wasted any time. "The topography of the search area," the scientists wrote, "is that of a gentle southward slope on the west side of a submarine canyon. This canyon is hereby named Alvin Canyon in honor of the submarine that was responsible for this study...."

Alvin was building a following and not just of scientists. In fact, most fans were not scientists. Letters came from Europe, South America and the United States, many from schoolchildren. An Attleboro, Massachusetts, youngster wrote:

> *Dear Sirs*
> Please send me all the imformation that you can on your deep diving vessels. I saw your show on channel 2 about Alvin and my interest was arroused in oceanography.... The thougt of growing things in the sea for human consumation has been in my head for a long time. I am in the first year of high school.
> *Nicholas W. Graetz*
>
> P.S.: Encosed are four 5 cent stamps to help pay for the cost of mailing.

Others wanted to know if they could come to Woods Hole and see this strange submarine. Samuel Griffith had something else in mind:

Dear Sirs

Enclosed is a copy of an idea for a two-man submarine which I have developed for use in discovery and research of the ocean floor. I would appreciate your opinion of my design.

I am 11 years old and in the fifth grade at Eastern State School, Trevose, Pa.

Sincerely,
Samuel Griffith

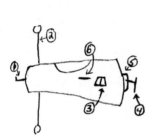

1. Retractable Laser Beam
2. Mechanical Arm
3. Retractable Propeller
4. Propeller
5. Rudder
6. Diving Plane

The Alvin Group, answered the boy's letter.

Dear Samuel

Your parents and teachers may be very proud of you for your active and creative mind....

Your design is very good and contains some excellent features. You will learn though, that between the sketch showing a design concept and the working drawings, which guide the mechanics in fabrication, there is a lot of hard work, called "Engineering...."

Best of luck in your endeavor; and thank you for writing to us.

There were several letters from Louise Hutchinson of the Chicago Tribune, who was determined to meet the little submarine that found the H-bomb. She got her chance in September. *Lulu* steamed a short distance from the dock, and Hutchinson, clutching a notebook, two pens and a bottle of pills, climbed into *Alvin*'s sail, becoming the second woman to dive. (The first woman was WHOI geophysicist Betty Bunce, who dropped to 45 feet in the murky harbor during *Alvin*'s trials in 1964.) Hutchinson's palms were wet; she didn't swim. *Alvin* backed off the cradle and began to sink amid the frothing bubbles from the air in the freeboard ballast. Out a porthole the reporter saw the barnacles on *Lulu*'s pontoons and water rising. "Mac!" she shouted. "We're under water!"

But Hutchinson soon forgot her bottle of pills, which wouldn't have helped anyway unless they had the power to make her unconscious. The only rough ride in any submarine is at the surface. Beneath the waves *Alvin* moved as if it were cutting through hot butter.

"It's like drifting through a dream," she said of her gentle fall to 118 feet. "The turquoise water turns a deep twilight blue. We pass through layers of plankton, the tiny animal and vegetable life that fish feed on. Who from a prairie state would ever believe plankton looked like this?...It's like falling through a snowstorm."

Alvin's passenger sphere, she wrote, was smaller than the Apollo space capsule. Three wasn't a crowd, it was a mob. She was not impressed with *Lulu:* "If you met her on a dark night coming down the Chicago River, you might never go near the Chicago River on a dark night again."

12

Lulu looked like a piece of the dock that had ripped off. She wasn't really a catamaran because her pontoons were almost entirely submerged. By now WHOI knew the consequences of making the forward arch lower than the rear arch. The arrangement was made initially for the sake of better visibility from the bridge, which was aft. But in heavy seas, water slapped against the lower arch and the entire ship would tremble and stop or just go backward. And *Lulu* did not have the power to save herself. She had been nearly run down, once by an aircraft carrier and again by a freighter; the men on *Lulu* frantically waved flashing lights until the big ships changed course.

The trailer that held the mess and galley was exchanged for permanent space built in over the port pontoon that held the engines. New air conditioning went into the starboard pontoon, which the crew called the "Tube of Doom."

Neither was the Drugstore anything to brag about. The corner building with the Victorian-like turret was falling apart. Crumbling plaster bled from the torn wallpaper. When the wind worked up a gale, as it often did, the building swayed and pencils rolled across desks. Bill Rainnie had warned Marge Stern, their new secretary from Louisiana, about it so she wouldn't be frightened. What bothered Stern most was the plumbing. Twice when she flushed the toilet the man from the drugstore downstairs rushed up the stairs and yelled at her for wetting his card display. But like *Lulu*, the Drugstore was home; Stern loved her job and the people. The fellows who went to sea were a little rough around the edges but they were good men, she said. They played hard, they drank hard and they worked damn hard. Swore hard, too.

In 1968 the Drugstore formally asked WHOI's director "for purposes of uniformity and morale" to officially christen *Alvin*'s mother ship RV *Lulu*, with the Research Vessel designation the same as it was for the other ships. The crew had already painted on her side "LuLu." Paul Fye did not argue.

Hopes of funding from the Office of Naval Research for a new mother ship had grown increasingly dim. The talk in Washington,

D.C., was not encouraging. ONR said the Vietnam War was bleeding its budget, everyone's budget.

Rainnie said *Lulu* was slowing them down. Some thirty civilian scientists dived in 1967, but almost half the available diving time or 120 possible dives were lost to bad weather or *Lulu*'s imperfections—her dinosaur-like speed in reaching dive sites and constant mechanical trouble.

Despite the mother ship's problems, it is unlikely that there would have been enough scientists interested in making the "lost" 120 dives; or that *Alvin*, still a prototype, could have accommodated them. Only eleven new scientists made dives in 1968. And it was funding constraints that kept *Alvin* and *Lulu* close to home in the North Atlantic for the entire 1968 season; of a total 67 dives, 43 were made by scientists. The balance was for trials, pilot training and the Navy.

That summer the Navy borrowed *Alvin* for five weeks to inspect seamounts on which it planned to erect listening towers off the Azores. The divers saw the detail their sound instruments had missed. One chosen seamount was far too narrow for a tower.

On that expedition, *Lulu*'s new cook, Big Ed Brodrick, befriended a Portuguese fisherman who gave him two thirteen-pound lobsters. These went into the biggest pot in the galley, a roasting pan, and mixed with Big Ed's stuffing, were served in the lobster shells. Hotel style, he called it.

Big Ed was widely regarded as the best thing that ever happened to the mother ship. He was thirty-one, five feet ten, and weighed more than 300 pounds. Some of his meals (like the roast beef, Yorkshire pudding and strawberry shortcake) were extraordinary enough to stick in memories for twenty years. Brodrick had little professional experience. In 1965 he became the cook on WHOI's ship *Gosnold*, and the following year enrolled at the Culinary Institute of America, then located in his home state of Connecticut, for a degree in chefing. After a year of la-di-da cooking at resorts and inns, he was eager to return to Woods Hole to cook for "real working people who were sober and hungry." His assignment was to feed the men on *Alvin*'s mother ship.

They loved Big Ed not for his meals alone; his heart was as big as his girth. He was a sweet young man, easy to please and unbothered by the quirks of the crew. When he served tomato soup, it was with Ritz crackers because the chief engineer demanded Ritz crackers with tomato soup. He didn't use any butter on the sandwiches for the dives because the pilots said butter for some reason tasted rancid 6000 feet down. The lunch he usually packed them included garlic-flavored baloney and cheese on white bread, a thermos of hot beef bouillon, apples, cookies and another thermos of water.

Another navy expedition in 1968 involved inspecting a World War II airplane discovered during an *Alvin* dive the previous year. The plane was an F6F Hellcat ditched by its pilot in 1944 about 110 miles east of Cape Cod. It sat upright in more than 5000 feet of water, wheels retracted and partially sunk in the mud. The navy's metallurgists were most surprised that the craft looked untouched after being submerged for 23 years. They wanted to raise the plane, hoping it would teach them something about the long-term behavior of metals in the corrosive ocean. Plans for retrieving the plane were never carried out, but *Lulu* and *Alvin* returned to the Hellcat area with naval officers, including WHOI's new Boston-based ONR project manager, Lieutenant Robert Ballard. The goal of this mission was to test experimental navigation gear developed by Skip Marquet by trying to relocate the Hellcat with it. They did, and on the first try.

Marquet also tested a new type of light—thallium iodide, the kind used in television studios. It glowed slightly greenish, and while not as bright as the standard quartz lights, it penetrated the perpetual midnight of the deep sea much farther than *Alvin*'s other lights.

Good progress had been made on sampling tools and safeguarding samples during *Alvin*'s ascent. Several different kinds of trays in front of the center porthole had been tried. The most enduring was a steel mesh basket which carried two milk crates that held several coffee cans for samples. Covers had been added to the emergency dump switches to make it more difficult to release an item accidentally. The sampling basket held the record; the pilots had inadvertently jettisoned it a half dozen times.

Alvin's arm was sent to its designer Bud Froehlich at PaR, a small company formed by ex–General Mills engineers. Experience had taught that the arm's two flat pincers were well designed for operating tools, but not for picking up nature's asymmetrical objects. The new pincers were curved. PaR charged WHOI $4700. A new arm, the engineers said, would have cost about $80,000.

Scientists who wanted special tools for their dives, as almost all did, discussed their needs with the submarine crew or the WHOI shop men. The person inspired first was the one who usually designed and built it.

At sea the Alvin Group used odds and ends for scores of jobs. Filled with oil, a length of old automobile radiator hose endured the pressure at 6000 feet. It was used to attach items to *Alvin*. A plastic cup over the gas-venting pipe in the batteries kept out seawater. That was McCamis's idea. He also built the bubble jug, a clear plastic container with valves used to detect air in the oil systems.

Lacking formal training, McCamis would describe the tool he envisioned to the Alvin Group draftsman, Tom Aldrich, who would draw the item and pass on the blueprint to the shop. McCamis's talent was considerable, but he was sensitive about lacking the education the engineers had. In one discussion about πr^2 he huffed off, saying: "All I know is pie is round and johnny cakes are square."

One of *Alvin*'s engineers used a condom to silence the pinger that was added to *Alvin*'s exterior in case a spot needed to be marked. The pinger seemed like a good idea, but the constant pinging drove the pilots crazy. A little oil in a condom fit over the electrodes on the end of the pinger kept it quiet, usually through the entire descent, before it burst. The condom worked so well that Clifford Winget, the engineer who thought of the idea, wrote a purchase order for a gross of condoms from the real drugstore under the *Alvin* group offices. WHOI's purchasing department, however, intercepted the purchase order, declaring that condoms could not be bought with federal research funds.

The tall ruddy-faced, white-bearded Winget had a deep chuckle that rumbled with mischief. As a test pilot he had "borrowed" an air force plane, not once but three times, and was court-martialed three times. When he joined the *Alvin* group in 1967, Broderson told him that he had to earn the distinction of wearing a red beret. "Are you kidding?" Winget said. "The garbage men in New York City wear those things."

Rainnie tried to convince him to be a pilot, but Winget declined. After hurtling through the air at 350 miles an hour, he said, he had no intention of drifting at a half knot in an eggshell. He was happy enough as *Alvin*'s chief seagoing engineer. "Those were fun times," he recalled. "You could let your mind go absolutely wild." And he did.

Many of the ideas for samplers, which filled a 180-page volume, were the products of Winget's creative mind. There was the ice-tong retriever, the clam rake, the fish scoop, several mud thermometers, and a stun gun for big fish. In answer to the biologists' need for something that could vacuum up fragile animals, Winget designed the slurper. It looked like the aspirator in a dentist's office. A funnel on one end slurped animals into a bag on the other end.

Of all Winget's inventions, his crowning achievement finally enabled geologists to take a piece of hard rock with one arm in a neutrally buoyant submarine. Winget's hard rock corer (a four-martini job, he said) looked like a caulking gun. It held a diamond-tipped drill, driven by a battery-powered electric motor bathed in pressure-compensating oil. It could take a three-inch core of hard rock. By reversing the drill direction, a suction was created to hold the core in the bit. Reversed again, the drill released the core into a coffee can in the milk crate. Geologists used it for the first time in 1968 to sample the wall of a canyon.

Allyn Vine's officemate, Bill Schevill, who had argued so hard in 1961 for a submarine with enough windows, finally got his chance to dive. Schevill's wartime curiosity about fish noise had developed into a full-time career that focused on cetaceans. He said he wanted to find out how they made a living. He and his assistant, Bill Watkins, presented the first direct evidence that the bottlenose dolphin echo-locates like a bat. They debunked early notions that a peculiar underwater sound was a whale's heartbeat or a communist trick. It was a stranded unhappy finback. They taped cetaceans clicking, wheezing, creaking and moaning, and analyzed the patterns and rhythms. Whales and porpoises contributed much to the natural cacophony of the sea, and what incredible sounds they were. Every sperm whale seemed to have its own unique speech.

Schevill and Watkins went to observe whales off Provincetown at the tip of Cape Cod in *Alvin*. For hours they looked, and finally the sonar glowed with a large target. It was so big that they thought it was a fishing boat, but surface control on *Lulu* said there were no boats around. So, they thought, it had to be a whale. Seeing nothing out the portholes, they watched the sonar scope. Their excitement turned to alarm. The huge thing kept coming. They braced for collision and suddenly the target disappeared off the sonar. They never learned what it was.

Near the end of another dive, the sonar picked up another target which *Alvin* approached. It was a can of O'Keefe's Vienna Ale.

In their four dives, Schevill and Watkins didn't see a single whale, although they heard them and wondered if the cetaceans were outsmarting the humans. As *Alvin* glided beneath the waves, three finbacks broke the surface and one moved between *Lulu*'s pontoons.

The Alexander's Acres group dived again, this time at twilight to see if lantern fish migrated, as many animals did, with the rhythm of dawn and dusk. But rough seas allowed them to make only three dives.

The benthic biologists were faced with the problem of quantifying what they had seen from *Alvin* the previous year. Just how abundant was the life? They had to know the submarine's precise angle and altitude, and the optical characteristics of the cameras, including the lens distortion, all difficult numbers to come by. Bob Hessler read up on aerial photogrammetry and with his colleagues, drew a grid on *Lulu*'s deck and photographed it from inside *Alvin*. With that information they worked out the lens characteristics. The grid would be superimposed on the new photographs which would include a tilt meter mounted on the submarine.

On hearing about the benthic biologists' need for a precise altitude, Bill Gallagher from the shop made a mold with one of his alpine skis and fashioned a fiberglass ski for *Alvin*. It was attached to the underside of the sampling basket. Not only did it maintain constant altitude; it left a track in the mud that remained undisturbed in some areas for months, making a kind of road map. The truncated ski also made driving in the mud easier by giving the pilots feedback. Now, when the seafloor sloped imperceptibly up or down, they felt it. And they knew with greater assurance when *Alvin* was down or still slightly hovering. And finally, the engineers liked the extra support the ski gave the sampling basket. So the ski stayed on for good.

George Hampson shared some of the animals he had collected from *Alvin* the year before with another WHOI biologist, George Grice. Hampson suspected that the creatures were copepods, a crustacean no bigger than a grain of rice. Except for the one-celled protozoans, there were more of these shrimplike creatures in the ocean than any other kind of animal, and that made them a vital link in the food chain and of considerable importance in the ocean's ecology. Grice, whose specialty was copepods, had never seen anything like Hampson's specimens. They were downright bizarre. Mostly colorless, they had strange spots on their bodies: sensory organs? to sense food? danger? a mate? Hampson couldn't be positive that the specimens came from the deep sea because his net didn't close properly.

But Grice was intrigued and he towed new nets from *Alvin* in the same area that Hampson had sampled. This time the doors stayed shut during the entire ascent and Grice could be certain that the animals came from the deep. He spent many hours hunched over a microscope before realizing that he had hit the jackpot: eighteen new species of copepods. The first names came easily—*alvinae, hampsoni*. The others were more descriptive: *distinctus, elongatus, minutus*.

John Schlee dived again with a colleague from the U.S. Geological Survey, this time in the Gulf of Maine, hoping to find bedrock in order to date the coastal plain that was once above water. But the boulders and cobbles left from glaciers hid the bedrock, and the weather allowed only three dives. One of *Lulu*'s engines died during a storm and until the crew could get it working again, holding position was impossible and *Lulu* nearly went parallel with a trough. McCamis fretted that *Alvin* would slide off the deck. It didn't, but on their return to Woods Hole, he told Rainnie that the airplane cargo straps used to tie *Alvin* down at the mid-deck cradle weren't sufficient. They needed four big posts at each corner.

As senior pilot and usually the expedition leader for *Alvin*'s second full season of science, McCamis was the one in charge at sea. He was by far the most experienced of the four pilots. Rainnie hadn't been in *Alvin* all year; he was busy trying to drum up funding in Washington. He agreed to two hitching posts. It wasn't enough, McCamis insisted. And that wasn't all. The steel cables that lifted *Alvin*'s cradle were worn and should be replaced immediately.

When McCamis got going, it was hard to calm him. Rainnie had seen him lose his temper at simply being ordered to abort a dive because of bad weather; and once in the Bahamas he boarded a plane and didn't return until days later when his temper cooled. McCamis's way was often, but not always, the right way.

The cables were three-quarters of an inch thick, rated for 47,600 pounds each. *Alvin* weighed about 31,557 pounds; the cradle weighed 7250 pounds. They were installed in April 1967 and scheduled to be replaced at the end of the 1968 season, which was almost over. There was only one more expedition to go. But McCamis was right, they were a mess, covered in rust and frayed steel whiskers.

"Cable like that don't like to bend over a small sheave." McCamis wouldn't let up. "Ain't much more that cable gonna take."

All his life McCamis's big mouth had gotten him into trouble; he was helpless to keep his voice from rising when he thought he knew better, and he could not keep his mouth shut now. Unless the cables were changed, McCamis said, he would not go out as pilot or expedition leader. It sounded like a threat to Rainnie, and it was. McCamis saw how angry his boss was but he refused to back down.

The last cruise was a six-day expedition about a day's sail from Woods Hole for the WHOI Buoy Group, a team of engineers who concentrated on a single challenge: to develop better moorings. Moorings typically lasted for only about two months and many disappeared altogether with their attached instruments and collected data. But often moorings were the only means for getting certain information, especially for the physical oceanographers who needed long-term records of water movement.

Instead of hemp lines which fish nibbled, the Buoy Group

was experimenting with nylon and dacron. To keep a mooring from sinking if separated from its anchor, they added large hollow glass balls encased in plastic hard hats for buoyancy. Some lines carried navigational pingers and flashing lights. One of the new moorings was the donut buoy—a six-foot-high tripod mounted on a big fat ring of fiberglass. Paul Stimson of the Buoy Group wanted to dive under the donut buoy Alpha, deployed about three weeks before, to photograph the line and the experimental anchor that he had designed. He also planned to add another line he had rigged as a fish-bite experiment.

At 9 o'clock in the morning on October 16, RV *Gosnold* escorted RV *Lulu* to the bright red Buoy Alpha. The weather was fair for *Alvin*'s 307th dive. McCamis was in an unfamiliar role behind the wheel of the small boat escorting the swimmers. Rainnie was on the bridge where McCamis should have been. Stimson and Roger Weaver, a pilot in training, climbed into the sub. Pilot Ed Bland took his place in the sail and watched the line handlers pay out rope as he backed out *Alvin* from between the pontoons. Bland ducked inside, shut the hatch, and the sub dropped through the frothing bubbles at the surface.

They were down only about fifteen minutes when they discovered a short circuit in an outside camera, and surfaced. The repairs didn't take long.

"Prepare to launch," Rainnie repeated from the bridge.

Broderson placed the ladder back into the passenger sphere for Stimson and Weaver. The ladder came out and Bland got back into the sail. As *Lulu*'s master held the catamaran in position against a 15-knot wind, the cradle rose and the chocks *Alvin* perched on were removed. Broderson looked to the line handlers for a nod. From the sail, Bland did the same and then signaled to start lowering the cradle.

The cradle dropped a foot, seven more feet to go, and suddenly *Alvin*'s nose pitched down and Rainnie saw the wispy tuft of Bland's white hair disappear into the sea.

13

Paul Fye was at a meeting across the street from the Drugstore when Frank Omohundro, the former Bureau of Ships engineer who became the Alvin Group's quality control man, tapped his shoulder and asked him to step outside. The director's only clue to why he was being interrupted was Omohundro's ashen face. They walked outside.

"I thought you should know," Omohundro began, breathed deeply. "That *Alvin* has been lost."

"What?"

Omohundro said it again as slowly as he delivered the line the first time. Fye repeated his line, too, impatient with Omohundro's words; they dripped from his mouth, one by one, one every second, it seemed. Omohundro began again. Fye was beside himself. "Was anybody hurt?" he asked. Omohundro's lips crept sideways.

Ed Bland saw it happen from the sail. He saw the cradle's forward port cable snap and then the forward starboard cable. When he felt the submarine slide, he instinctively crouched inside the sail.

Alvin's aft propeller struck the cradle as it plunged into the water. Bland's head hurt, he couldn't breathe.

Roger Weaver, whose face only seconds before was nuzzled up to the center porthole, was trying to speak. He saw Paul Stimson's eyes and no more, and suddenly it was dark and he was pressed against the sphere by onrushing water. Water entered his mouth when he tried to speak; he was trying desperately to get out the words, *Let's get the hell out of here.*

Bland, still straddling the open hatch, gulped for air as the buoyant submarine bobbed back to the surface. He saw that one of the windows in the sail was shattered—by his head? Water poured into the submarine, taking away all that buoyancy. *Got to get out.*

Lulu's skipper could see that if Bland wasn't going to drown, he would be crushed against the underside of the deck by the big drums wrapped with the cables. He quickly jerked the ship forward in an effort to center *Alvin* between the pontoons. The submarine slid into the trough of a wave, tilted, and Bland jumped from the sail onto a pontoon.

The six ropes attached to the submarine whipped one by one out of the line handlers' hands. A hose in the hydraulic lift system burst in a spray of oil.

Weaver pulled himself up into the sail as the submarine bobbed beneath one of the steel drums wound with cable. Bland quickly pushed him back down. Weaver would be killed for sure if his head cracked on that drum. Weaver tried again, the sail bobbed at the drum, and Bland pushed him back again. Weaver looked down and saw Stimson between his legs and water rising in the sphere. He stood again and in one frantic heave threw himself into Broderson's outstretched arms. Stimson's drenched body might well have been attached to Weaver's feet. He was right behind. And the last of *Alvin* disappeared.

It took about 60 seconds from the time of the first cable parting to the sinking of the submarine. The three men escaped with only bruises and scrapes; Bland was limping. Broderson had an arm around Weaver, the unflappable guy who would talk to him while standing on his head; he said the yoga posture was refreshing. "If you had removed the ladder, Brody, I never would've gotten out," Weaver said. Broderson nodded silently; of course he had removed the ladder, as he always did after the last passenger got into the sub. The "steps" Weaver had climbed were made of thin air.

The seconds of grisly panic were replaced with relief and then another kind of panic. In a frenzy, everyone began to throw objects overboard to mark the spot. All six aluminum lawn chairs were heaved, scrap pieces of steel, a 55-gallon barrel, Stimson's untried fish-bite experiment, all of it went over the side, anything with metal that could be lifted and might be picked up by sonar.

In 5000 feet of water, the sonar at the surface probably would not see the articles. Even *Alvin* might be missed. Oceanographers learned some of the limitations of their sound instruments in an experiment they performed during the *Thresher* search in 1963. They brought a junked automobile aboard one of the search ships (much to the dismay of the captain) and pushed it overboard. It disappeared from the echo sounder at a thousand feet.

Luckily Buoy Alpha was there to mark the spot. But the donut buoy was experimental, it had an untried anchor, and it was hurricane season. They were gripped with the awful reality of the situation.

Lulu called her escort vessel, *Gosnold*, which was about two hours away. WHOI's ship *Chain*, which was headed home, changed course and steamed to *Alvin*'s mother ship. A Coast Guard plane and WHOI's small seaplane flew over the area and took bearings.

When *Gosnold* arrived, it immediately dropped a new mooring and with *Lulu* began to sweep an eighteen-square-mile area with sonars. The sonar images and samples of the seafloor were brought back to WHOI's geologists who spent the weekend piecing together a bathymetric map and wringing every possible bit of information they could from what they had.

They had the coordinates: 39° 52.1' north and 69° 11.9' west, 135 miles southeast of Woods Hole. It was like saying *Alvin* lay somewhere between 42nd and 79th Streets in New York City. But in this place there were no buildings, no street signs, no nothing except a muddy bottom and 5000 feet of water above it. The satellite fix, it turned out, was useless because of a two-mile error.

The echo sounder images indicated that the seafloor sloped slightly to the south. There were no boulders that might have chewed up *Alvin*. The bottom was clayey mud, not soft enough to cover the submarine completely. The geologists predicted that *Alvin* was sunk in up to three feet of the mud.

At least one ship remained on site. When a Russian trawler moved in, two Coast Guard cutters arrived and stood guard. The Russians jammed the radio waves, making communications difficult and sometimes impossible. WHOI's ships dropped two more moorings. The navy sent Woods Hole 20,000 feet of nylon line.

What they needed was another deep-diving sub.

More than two dozen small *Alvin*-like vessels, variously called submarinos, manned submersibles and minisubs, existed. All but three were shallow divers. Lockheed's *Deep Quest* was designed to operate at 8000 feet, and General Motors' *DOWB* or Deep Ocean Work Boat, to 6500 feet. But both were launched in 1968, and infancy was a definite disadvantage. *Aluminaut*, which could dive to 6500 feet, was by far the more experienced. At $7000 a day, it was also the most expensive. General Motors was willing to charge about half that amount to rent *DOWB*. It was *DOWB* that WHOI chose.

Filled with water, *Alvin* had to weigh about 10,000 pounds. Nothing so heavy that lay so deep had ever been salvaged. No one knew if *Alvin* was crushed, upright or on its side, or whether it could be raised. In fact, no one knew where *Alvin* was. In a week of towing cameras, not a single photograph showed any sign of the sub or related debris. Hurricane Gladys forced *Gosnold* and *Chain* back to port; when the ships returned to the loss site October 22, Buoy Alpha was gone.

On October 23, an air force cargo plane whisked *DOWB* from Santa Barbara, California, to Cape Cod. Accompanying the submersible was the director of General Motors' deep submergence project, retired navy Captain Swede Momsen.

It took more than a week to outfit *Lulu*, test the quickly assembled launch system, and prepare the bright orange *DOWB*, which had virtually no instrumentation and no windows. *DOWB* used a system of periscopes.

Momsen was making last-minute checks to the hatch in the freezing rain when he slipped and fell twelve feet onto the cement pier. He was rushed to Falmouth Hospital with a broken back.

The mood that Halloween night at the Captain Kidd bar in Woods Hole was somber, but not sober. In the morning a big jack o'lantern was waiting at the WHOI dock. It was the rotund orange *DOWB* painted with the mask of a smiling moustachioed pumpkin.

Nobody would admit to the deed, and there were many suspects, not least the shop men who had secretly placed a statue of a nude woman out by the docks to protest WHOI's decision to grow grass there rather than build them a new workshop. McCamis was considered too serious to be a suspect, but it was he and his girlfriend who did the paint job.

The General Motors men quietly removed the black paint while the *Alvin* crew and the shop men laughed and pointed accusing fingers at each other. They needed to laugh. It was pouring and cold; the towed cameras still hadn't found *Alvin*.

The Coast Guard cutter *Sassafras*, *Lulu* with *DOWB* aboard, and *Chain* sat out two days of miserable weather. Seas were even rougher the third day, November 3, but by midday the searchers were out of patience and headed for the site anyway. *Chain* steamed at top speed in circles around *Lulu* to create a lee so they could launch *DOWB*.

"You can't imagine what it was like, with the *DOWB* going down, down, and the tension building," Bobby Weeks, WHOI's aircraft pilot and scuba diver, said. "You could hear a pin drop. That damn telescoping thing going down, down, and suddenly . . . ah!"

DOWB wasn't in the water an hour when things began to go bad. The sonar didn't work and neither did a propeller. But worse, the optical system didn't work. Nobody could *see* anything from *DOWB*. Except for *Chain*, which remained to tow a camera sled, the ships returned to Woods Hole.

The nurses at Falmouth Hospital had given up trying to keep Momsen in bed in traction, which seemed to him beside the point. He knew he should have been out there with them; if he had been, they wouldn't have come back in, they could have made the repairs at sea. In a brace and harness, Momsen walked out of the hospital, which was difficult enough, and managed to implant himself in a borrowed car. And arms fully extended on top of the steering wheel, head stiffly fixed forward, he drove to the WHOI dock.

But he was in no condition to join them when the fleet left again at the next break in the weather on November 9. By afternoon the weather began to deteriorate and the radio reports coming in sounded ominous. *Lulu* and *Sassafras* never reached the loss site.

> SASSAFRAS WITH LULU TRYING TO FIND A LEE IN MARTHA'S VINEYARD AREA.... REPORTED NE WINDS UP TO 90 KNOTS.... [STARBOARD] TRUNK DISHED IN BY WAVE. CRACKED WELDS OBSERVED. BULKHEAD SHORED UP....

They were hanging on for their lives. *Lulu* was no match for the forces of an open ocean and hurricane-strength winds. *Chain* took green water over her bridge which rose thirty feet above the surface. It was too late for *Lulu* to make it back home; she had no choice but to try to ride out the violent seas. Faithful *Sassafras* stayed close by.

The waves had smashed in the bulkhead doors on *Lulu* and the crew had to use the emergency scuttle, smaller than a manhole, to reach the safety of the starboard pontoon. Big Ed Brodrick wasn't sure he could fit through it. But even if he could somehow scrunch down through that little scuttle, Brodrick worried that he might not get back out. The thought of being stuck in the Tube of Doom was more terrifying than getting swept off deck. Brodrick made his way to the galley on top of the pontoon, where to his surprise, he found *Alvin*'s electrician, Bill Page, also a big man. They were soon joined by most of the crew who shared their sensibilities about the worst of two evils.

Lulu and *Sassafras* limped home three days later, both in need of repairs.

At the next break in the weather four days later, they tried again. For a week they rode out more bad weather and on November 22, Woods Hole gave in to Mother Nature. Not a single photograph showed any signs of *Alvin*. Then a cable arrived.

> NORWEGIAN BENOIL FROM CURACO [Curacao] ... PICKED UP DRIFTING CIRCULAR ORANGE DAMAGE BUOY STOP. ... THE BUOY IS MARKED W.H.O.I.

A passing oil tanker, which sent the cable, picked up Buoy Alpha 80 miles southwest of its anchorage.

14

We have got to get it back.
    Paul Fye, Director of WHOI

Bill Rainnie went to Washington, D.C., this time to the Naval Research Laboratory which operated *Mizar*, the navy's ship used in the Mediterranean search for the H-bomb. *Mizar*'s computer-aided camera-tracking system was still the most sophisticated in the world. But it was busy. In November 1968 *Mizar* had found the nuclear submarine *Scorpion* which had mysteriously disappeared with all hands southwest of the Azores. Before taking on the next assignment, *Mizar* was getting her gear refurbished. *Alvin* was not a navy priority, but NRL wanted to help Woods Hole.

After months of frustrating waiting for Woods Hole, the navy agreed to give *Mizar* two weeks to try to find *Alvin*. On June 4, 1969, more than seven months after *Alvin* disappeared, *Mizar* arrived in Cape Cod waters and began ten-hour-long runs of towing back and forth across empty ocean.

Days passed with no sign of the sub. On the fourth run, the cameras caught a section of nylon line. No one could be sure it was from *Alvin*, but it was the only hope in almost a week of towing. Five days later, a fraction of *Alvin* showed in a corner of one photograph. Time was up on June 14. This was *Mizar*'s fourteenth and final run, and as it turned out, the best. The photographs fully captured the sub 5200 feet down. "It was a thrilling moment," Skip Marquet said. "Especially after all that wild speculation about whether we'd ever find it, and the state it would be in if we did find it. And there it was, pretty as you please."

Alvin sat upright, sprinkled with a little sediment, intact except for the aft propeller which was ripped off but still attached by hydraulic lines. The lid on the sail was open. They couldn't see inside but hoped the hatch was also open. The only practical way to lift the sub was to insert a long bar into the sphere. But if that was not possible, they would think of something else. It didn't occur to the Alvin Group that *Alvin* might not be salvaged.

Paul Fye admitted that he had his doubts. So did the navy. The Chief of Naval Research, who called a meeting to discuss the matter, said he wanted to leave *that damn toy* on the bottom of the ocean. The eight people at the chief's meeting were about evenly divided over whether to salvage *Alvin*. Their heated discussions ranged wide, from money and compassion to national security and pride. What if the Russians salvaged it? Nobody knew what it would cost to rescue the submarine; but for sure, it would take more money to repair it.

The pro forces tried to focus on the issue of salvage and avoid the matter of repair costs. WHOI's ONR project manager, Jack Donnelly, delivered an impassioned plea to save *Alvin*. He heard himself repeating the same words. *We just can't leave it there. Even if we recover it to put it in a museum. We just can't leave it there.*

The vote was six to two to salvage.

The Alvin Group's at-sea crew was keeping a low profile. They worked on *Lulu,* attending to anything that needed attention, and when there was no more to do, they painted over fresh paint. Some of them did nothing, and some tried to drown their depression with booze. It was a hard time for a submarine crew with no submarine and no certainty about their future. When word came that the navy would fund the salvage, they rejoiced.

This time WHOI chartered *Aluminaut.*

In mid-August 1969 a fleet of ships gathered off the elbow of Cape Cod. Mindful that Hurricane Camille was closing in, the search teams worked around the clock. It took three days for *Mizar*'s cameras to relocate *Alvin*.

The plan called for *Aluminaut,* outfitted with two mechanical arms, to dive with a reel of line mounted on its bow. At the end of the 7000 feet of double-braided nylon line was a long aluminum toggle bar padded with syntactic foam. Winget, who made it, painted it bright yellow with black lettering: "Fickle Finger." If *Alvin*'s hatch was open, the toggle bar would be dangled into the sphere; a sharp tug would snap it into horizontal position, securing it inside the sphere across the hatch opening. Then the red submarine would pay out the line as it ascended; at the surface, the reel would be transferred to one of the ships so *Alvin* could be winched up.

It only sounded simple.

Efforts to secure the reel to *Aluminaut*'s bow were abandoned after several frustrating hours. *Mizar* would hold the reel.

Aluminaut dived through the eight-foot waves with McCamis and the Reynolds pilots. At 2000 feet, the sonar screen went blank. *Aluminaut* was totally helpless to navigate. *Mizar* directed her from the surface, hoping to steer her to visual contact. It was a long shot.

Aluminaut sank slowly. Two hours passed, three hours, and the submarine reached bottom at 5000 feet. Another agonizing hour went by, and another.

"We have *Alvin* in sight...."

The news brought a roaring whoop from those at the surface.

Alvin looked pretty good, pilot Bob Canary reported. Now they had to locate the lift line which held the toggle bar. It floated somewhere nearby. Seemingly trivial tasks were big victories under water. It took two hours to find the line and return to *Alvin* with the bright yellow Fickle Finger.

Getting it through *Alvin*'s hatch was another matter. Canary said it was like trying to thread a wet noodle into a soda bottle in a half-knot current. One of *Aluminaut*'s propellers stopped, making the task all the more difficult. There were too many people aboard for the carbon dioxide scrubbers; McCamis had a headache that wouldn't quit. After twelve hours, they gave up.

When *Aluminaut* surfaced, the seas were too rough to make repairs and recharge the batteries without taking in water through the hatch. Reluctantly, the fleet returned to Woods Hole.

On August 27, they went out again. *Aluminaut* had been submerged almost fourteen hours, working at the wet noodle trick. Canary was trying to be careful. Every time he tried to insert the toggle, he tore off a part of *Alvin*. There wasn't much left of the fiberglass sail, and both amidships propellers had been ripped off. One of *Aluminaut*'s batteries was leaking and the pilots worried about an explosion. It was 3 A.M. McCamis couldn't stand it.

"Look, all you're doin' is tearing *Alvin* apart," he said.

McCamis said he grabbed the controls and drove *Aluminaut* up onto *Alvin* so Canary could insert the toggle bar. Canary said he alone handled all the controls. In any case, the Fickle Finger was inside. *Mizar*'s winch turned and *Alvin* rose.

At about 85 feet below the surface scuba divers trussed the arm and the props, and wrapped a net around *Alvin*. Three large air-filled pouches were attached, each providing more than eight tons of buoyancy. The redundant lift force was a good idea. One of the pouches burst and was quickly replaced.

Slowly *Mizar* towed the submerged *Alvin* to Menemsha Bight off Martha's Vineyard where Dan Clark waited with his barge and crane. *Alvin* broke the surface on September 1, 1969. Bobby Weeks jumped in with the end of a hose to pump the water from the passenger sphere. Something was in the way. A jacket floated out of the sail. Weeks tossed it onto the barge. The lunch bag floated out. He threw that, too, and pushed in the hose. With the water out, *Alvin* was lifted onto the barge and doused with fresh water.

Clark hoisted a large bright purple and orange flag that Adelaide Vine had made for this occasion. In the center of the flag was a patch of white that was the unmistakable silhouette of *Alvin*. The flag waved from the stern as the injured sub headed home from its long, long journey up from the deep sea.

Planes, helicopters and scores of small boats, even a canoe, met *Alvin* in Vineyard Sound and escorted it to the Woods Hole dock. The innards of the passenger sphere were spread out on Clark's barge: the depth recorders, fire extinguisher, flashlights, tape recorder, cameras, compasses, three life jackets, three pairs of long johns and imploded cans of emergency drinking water.

The biologist Howard Sanders walked among the scattered debris, shaking his head. What a sorry sight.

"Hey, Howard, look at *this*." Winget held up a baloney sandwich which he had taken from the bag Weeks tossed onto the barge.

"Looks good enough to eat, doesn't it?" said Winget.

"How's it taste?"

"Salty but still tastes like baloney."

Surely, Sanders thought, Winget was joking. The engineer swore he wasn't; he showed Sanders the other sandwiches and the three apples; all looked fresh. But how could that be after being at the bottom of the ocean for ten months?

Sanders took the lunches back to his laboratory and called WHOI's senior microbiologist.

15

We are still hopeful ALVIN can go back into action in 1970, but don't have the green stuff in hand yet.
    Bill Rainnie, November 6, 1969

The Alvin Group did not wait for word from the Office of Naval Research, which was trying to decide how to divvy up its research dollars and how much it could give *Alvin*. They carried their crippled sub to a rented warehouse a few miles from Woods Hole and immediately got to work.

They kept telling themselves how good *Alvin* looked. The mechanical arm was fine. With a new paint job, the passenger sphere would be perfect. The fifteen titanium spheres in the ballast systems were untouched by corrosion, although they were covered in a strange cementlike film. Two batteries had full voltage, some lights lit, and the motors ran.

But everything inside the passenger sphere, including the expensive electronics, suffered the effects of saltwater and pressure. The navy sent technicians and ultrasound tanks to clean the gear. The circuits looked new again, but the capacitors and transistors were crushed. The aft prop and the conning tower were beyond repair. The side cavities of the freeboard ballast system were crushed, and corrosion had eaten through large sections of the aluminum frame.

Alvin's engineers called and visited dozens of companies for cost estimates, which were all much higher than they expected. WHOI didn't have the equipment for making all the needed parts; none of them had ever worked with fiberglass. But this was a can-do group. They bought surplus equipment and used the odds and ends from their home basements and toolboxes, and they learned to do things they had never done before.

"These guys were smart, conscientious," Winget said. "They were proud of *Alvin*. When the submarine was diving and surfaced with one of its systems on the verge of failure, they would bust their chops all night long to get that sub ready to go out the next morning. They were more than dedicated. I told them what had to be done and I told them what should not be done. Then I said: Go to it, create."

Alvin's technicians went to Bill Gallagher, who had fashioned *Alvin*'s ski, to learn about fiberglassing. Gallagher rode a unicycle when he was working to keep his arthritic joints moving. At the warehouse, he would pedal up to the coffeepot to fill his mug and begin his lessons, instructing *Alvin*'s mechanics, electricians and electronics technicians in the fine and messy art of fiberglassing.

Alvin's new sail looked like the old one but had extra features: behind it was a handle and a slightly elevated platform for the escort swimmers. Instead of being part of one continuous section of fiberglass that fit over the entire top half of the submarine, the new sail was separate. Now when the motors for the amidships props had to be serviced, the whole forebody didn't have to be lifted, just the sail.

The crew was proud of that sail. When a visiting engineer criticized it for being too weak, Bill Page kicked over the sail and jumped on it. "There!" Page said. "It's strong enough."

For advice about getting rid of the carbonate film on the titanium spheres without harming the metal, the Alvin Group spoke to navy labs and the aircraft industry, the primary user of titanium. Nobody could tell them how to clean the spheres, but they got an earful about what not to do. One aircraft firm warned not to get near titanium with a Magic Marker. It turned out that when a pit periodically appeared in the titanium shroud around a jet engine, an inspector would circle it with a felt pen for the repair crew that followed. But months later, the inspector would find a large hole in the same spot. After much tribulation the company realized that some chemical in the ink corroded titanium.

Winget went to his newly hired engineer, Barrie Walden. "Listen, kid, we gotta clean these things," Winget said. "Don't worry, no one else knows what to do either."

Walden chose his tools: a toothbrush and a bunch of wooden tongue depressors. He dribbled a weak acid onto a sphere and started to pick away at the cementlike calcium carbonate. It soon became apparent that this job was going to take forever. Walden then cut a barrel in half lengthwise and filled it with the acid. He put the sphere on a spit so it was partially immersed in the acid; with one hand he cranked the spit, and with the other he worked the toothbrush and tongue depressors. It occurred to Walden that this, too, might take forever, but he couldn't think of a better way. Neither could anybody else. It took him several weeks to clean the spheres, but it worked and the titanium was intact.

The engineers realized it was "ludicrous," as Walden put it, to depend on scuba gear in an emergency. Even if the hatch could be opened under water, which was extremely unlikely, the exhalations of carbon dioxide would pressurize the sphere and possibly blow out the hatch and plastic viewports. Walden installed rebreathers, closed-circuit systems that delivered oxygen and returned it purified. They looked like gas masks.

To make the mercury trim system more efficient, the three fiberglass balls were exchanged with four stainless steel balls with rubber diaphrams. Winget got them for a good price as surplus; they had been used as hydraulic reservoirs in World War II fighter planes.

The only parts of *Alvin*'s structure made outside Woods Hole were the aluminum frame and aft propeller. Alcoa, which made the original frame, did the job for a greatly reduced price.

Alvin did not return to action in 1970. Federal budgets were strained by the longest war in United States history, a race with the Russians in outer space, and the government's inability to keep the costs of its own submersible construction programs in check. There wasn't going to be a "wet NASA" after all.

The navy was building the first of several rescue submersibles, as recommended by the special panel appointed after the *Thresher* disaster. The estimated cost of these vessels was based on *Alvin*'s price, and reality was setting in thick. The first of the rescue submersibles, estimated at $3 million, was delivered in 1970. It cost $43 million. It could rescue twenty-four people at a time and dive at most to 3500 feet. One more that could dive to 5000 feet would be built.

Also under construction were two *Alvin* look-alikes made with HY100 spheres No. 1 and 3. The navy had planned to let WHOI operate one and a naval lab run the other. The sub marked for Woods Hole was called *Columbus,* the winning entry in a name contest held by WHOI. The Drugstore had even appointed *Columbus* pilots, who oversaw construction at Electric Boat. But the plans suddenly changed. The navy could not afford to operate the look-alikes, which had grown increasingly unlike the 14-ton *Alvin*. Each weighed about 24 tons, and the pilots said they cost at least $3.5 million each, not including the passenger spheres. In December 1968 they were ceremoniously dunked in the water, christened *Turtle* and *Sea Cliff,* and moved back into the workshop to be finished. When they finally left the Groton shipyard, the look-alikes were put in dry dock with *Trieste* in San Diego until the navy could afford to operate them.

The dozen-odd companies that built minisubs were struggling to find work for them. Westinghouse, which had joined forces with the French to build *Deepstar 4000* and *Deepstar 2000* (the numbers signified depth limit), concluded that as "sophisticated as [submersibles] may seem, their ability to do economically justifiable tasks in the sea is very unsophisticated."

By 1971 *DOWB*, the *Deepstars* and *Aluminaut* were retired. General Motors gave *DOWB* to a local college, but the school could not afford to operate it either, and the orange submersible with no windows was moved to a field overgrown with weeds.

The Alvin Group's future was unclear. The Drugstore's top two floors had been condemned as unsafe, and the Alvin Group was moved to trailers.

WHOI's ONR liaisons, Jack Donnelly and Bob Ballard, found a depressed Bill Rainnie in a small trailer in a parking lot a few blocks from the WHOI dock. The message they brought to Rainnie was one of hope but not without drastic change. They said he had to start charging scientists to use *Alvin*.

It sounded blasphemous. Charge *scientists*, the constituency he had spent years coddling and cajoling to take a dive for nothing? At least charge them a dollar, Ballard said. It was the principle of the thing. And forget the navy, he had to look elsewhere for funding. But he had looked, Rainnie protested. Then he had looked in the wrong places, Ballard said. They needed a long-term project for continuous funding, like exploring the entire continental shelf off the east coast, Ballard said. No, what they really needed was another H-bomb lost in some ocean so *Alvin* could find it and become a hero again. The *Titanic!* Find the *Titanic* with *Alvin*. Now that would really be a feat, the publicity would be terrific.

The ideas seemed to gush from the twenty-eight-year old Ballard. He was brash and cocky but serious and convincing. When Ballard was in college, he got a summer job to come up with missions, *justifications* for deep-diving submarines like *Alvin*. Ballard had the job with North American Aviation in 1962, the year the company was bidding on a contract to build the *Seapup*, the one a confident Andreas Rechnitzer had named *Andrea*.

Rainnie challenged Ballard to put his money where his mouth was. Ballard went straight to WHOI's personnel office and filled out an application for employment. Before his Navy commission expired on the last day of December 1970, Ballard was well into his new assignment as Rainnie's generator of ideas for funding sources and missions for *Alvin*.

The 1971 dive schedule was an unprecedented mishmash of funders, barter and IOUs—much of it Ballard's handiwork. WHOI allowed him to solicit customers from outside Woods Hole and charge scientists for dives. ONR came through with $600,000 and warned that funding for the following year looked even "darker." Other sources contributed another $600,000. The Advanced Research Projects Agency or ARPA, a little-known cog within the Department of Defense that funded high-risk, innovative research, sent $70,000. The recently created United States marine research agency NOAA (National Oceanic and Atmospheric Administration) bought several weeks of dive time worth $60,000. The University of Miami paid for a few dives. A group from the Bedford Institute of Oceanography in Nova Scotia and Lehigh University in Pennsylvania promised to kick in funding the following year if they could dive for nothing in 1971. A U.S. Geological Survey geologist was allowed to dive in exchange for lending Woods Hole a survey instrument which Ballard planned to use for his own dives to conduct research for his Ph.D. thesis in geology.

That degree had eluded Ballard. Scripps had turned him down for graduate school, so he went to the University of Hawaii for a year and a half. To support himself and his wife, he trained porpoises at Sea Life Park in Honolulu, and after navy duty in Boston as liaison to ONR-funded scientists in New England, he applied to WHOI for graduate school. He was turned down because, the front office said, he was an employee, Rainnie's salesman. Ballard suspected he was rejected because his grades were too low. But he applied to the University of Rhode Island and was accepted. Another of his accomplishments was gaining his own funding from the National Science Foundation for his graduate research with *Alvin*.

Ballard was a great salesman and savvy enough to call on well-funded scientists, such as Ruth Turner, a Harvard biologist who was an expert on wood-eating molluscs. Turner knew Ballard from his ONR days when she nagged him to get that Ph.D. Ballard knew she dropped boards in the ocean to catch her prey; he asked her if she would like *Alvin* to take down some of her wood planks. She said sure. It certainly would be interesting. There was no direct proof that wood borers lived in the deep sea.

Ballard called WHOI biologist Fred Grassle who would be responsible for taking down Turner's wood. But why couldn't Turner herself dive? Grassle and Ballard went to Paul Fye to argue the point. WHOI would not allow women to live aboard *Lulu* and was uneasy about a woman urinating into a plastic cup in front of two men in *Alvin*. No female scientist had done any work with *Alvin*. As it was, no lone woman was allowed on WHOI's ships; there had to be at least two women aboard simultaneously or none at all. (When WHOI's biology department chairman asked why, the front office said that if a woman got sick, only another woman could tend to her.) Ballard and Grassle eventually prevailed and Turner was allowed to dive, but only if another woman accompanied her on the expeditions.

Ballard's touch extended also to publicity. Fye appreciated the importance of blowing the institution's horn, but *Alvin* didn't have much of a horn to blow. The investigations into the accident had concluded that the cables on the cradle should have been replaced; they broke from the stress of wear at sea, where the prime and insidious culprit is saltwater. An outraged McCamis was calmed by Earl Hays who urged him to say nothing of his fight with Rainnie over the cables. If he talked, Hays warned, the navy would probably stop funding the project altogether or take *Alvin* away from Woods Hole. Neither McCamus nor the other crew members who had also expressed concern about the cables said a word. The unspoken threat of losing *Alvin* was real to all of them. The Alvin Group pointed a tacit finger at the Marine Department, under whose jurisdiction WHOI recently had placed *Lulu*.

McCamis stayed out of the heated, bitter arguments that arose when ONR announced that it could afford only one submersible, *Alvin*. Val Wilson, who had been appointed senior pilot of the look-alike *Columbus*, a.k.a. *Sea Cliff,* was made senior pilot of *Alvin*, and McCamis moved to the warehouse to build sampling tools. He would make only eight more dives, and in a few years, leave Woods Hole for good.

Jack Donnelly, whose ONR commission was up, accepted Rainnie's offer to work for the Alvin Group. He hired on as an engineer but almost immediately began pilot training.

The trade newsletter *Ocean Science News* had declared WHOI the "boob of the year" for committing the "goof-of-the-year" by choosing *DOWB* to salvage *Alvin* in 1968. Rainnie wrote the editor: "Are you going to take it back now that she's back and ready to go?" The editors did in their May 21, 1971, issue: "About that "boob of the year award," we take it back, we take it back with real pleasure."

Alvin's fan club of people unschooled in science and technology continued to write. Friends of *Alvin,* the Charles Kavaloski family, from Cheney, Washington, wrote:

To Whom it Concerns
Two young children plus a pair of interested adults are concerned about the mini-sub "Alvin" and wonder if he's been recovered yet....

The Alvin Independent School District in Alvin, Texas, wrote and so did another Alvin.

Dear Sir
Through life I have carried the name Alvin and always disliked it since no one ever called me Al until late in life....

Water Baby

And when that lug put forth that Alvin The Chipmunk Song that almost collapsed me!

For some time I swore I was going to overtake him some day and... try to relocate his beak at a point in rear of his left ear, this with a sharp left hook....

But now having avidly kept track of Alvin The Sub I feel that he (Hell's Bells that she does not sound right—yet a ship is, after all, a she) has finally redeemed the name.

I shall be [on Cape Cod] during my vacation. Might I be permitted to plant a kiss on her bow.... I would hugely appreciate it....

> Respectfully yours,
> *Alvin E. Foss*

Alvin returned to sea in the spring of 1971. Its first port of call was the New England Aquarium in Boston where the Sea Rovers, a national scuba divers' club to which Ballard and Ruth Turner belonged, was holding its annual convention. Ballard arranged to have the Sea Rovers conferees bused to the aquarium. Before the museum opened Sunday morning, there were 200 people waiting in the rain to get a chance to board *Lulu* and see the small white submersible. More than 4000 people came to see *Alvin* during its two-day stay.

In mid-June *Alvin* headed for its first dive site, the bit of ocean that had swallowed it whole twenty months before. The sub was going to set another precedent. And another and another and another....

III Wonderland, 1971–1982

"It's something very like learning geography," thought Alice, as she stood on tiptoe in hopes of being able to see a little further. "Principal rivers—there are none. Principal mountains—I'm on the only one, but I don't think it's got any name. Principal towns—why, what are those creatures . . ."
      Lewis Carroll, *Through the Looking Glass*

Suddenly we realized that we had shrunk down to the point where we had become the microorganisms and Alvin *was a ladybug. It was literally like Alice in Wonderland.*
      Howard Sanders, biological oceanographer

16

The incredibly fresh-looking sandwiches and apples that Howard Sanders carried to his laboratory attracted much attention. The microbiologists could not explain how after ten months at the bottom of the sea the lunches could still look fresh and, in fact, be fresh, untouched by decay. Could it be? No. Perhaps, someone suggested, the food had sat in a pool of battery acid.

The scientists photographed, poked and prodded the three waterlogged apples and three baloney and mayonnaise sandwiches. The apples tasted like apple, even smelled like apple. The concentration of enzymes in the fruit was equivalent to that of fresh apples. The baloney was still pink. Seawater had seeped into the crushed thermos bottle of bouillon but it still tasted like perfectly good broth. The usual amount of bacteria was present in all the food.

Another puzzle was the healthy state of the bacteria. Like most life, bacteria brought up from the deep ocean were usually dead at the surface from the drastic pressure and temperature changes. In the zippered lunch bag, the food had been protected from scavengers, preserved by a combination of high pressure and cold temperature. In the biologists' refrigerator, all the food spoiled in a few days.

Alvin's misfortune immediately sparked a new field of study. While the submarine was being repaired at the warehouse, WHOI's microbiologists tried to duplicate the unintentional experiment. They packed the essence of the same lunch—starch, sugar, protein and lipids—in containers with tiny holes to keep out large animals and, with the help of the engineers in the Buoy Group, fastened the abbreviated lunch packs onto the lines of moorings used by the physical oceanographers for other experiments. When the organic material was retrieved with the buoys several months later, it was in excellent condition, proving that the preserved state of the *Alvin* lunches was no fluke. The metabolism of the bacteria was as much as a hundred times slower in the deep sea.

"The implications of the *Alvin* lunch experiment are obvious," microbiologists Holger Jannasch and Carl Wirsen wrote. "The deep sea is not a suitable environment for dumping solid organic wastes."

The benthic biologists proposed a series of experiments in the same area over several years. They wanted to know, as Bill Schevill put it, how the creatures in and on the muddy seafloor made a living. What did they eat? How long did they live? How would they react to a change in their environment caused by a spill of oil or wastes? Would they stay put and die or move to a better spot? And did their life processes operate at the same numbing slowness of the bacteria?

Alvin's first job in 1971 was to set up Deep Ocean Station or DOS No. 1, the first permanent laboratory in the deep ocean. The site for DOS No. 1 was an area where the benthic biologists had dredged for six years; it also happened to be the general area of *Alvin*'s accidental sinking.

In the next two years, biologists made six trips to DOS No. 1. Fred Grassle disturbed small patches of occupied seafloor with No. 2 fuel oil and fertilizer, and left trays of sterilized, uninhabited mud. The microbiologists injected organic material into the seafloor. Others put down small instrumented jars to measure respiration and found that deep-sea animals needed ten to a hundred times less oxygen than their shallow-water counterparts.

Gilbert Rowe went to considerably more trouble; he had twenty tons of garbage shipped from San Diego for his experiment. Not that California garbage was special. Rowe had done research for a man in San Diego who had patented a garbage shredder. The Californian wanted to know if the ocean was a safe dumping place for regular household waste; if it was, his shredder would be more saleable, especially when backed by scientific evidence. At Rowe's request, he sent WHOI free of charge ten two-ton bales of garbage.

Rowe and his graduate students dumped four of the bales in shallow water (in reach of scuba) off the Woods Hole Yacht Club, and once a month for about a year monitored the artificial reef. Although the garbage was shredded, they could make out portions of tin cans, sneakers, plastic bottles and other unwanted items. As expected, the clear plastic wrapping eventually developed holes and the garbage, like that in any town dump, produced methane.

Next Rowe pushed a bale off *Lulu* at DOS No. 1. During two years of monitoring the San Diego garbage, he was most puzzled by what did not happen. Unlike the land and shallow-water refuse, the deep-sea garbage did not emit methane. He never determined why.

The remaining bales of garbage stacked in a WHOI parking lot behaved as expected in the bacterial nirvana of sea level pressure and temperature, and WHOI's front office asked Rowe to remove them immediately.

After only three months Ruth Turner's wooden planks at DOS No. 1 were riddled with holes and wood borers. The results of her experiments contradicted all the other DOS No. 1 work which indicated that deep-sea life processes transpire extremely slowly. It would take Turner more than a decade to understand why.

Returning to DOS No. 1 was never easy. The tall stakes, topped with paddlelike plates of aluminum, were good sonar targets, but large rocks and hills also produced strong signals, and dives ended often at a rock. It wasn't unusual to spend half a day looking for the station. On a few expeditions, it was never found. Rowe tried to make the search easier by fastening an old bicycle wheel with an odometer to *Alvin*'s sampling basket. Wheel on the bottom, they "hover-drove" in the mud to DOS No. 1. Knowing the precise distance was useful, but only if *Alvin* was traveling in the right direction.

With the lessons of *Alvin*'s accident and perhaps the hindsight of the jinxed Civil War vessel *Hunley* and so many other early combat submarines, *Alvin* was launched now with its hatch closed; passengers did not step inside until *Alvin* and the cradle were in the water. A second pilot stood on the closed hatch in the sail to motor *Alvin* away, and he was returned to the mother ship with the escort swimmers by a Boston whaler.

Changes had been made to *Lulu*, too. The forward tips of the 98-foot-long pontoons were tapered like the bow of a monohull, adding seven more feet. Because she displaced more water, *Lulu* sat slightly higher in the water and better handled her weight load. Instead of cables, the cradle used enormous anchor chains tested to lift more than 116,000 pounds, about five times stronger than necessary. And Big Ed Brodrick got an ice cream maker and a dough hook for the galley.

Rowe hated the lack of privacy on *Lulu*. He hated to go to the mess to watch movies with the crew because he couldn't stand the cigarette smoke. He was thankful that his bunk was one of the four in the new portable van on deck. His assistant, a lowly graduate student, slept below in the long starboard tube lined from stem to stern with bunks stacked three high. The young man didn't mind it so much, but he had some difficulty with the crew. Someone had put two rubber boots filled with hardened cement in his bunk. The senior biologist, who knew it was a message to malodorous passengers, told his student to take a shower.

In keeping with WHOI's policy of never-one-woman, there was at least one other female on Ruth Turner's expeditions to DOS

No. 1. It had taken a few weeks for Bob Ballard and Fred Grassle to convince the front office to allow Turner and Tracy McLellan, a WHOI graduate student, to dive. McLellan thought she was lucky to be allowed to dive at all and luckier that her allotted dive was unsuccessful in finding DOS No. 1; that meant another dive and more time to look around. She counted 263 fish. "For once I finally saw these things that I had only seen pickled," she said. "There it was alive and going by. It was wonderful."

The young biologist positioned a cage on the seafloor, hoping to learn what would happen to the creatures under it once they could not be eaten by larger animals. On her return to DOS No. 1, McLellan was surprised to find that in a year there were not more but fewer creatures under the cage. She concluded that the cage must have interfered with the current, the one that was not supposed to exist in the deep sea, and that the animals in the mud beneath her trap depended on the nourishment of marine snow.

Most of the biologists had no funding for their dives to DOS No. 1 in 1971; the Alvin Group paid for them out of the pot of money from the Defense Department agency to develop instrumentation. The following year, however, the National Science Foundation was impressed enough with the DOS No. 1 work to fund them. When the funding didn't stretch far enough, the Alvin Group let them dive anyway, and if anyone asked, the administrators said that the scientist happened to be along on a dive to test a new instrument.

Alvin also dived off Florida and the Bahamas in 1971 and 1972. In the Florida Straits at about 2000 feet, scientists from NOAA and the University of Miami discovered strange elongated mounds, one as high as 125 feet. Geologist Conrad Neumann called them lithoherms.

Trying to sample one with *Alvin*'s pincers, Neumann said, was like trying to pick up a piece of Interstate 95. He did not have Cliff Winget's hard rock drill; most pilots preferred the quicker procedure of prying loose bits of rocks with *Alvin*'s pincers. Neumann eventually got a sample, but it wasn't until many hours later that he realized what it was.

It was about 4 A.M. and he was reading the notes from his dive. *Alvin* had landed atop one of the huge domes amid yellow sponges as big as catchers' mitts and crinoids which resembled skinny starfish perched on thin stalks. At least he thought they were crinoids. The only crinoid Neumann had ever seen was a fossil. He also saw a feathery creature that reminded him of a fossil from the Paleozoic Era, some 600 million years old. Like crinoids. How, he wondered, could a current, even the mighty Gulf Stream, scour hard rock to form these domes? Suddenly it dawned on him that it wasn't currents but animals which created the lithoherms.

"It was a big day for me," he recalled. "It was like suddenly the movie began to run in my mind and all this came together. I had to tell somebody. Everybody else was asleep. I went to the galley and the cook was kneading bread, and the ashes from his cigar kept falling into the dough. I told him the whole story. He listened, too."

Lithoherms are made of many organisms which attach themselves to hard seafloor, forming a cluster which traps sediment, hardens, and eventually turns to rock. More animals attach themselves, and the mound grows like the concentric layers of skin in an onion.

Two years later, Neumann found more lithoherms, not in the Atlantic but in Kansas, in a place once covered by an ancient ocean. They, too, were covered with similar organisms, but these were fossils.

In other dives off Florida, geologists found a 200-foot-wide crater and concluded it was a sinkhole. The pilot lowered *Alvin* about a hundred feet into the hole, but not daring to go further, brought it back up. *Alvin*'s resurrection apparently surprised an eight-foot marlin which charged the sub. "It was very deliberate," pilot Ed Bland said. "There were actually two blue marlin. They slowly circled the submarine, maybe twice, and suddenly one of them just came charging right at the two lights on top of the sail and destroyed them thoroughly." The lights went out with a loud crash and the fish fell in a woozy dance to the bottom, dead.

Following that episode, WHOI's engineers thought it prudent to try to answer those questions about plastic portholes that first arose in 1967 when *Alvin* was attacked by a swordfish. To simulate an attack in the laboratory, they shot a half dozen fish swords with an air gun at one of *Alvin*'s spare portholes.

An ingeniously engineered weapon, the fish sword is a honeycomblike structure about three feet long, filled with fluid to cushion the shock of a hit. Charging at 60 miles an hour, it can do a lot of damage. Reports of swords penetrating wooden keels more than a foot thick were not uncommon. But as strong as the swords were, the methylmethacrylate was stronger. The swords bent, bowed and broke; the porthole suffered only a few scratches. Knowing that a fish sword moving at top speed couldn't even chip Plexiglas, the engineers were confident that *Alvin*'s three-and-a-half inch thick windows were safe from curious marine life.

Not everyone shared the same logic of what constituted safe. For some scientists, there was no *safe* in a sealed seven-foot sphere in deep-sea pressure. While this reasoning might not be logical in the least, it was preferable to the death grapple of claustrophobia. Few admitted the fear which kept them out of *Alvin*. One scientist, who as a child had been trapped in a cave, tried. He got as far as the sail. Elazar Uchupi didn't get too far either.

Uchupi was a senior WHOI geologist whose Basque roots were easily discernible in his accent. He had become accustomed to Bob Ballard banging on his door at midnight with a Coca-Cola, a rolled-up seismic map and a question. He enjoyed the graduate student's inquisitiveness and sharp intelligence. And he accompanied Ballard on several expeditions to the Gulf of Maine in 1971 and 1972.

Ballard tried everything he could think of to get his mentor into the submarine; he even told the pilots to forego the predive briefing about what to do if, in the most improbable of events, the pilot had a heart attack. But Uchupi would only look up through his bushy eyebrows and wag a finger in Ballard's face. Ballard made his first *Alvin* dive in July 1971, and by the end of the 1972 season, he made twenty-three more, holding the record for the scientist with the most *Alvin* dives.

The Gulf of Maine dives were part of Ballard's thesis research on plate tectonics, the theory that the continents ride on slow-moving blocks of the earth's crust. The idea of continental drift was championed in 1912 by German meteorologist Alfred Wegener, who suggested that our present five continents were once part of a single primordial landmass called Pangaea which split into five parts. It seemed logical from looking at a map: South America's knee bend fit neatly into Africa's armpit. But the theory caused great controversy and Wegener was ridiculed by many.

A major problem was that Wegener didn't have any idea about how ocean floors were made and destroyed as the continents drifted about. Indeed, nobody understood.

Geologists remembered Wegener a few decades later when their echo sounders identified mountains in the deep ocean—first in the Atlantic, then in the eastern Pacific and Indian oceans. In the 1950s, soundings showed that these ranges were actually part of one colossal chain of mountains girding the entire planet like the seams on a baseball. This 40,000-mile-long mountain chain was named the midocean ridge. In some places, the tips of the mountains surfaced as islands, such as Iceland. Running down the center along portions of the ridge was a massive canyon or rift; the scientists called it a rift valley, about as deep and as high as the Grand Canyon. In the 1950s, Harry Hess of Princeton suggested that this rifted mountain range was the site of seafloor spreading. As the continents drift, the rifts widen and are filled in with lava which hardens and becomes new seafloor.

But dramatic proof of this phenomenon came from measurements of magnetism in the seafloor. As hot lava cools and hardens, its magnetite crystals align themselves with the earth's magnetic field, remaining forever locked in that direction. From time to time, the earth's magnetic poles switch, reversing the direction of the magnetic field. Geophysicists learn much from these stripes of new seafloor: their alignment indicates polarity; their width and the pattern of wiggles indicate age.

However, these squiggles were not understood until the Vine-Matthews Hypothesis and Fred Vine's paper of 1966, which set geophysicists afire. The "Vine Bible" proved that there were distinct patterns to magnetic recordings of the seafloor. The patterns on the edges of two separated plates matched up perfectly.

Continental drift seemed after all to explain the earth's most prominent geological feature, and Wegener's ideas evolved into a unified theory of plate tectonics. Geologists counted seven major plates. Driven by the process of convection caused by the heating in the earth's interior, the plates pull away or push into each other, grind against or slide beneath one another. Where two plates spread apart, molten earth from deep beneath the ridge erupts and hardens to fill the crack, adding more real estate onto each of the drifting blocks. Where plates collide in the ocean, deep trenches and arcs of volcanic islands such as the Marianas are formed. Mount Everest, if placed in the 35,800-foot-deep Mariana Trench, would still have more than 6800 feet of water over it. On land, clashing plates create mountains. When India rammed into Asia, the Himalayas were formed.

If the two plates in the North Atlantic separated, Ballard reasoned, there should be evidence of it in the continental shelves. At about the time the continents were thought to have pulled apart in the North Atlantic, a structural rock formation unique to the separated plates developed. This formation, called the Newark System, had been found in the Appalachian Mountains. Ballard determined to look for the Newark System in the Gulf of Maine, where the Appalachians continued but submerged.

Ballard did find pieces of the Newark System, rocks that could not have been found blindly with a dredge from a ship, because these were lying beneath other kinds of rocks. He used the land geologists' method—a combination of mapping, sampling and observation. It didn't come naturally to most marine geologists; in fact, many told Ballard it was impossible in the ocean, even with *Alvin*.

But those who used the submarine were learning otherwise. When Lamont marine geologist Bill Ryan dived with land geologists, he was amazed at their proficiency. "These colleagues I invited had never been to sea," Ryan said. "They did their work in the Alps and the Dolomites. But even on their very first dive, the field geologist is a far better *Alvin* user than the marine geologist. They know what to look for and what to describe. They're very good at figuring out where they are. Maybe because they're used to working in the fog and clouds."

Ballard's biggest challenge was mapping—knowing precisely where a sample had come from. He achieved moderate success with the help of Skip Marquet's navigation system, which was still experimental. The pilots didn't put up much of an argument when Ballard insisted on operating the mechanical arm himself to collect samples. On the early dives most scientists became so impatient with their efforts—understandably, they hadn't practiced—that operating the arm had become another of the pilots' jobs.

The crew called Ballard the White Tornado, after a TV commercial that likened a cleaning agent's powers to a whirlwind. They wondered if he slept. It wasn't easy keeping up with Ballard, especially in a place like the Gulf of Maine. The weather turned violent and it was always cold. Those on *Lulu* slept in their clothes. Despite wetsuits, the escort swimmers dashed to the head after every launch and retrieval, and stood under the steaming hot shower which scalded their tongues, the only portion of their anatomy that wasn't numb.

Each night after the dive when *Alvin*'s batteries were being recharged and all but the night watch slept, Ballard conducted seismics from *Lulu*. This involved towing hydrophones and a large air gun which went off every thirty seconds. Recorders on the mother ship graphically captured the echoes of the booms. It was a lot safer than throwing TNT over the side, but still noisy. The men trying to sleep in the Tube of Doom felt as if they were in a submerged belfry with sonic bells. *Boom! Boom!* every half-minute. One of *Lulu*'s sailors was particularly annoyed. The man had that pasty complexion ships' oilers get from working in a sunless engine room. So vexed was the sailor that one night, after swiping a bottle of cooking sherry from the galley, he confronted Ballard and threatened to kill him. George Broderson, the ever-present hunting knife on his belt, rushed to the commotion. Ballard refused to stop the seismics. "I quit!" the sailor declared. Broderson escorted him below deck and Ballard went back to work.

He hooked up a speaker to the sonic gear in the mess and stretched out on a bench. If something went wrong with the seismics, Ballard reasoned, the absence of booms from the speaker by his ear would awaken him. He sank his head on a pillow and suddenly sensed another presence in the room. This time, the sailor had a knife. But Broderson was on his heels and wrestled the weapon away from him.

Ballard was too riled up to continue working. He went to the small bunk van on deck. "That loony tried to kill me," he told his technician, Earl Young.

"Yeah?"

"Yeah."

"Go to sleep."

Young got up to relieve himself. He had a leg out the van when he saw there was no deck to step on. The waves had pushed up the plywood deck planks. He stopped his fall by grabbing the door and, cursing loudly, hung on as he swung out over the churning black ocean.

17

The Alvin Group wanted to go deeper because there was deeper to go. Through a combination of navy red tape and Paul Fye's conservatism, *Alvin*'s depth limit was kept at 6000 feet, but the group knew their sub could easily reach 10,000 feet. There was some doubt, however, about whether the pump in the oil ballast system would work efficiently at that depth. But the Drugstore thought it certainly could handle 8000 feet. After years of trying, Bill Rainnie convinced Fye in 1967 to agree to extend the depth to 7500 feet; and the Office of Naval Research formally issued the request to higher-ups in the navy. But when all parties met to review certification standards, an ongoing saga, the depth extension was not discussed. Why, the navy wanted to know, had *Alvin* exceeded its depth limit by a few hundred feet, not once but three times? Bill Rainnie said it was unintentional, the depth recorders blinked out.

"It wasn't dangerous," Skip Marquet said. "These were like excursions beyond the speed limit and the rule was, you were supposed to report yourself for speeding. If you didn't, they would remove your certification and put you out of business."

The same year the Alvin Group sent ONR a proposal for funding to build a craft for 20,000 feet. While this depth did not demand titanium, it was the obvious metal of choice because of its light weight. Titanium was 40 percent lighter than steel; thicker walls to withstand the greater pressure would not add as much weight as a steel sphere. ONR rejected the proposal, but the *Alvin* engineers, still hopeful of finding a backer, visited various naval labs in Washington, D.C., to talk up the idea. They won over an old friend, the Naval Applied Science Laboratory, which had tested the purity of the HY100 used to make the spheres. Making *Alvin* a titanium passenger sphere would be a kind of ultimate experiment capping the lab's years of research on titanium. In January 1968, the lab told Rainnie it had the go-ahead to order titanium alloy 621.

To Fye, the major advantage of a titanium hull was an increase of a few thousand pounds in payload, not depth. A titanium sphere identical to the HY100 hull would not increase *Alvin*'s depth; in fact, it wouldn't be as strong as the original sphere. But greater depth was unquestionably what fired the Alvin Group and the naval engineers in cahoots with them.

That spring the Alvin Group and navy engineers put the specifics on paper. It was feasible, they wrote, for a titanium sphere with slightly thicker walls to have a collapse depth of 18,000 feet, which would double *Alvin*'s depth. However, they continued, since less was known about the behavior of titanium than steel, and even less about a titanium sphere under great external pressure, "prudence would require a more conservative approach in establishing an operating depth for the titanium." They suggested a 9000-foot depth, and to confirm their numbers, pressure-testing a second titanium sphere to destruction. The cost of building and testing both spheres they greatly underestimated at $1.8 million. Thus, the navy's Project Titanes was born.

First, the navy made *Alvin* a new pump for the major ballast system, which was installed in the spring of 1972. The oil in the six small titanium balls was drained for good. By moving from the balls to rubber-sided compartments, the oil had changed *Alvin*'s displacement. The new system used seawater, let in and out by an ingenious ceramic pump to vary weight. Unlike any other pump, its pistons were lined with titanium dioxide ceramic, allowing it to move seawater under pressure without the gears being plugged up or scoured by the millions of gritty particles that are everywhere in the ocean. While it would require several years to perfect, the pump even as a prototype was a marvel.

Navy engineers also designed for *Alvin* what they called "bells and whistles," such as special valves and meters. But the Alvin Group distrusted any unnecessary item, especially unnecessary experimental gadgets, and to the surprise of the naval engineers, rejected the bonus parts. "They were very protective of their submersible," said John Sasse, one of the navy project engineers. "They wouldn't let anybody put any trash on it."

The big billets for the spheres were rolled at Lukens Steel and pressed into four hemispheres, which went to the navy's Mare Island Shipyard in San Francisco Bay to be welded together into two spheres with the same outside diameter as *Alvin*'s steel hull; the inside diameter was sightly less because the titanium sphere's walls were .60 inches thicker. As far as anyone knew, they were the largest titanium objects in the world.

The depth to which the new sphere would go remained a matter of debate. The navy and Alvin Group engineers now said it was designed for 12,000 feet, based on a 1.5 safety factor. Fye wanted to keep the 1.8 safety factor, which would limit the depth to 7200 feet. But he compromised. Pending the pressure tests, he instructed his public relations man to refer to the new depth as "about two miles" or "about 10,000 feet."

During the winter and spring of 1972–73, *Alvin* was overhauled in Woods Hole and fitted with its new sphere; and housings for lights, cameras and other outside gear were strengthened for the pressure at 12,000 feet. In July the sub and its crew arrived at the navy's Annapolis lab. Before it was shipped to Woods Hole, the naked titanium sphere had endured unscathed the simulated pressure at 13,200 feet. This was its second trial in a navy pressure chamber, and this test would involve passengers. Rainnie, Sasse and another navy engineer had planned it over drinks one night on the Cape.

The 27-foot-long pressure chamber, still the world's largest, lay on its side; and the 100-ton lid of HY100 steel rested nearby on rails, ready to slide into place. But the ten-foot-wide chamber was not quite big enough to fit *Alvin* even with the sail off, so the power sanders came out and the syntactic foam on *Alvin*'s sides was whittled until many inches of the submarine's girth lay in piles of powdery dust. Slightly slenderized, *Alvin* slid nose-first into the tank and twice endured the pressure at 12,000 feet. For the third cycle, pilots Val Wilson and Ed Bland climbed into *Alvin*. The naval officer scheduled to join them developed heart palpitations. Another navy man was sent for. When he finally arrived and climbed in, the two-foot-thick lid was shut and the tank was filled with water. If at any time communication with the passengers was lost, the test would be aborted immediately. As the pressure climbed to 12,000 feet, the tremendous *thud* of an implosion inside the chamber set everyone's heart to galloping. "Everybody was tense anyway and it certainly was unnerving," Sasse recalled. "But obviously nothing major had happened. The tank didn't blow apart and we were still talking to the people inside."

They would not learn until the end of the test ten hours later— too many hours, as far as the passengers were concerned—that one of *Alvin*'s lights had imploded.

Based on the third cycle with people aboard, *Alvin* was navy-certified for 12,000 feet. But *Alvin* would not go to that depth in the ocean for almost two years. Fye forbade dives deeper than 1200 feet.

His action was prompted in part by three underwater accidents, all the summer of 1973. In June two men died in the *Johnson-Sea-Link* a few miles off Florida when the submersible got hung up on a junked destroyer at 350 feet. Two months later another minisub, *Pisces III*, flooded because of an open hatch and sank in 1575 feet of water off Ireland. The two passengers were rescued unharmed. The same summer a fire broke out in France's bathyscaph *Archimede*.

Fye was also heeding the advice of Jim Mavor. On returning from sabbatical and writing a book about Atlantis (Mavor said the legendary city was in the Aegean; Val Wilson swore it was in the Atlantic because on an *Alvin* dive he had seen monstrous columns off the Azores), Mavor accepted a new job as WHOI's first safety officer. The summer's accidents, he told Fye, proved the dangers of deep-diving and *Alvin*'s passengers were "not immune to becoming such statistics. In my judgment, deep diving by *Alvin* is not safe under present and planned conditions...."

Mavor itemized nineteen wide-ranging concerns. During launch and retrieval the crew may take "occasional unwarranted chances... due to over-confidence." He said the freeboard was inadequate; in high seas water splashed inside the sub. But Mavor's biggest worry was *Alvin*'s new penetrators.

The twelve penetrators held the wires that led from the batteries in *Alvin*'s belly to small holes around the portholes. The new penetrators were virtually identical to the old ones but they were of titanium, not steel, and there were more of them. There were no problems with the old penetrators, and the new ones did not leak in any of the pressure tests. But Mavor worried that at twice the old depth, the penetrators might pop out of their holes.

In technical terms, for a penetrator to pop out, its coefficient of friction has to be close to zero—as it is when you step on your car brakes while driving on ice and the car keeps going. Friction is what makes the brakes work and friction is what keeps *Alvin*'s penetrators from slipping out.

Pressure constantly pushes on the sphere, compressing ever so slightly those holes the tapered penetrators thread through. Pressure also pushes against the penetrators, making their fit snugger. Like every external part of *Alvin*, they have to be able to stretch and constrict repeatedly with changing pressure. The outside nut and rubber gasket on each penetrator are only weak seals meant for shallow-water pressure. The high-pressure nuts are inside the sphere. If the nuts are too tight, they might fight back, and something might break or perhaps the threads would erode. Even a microscopic leak could lead to catastrophe; a fine spray of seawater could enter the sphere with enough force to decapitate a passenger.

Exactly how much elasticity *Alvin*'s components needed was one of those nagging questions that want hanging. In earlier days, it was the taxi mechanic, George Broderson, who through default decided how much to tighten the inside nuts on the penetrators. He whacked the wrench with a sledgehammer, using a force that depended on his gut sense of what felt right, and what felt right to him was getting those suckers just as tight as he could. When a BuShips officer, bristling with authority, asked how tightly the nuts were torqued, Broderson gravely intoned *125 foot-pounds, sir.* It became the standard, not because Broderson said so, but because it worked. Knowledge gained from trial and error was better than no knowledge. The new penetrator nuts enjoyed the same torque, arrived at from experience, not an equation. And now a torque wrench, which had a dial to indicate pounds of force, was used.

As far as the Navy was concerned, *Alvin* had passed certification and could dive to 12,000 feet. The arguing and searching for answers was a Woods Hole affair that would drag on for months as the engineers tried to answer Mavor's question. *Alvin* made only sixteen dives in 1973, none of them deep, and the following winter and spring spent more time on the dock than in the water. The engineers experimented in the laboratory and in *Alvin* during its ocean dives—to 10, 15, 132 feet—one dive at a time, and each authorized by Fye and instrumented with dials to record how much the penetrators moved. If they moved more than five thousandths of an inch, Fye said, the sub had to surface.

The penetrators did not move that much, but they did move and their movement was confounding. Rather than move relative to the changing amounts of pressure, as expected, they moved erratically, and the engineers could not explain the numbers spread all over their charts.

They tried different kinds of lubricants. Without some grease, the metal-to-metal contact between the titanium cone and its matching hole could actually cause the parts to weld together. But the various lubricants made no difference.

The engineers stretched their imaginations. What would happen if someone accidentally used the wrong grease or spilled something into the right grease? They mixed the lubricants with fresh water, saltwater, Hoover oil and the other fluids aboard *Lulu,* and slathered the concoctions onto the penetrators. Nothing made a difference.

They tried a different kind of inside washer on the penetrator. Now they did have a problem. The new washers didn't make any difference or they leaked, and the engineers were at a loss to explain why—why didn't they always leak or why didn't they always form a tight seal?

Everybody was frustrated and impatient. *Alvin's* deepest ocean dive in 1973 was to only 3822 feet. The easy-going, mild-mannered Bill Rainnie, who had been snapping pencils in two during meetings on the "penetrator problem," left the Alvin Group for good. According to those who worked with Rainnie, *Alvin's* first pilot was burned out, weary of the battle for funding and the more recent battle to prove that the sub was safe. Before leaving, he hired Larry Shumaker as his replacement.

Shumaker, who had been in the navy since he was twenty-two, had piloted the *Trieste* and for the past seven years, Lockheed's *Deep Quest,* which had fallen on the same hard times as its kindred craft and was permanently shelved for lack of customers. Shumaker, forty-one, missed piloting under water. He took a sizable salary cut to be the Alvin Group's new manager and eagerly looked forward to the challenges ahead.

They were not quite what he expected. Not only was his authority to approve dives and ensure *Alvin*'s safety usurped, he faced the imminent cessation of all ONR funding. The indigent *Alvin* had not yet made a deep ocean dive with its shiny new titanium sphere, worth more than $7 million.

Soon after his arrival in April 1973, Shumaker called together the Alvin Group—men he hardly knew, but this made his unpleasant duty slightly easier. He urged them to look for other jobs. A few days later, Shumaker excitedly called in his staff again. They were back in business, at least temporarily. The government had decided to fund *Alvin* for an international expedition to study the range of mountains at the bottom of the Atlantic.

But the Mid-Atlantic Ridge was about 10,000 feet down, and *Alvin* hadn't even approached that depth. Fye held a meeting of engineers and aides to discuss extending the depth to 8000 feet. "Unprecedented and dangerous," Mavor said. "Nothing less than Russian roulette." But only Mavor dissented, and on January 29, 1974, Fye authorized dives to 8000 feet.

The months of destructive testing of the second titanium hull in a navy pressure chamber did not begin until February, and the results would not be in until that summer. Perhaps if these tests had been available earlier, *Alvin* would not have had to endure more than six months of diving in 500-foot increments with penetrators instrumented to record their infinitesimal creeps.

According to the calculations, the sacrificial titanium sphere would implode at 19,035 feet. But like the steel HY100 ball No. 1, it refused to implode. It didn't even crack, not even when the engineers drilled tiny holes and made tiny cuts in the welded seams. Hull II endured lifetimes of brutal pressure. In the test designed to destroy it, Hull II was held for seven days between the pressure equivalents at 18,000 feet and 22,500 feet, the chamber's limit. Seven threaded studs holding in the portholes blew out, and a couple of penetrators, the ones with the new washers WHOI was experimenting with, leaked.

In February, Shumaker moved to the offensive. *Alvin* had made four dives to 8000 feet, he argued, and no penetrator had moved more than .0055 inches. If *Alvin* didn't get to at least 10,000 feet once, how could it join the project at the Mid-Atlantic Ridge? He didn't need to add that if *Alvin* did not participate, it probably would not dive again, at least not soon and not with Woods Hole pilots.

Actually *Alvin* had made three dives to 8200 feet, one to 7700 feet and three to 7200 feet; the rest were shallower, and two of them were aborted because of penetrator "weeping." But Fye finally agreed to extend *Alvin*'s depth limit to 10,000 feet.

The following day, March 5, 1974, pilot Val Wilson shook Barrie Walden in his bunk on *Lulu*. For the past half year Walden had charted the penetrators' creeping with various greases. None of the graphs had ever made any sense. Walden thought Mavor's question was valid, but by now he had concluded that they were not going to meet Mavor's demand for "a thoroughly understood and permanent solution to the penetrator problem." Engineers Arnold Sharp and Skip Marquet agreed. "Those things were overdesigned by yin yang," Marquet said. "There was no question in my mind about their safety."

But the pilots understood less than the engineers, and that morning before the Bahamian sun peered above the horizon, Wilson was at Walden's side pushing against his shoulder.

"Come on, get up. Wake up!"

"What for?"

"You're goin' "

"Where?"

He was going to 10,000 feet.

Walden spent a few moments recalling his calculations, and without argument, made the deep dive.

There were no leaks at the penetrators.

18

Alvin's lucky opportunity, regarded by WHOI as the submarine's last chance, was decided at the highest levels of diplomacy, not science. The United States was laying the diplomatic groundwork for the second international Law of the Sea Conference.

The nations of the world first gathered in 1958 to begin to define one another's claims to the ocean, such as rights of fishing and minerals. In 1967 President Lyndon Johnson's Marine Council, a cabinet-level advisory board, recommended a second Law of the Sea Conference. In an effort to ensure that a spirit of cooperation would reign, the Marine Council suggested initiating joint oceanographic projects with other countries.

The council's executive secretary was Ed Wenk, who had designed the *Aluminaut*. Wenk now became a traveling salesman bearing goodwill toward other nations in their mutual stake in the ocean. In France, he found a kindred spirit in Yves LaPrairie, director of the government's oceanographic agency CNEXO or Centre National pour L'Exploitation des Oceans. They talked about deep-diving submarines, and in the course of several visits came up with the idea for a joint U.S.-French project to explore a portion of the Mid-Atlantic Ridge. In 1968 Charles DeGaulle and Richard Nixon discussed the idea, and the proposal gained real momentum.

CNEXO's chief scientist Xavier Le Pichon was not convinced that submersibles were any good for oceanography. He wrote WHOI geologist K. O. Emery for an informed opinion. Emery asked graduate student Bob Ballard to respond and convince the skeptic Le Pichon. Emery signed Ballard's enthusiastic letter.

Nobody had ever been to any underwater portion of the ridge. At least two WHOI geologists had suggested it to Paul Fye years before when they learned that *Alvin*'s depth (with the HY100 sphere) might be extended enough to reach it. Such an adventure tantalized many oceanographers, and word spread about the proposed French-American Mid-Ocean Undersea Study, or FAMOUS.

To gather formal consensus the National Academy of Sciences hosted a symposium at Princeton University in January 1972. Ballard gave an invited presentation on *Alvin*. He spoke of his success at finding the Newark System and more ancient rocks in the Gulf of Maine. It was proof, he said, that classic geology—mapping, observation and sampling—could be done under water. He himself had done it, what so many of them said could not be done.

When Ballard finished his talk, Frank Press, a geophysicist at the Massachusetts Institute of Technology, who would eventually become the president of the National Academy of Sciences, challenged Ballard to name one significant piece of science that had ever come from using a submersible.

Ballard heard another geologist reply: "There isn't any because nobody has tried to do any significant science from a submersible."

It wasn't true, but the record set forth in peer-reviewed journals, the only record that counted in science, was mediocre and incomplete. There were only about a half dozen papers reporting on science conducted with *Alvin*—on submerged canyons, new species of copepods, the continental slope, and the sound-scattering layer Alexander's Acres. The discoveries made by more recent divers were not yet published. The only paper Allyn Vine ever wrote on his *Alvin* dives (on acoustics) was classified by the navy.

Besides the written record, the scientific community was still extremely skeptical about the need to get inside the deep ocean. But the numbers of supporters had grown and those at the Princeton gathering voted to participate in FAMOUS with *Alvin*. Only a few dozen of them would actually join the expedition. They were voting not on personal involvement but on how the shrinking pot of federal research dollars should be spent. Maurice "Doc" Ewing of Columbia's Lamont-Doherty Geological Observatory shook a finger in Ballard's face and threatened to melt *Alvin* down into titanium paper clips if FAMOUS wasn't worthwhile.

The federal funding agencies did not immediately approve the funding, and more than once the American scientists wondered if there would be a FAMOUS. The agencies grilled Woods Hole repeatedly about the qualifications of the diving scientists who were selected by Ballard, still a graduate student, and WHOI geologist Bill Bryan. In addition to themselves, they picked Jim Moore of the U.S. Geological Survey in California; Tjeerd van Andel from Oregon State University, and George Keller of NOAA. But the government finally let them have their way and put WHOI geologist Jim Heirtzler in charge. On July 4, 1972, both countries made it official. The United States would use *Alvin* and France would bring the government-owned bathyscaph *Archimede* and the small deep-diving sub *SP-3000*, renamed *Cyana* and upgraded to reach a depth of 9843 feet.

Based on the likelihood of clement weather and the proximity of land (needed for mechanical problems and other surprises), the teams marked off a presumably typical section of the ridge about 60 miles square, some 400 miles southwest of the Azores. Then they learned everything they could about it from the surface.

Ballard convinced the navy to sweep the area with its classified sonar; aircraft carrying magnetic sensors flew over the site; ships surveyed with other gear that detected scores of small but almost continuous earthquakes. *Mizar's* cameras took 5000 pictures. WHOI's camera sled ANGUS shot thousands more.

ANGUS (Acoustically Navigated Geological Undersea Surveyor) was built for FAMOUS to carry two deep-sea cameras that geophysicist Joe Philips found languishing in a basement at WHOI. The cameras and pinger were housed in a framework of welded pipes which after a few towings over the FAMOUS area sagged at both ends, and a camera dropped off. But ANGUS' photographs were so good that nobody wanted to give up on it. Philips and Bryan dragged it to a welder in Ponta Delgada. In two days he presented them with a new frame of thicker pipes that was much stronger and more elegant. The welder put a curved pipe at one end and instead of plywood used metal for the small fins at the stern. The Portuguese ANGUS frame would endure about a decade of being bashed against rocks.

From these sources and others, the French and Americans had an enormous amount of information and they hadn't even gotten under water yet. They had the finest bathymetric map possible, the most detailed map ever of any part of the ridge and more than enough photographs to fill a gymnasium.

The pictures were laid out in a mosaic on the floor of a naval gym in Washington, D.C., and the scientists spent days walking back and forth among the photographs to familiarize themselves with the terrain that awaited them in the Atlantic. Pilots and scientists went to Hawaii and Iceland to climb volcanoes, what they expected at the Ridge. Those who had never been under water dived now. Only George Keller and Ballard had been in *Alvin*.

On Bryan's training dive to 918 feet off the Bahamas, he noticed water dribbling from a penetrator. "Do you usually have water coming inside?" he asked pilot Jack Donnelly. "No," Donnelly said.

Because there was always condensation inside the sphere, it was almost impossible to tell by looking whether a penetrator was actually leaking. Bryan ran his finger along the dribble and tasted it. "Well, you do now," he said. "That's saltwater!" Donnelly calmly dropped the remaining squares of steel on *Alvin*'s sides to ascend.

Neither did volcanologist Jim Moore during his training dive in the fall of 1973 find a well-running submarine. Moore scribbled in his notebook after his dive to 132 feet:

> Aborted. Leak in system. Realization that many things need debugging on the submarine. Main problems with *Alvin* sensors and [the new ceramic pump] system not working. Insufficient freeboard to prevent splash into hatch. Crosstalk among electronic gear causing poor operation of underwater telephone, sonar and television screen....

The French grew antsy waiting on *Alvin*. The summer of 1973, they invited Ballard to join in the first dives at the FAMOUS site in their bathyscaph *Archimede*. On the second dive, the one with Ballard, a fire broke out and the passenger sphere filled with smoke. All three passengers put on oxygen masks. When Ballard tried to take his off—there was no oxygen coming through it—the French thought he was panicking and pushed it back onto his face. The more he struggled, the more the French pushed. In a last desperate attempt to make himself understood, Ballard grabbed his throat and the French pilot turned on his oxygen.

Archimede made five more dives and withdrew until the following summer, when it would dive with *Alvin* and *Cyana*.

Alvin made it with no time to spare. Following its deepest dive to about 10,000 feet on March 5, 1974, the sub made only four science dives, all with the Deep Ocean Station biologists, to retrieve experiments. And after five more test dives it was time to go. On June 6, WHOI's new ship RV *Knorr*, which carried *Alvin* and towed *Lulu*, headed for the Azores.

The day before leaving, Ballard successfully defended his thesis. Now he was Dr. Ballard.

Knorr carried a full twenty-four-person complement of scientists, graduate students, technicians, and two members of the press—a reporter from the *New York Times* and Emory Kristof, a photographer from the National Geographic Society, to document the first human probe to an underwater seam of the planet.

The FAMOUS site was twenty miles long and ranged in width from about a half mile to two miles. It was marked at each end by a volcano, subsequently called Mount Pluto and Mount Venus. The French team dived at the latter, while the Americans explored around 700-foot-high Mount Pluto.

For this mission Cliff Winget concocted a water sampler from the suction cups of two toilet plungers. When triggered, the cups clapped together to enclose a tube that collected water. The various names for the gadget would not be allowed on prime-time television, but it worked.

Alvin also carried Skip Marquet's Data Logger, a small box of electronics that automatically recorded depth, altitude, gyro heading and time. With the Data Logger and Marquet's navigation system, the scientists knew with more precision than ever before where *Alvin* was.

The new navigation system used transponders. Three of these sophisticated pingers were lowered to tread slightly above the seafloor so that fixes could be made from the triangular network. Instead of constantly pinging, the transponders were programmed when to ping and how to ping. Their built-in clocks were synchronized with clocks in the transponders on *Alvin* and *Lulu*. By the fourth dive, *Alvin* was landing consistently to within fifteen to thirty feet of the designated spot. The geologists took the bathymetric map with them on their dives and plotted each sampled spot. For the first time, they could tell where a rock came from relative to another rock. "We were really proud of ourselves," Marquet said.

The pilots were proud, too, as evidenced from the comments jotted on the dive log: "highly successful" for the first dive; and for the second, "very successful—no penetrator leaks at all."

Maintaining the navigation network was hard work. Marquet plugged in the coordinates with a borrowed hand calculator because the Alvin Group couldn't afford a computer. Porpoises, wonderful mimics, played havoc with the transponders. Once they figured out a transponder's unique pattern of pings they could talk it to death, that is, exhaust its batteries. A transponder lowered near a high shelf or outcrop was ineffectual; its sound pulses bounced off the geology in its way.

As the submersibles dived, the surface teams dredged rocks and towed temperature sensors and other instruments. One day the dredges were lowered seven times from *Knorr* and came back with only two rocks.

Because of the earth-birthing process—magma belching upward from deep within the planet's hot interior and producing new strips of seafloor—the oceanographers expected to find slightly higher than normal water temperatures. But nobody found anomalous temperatures or fire or brimstone. Creeping along a seam of the earth was like driving in first gear through a light snowfall at night. The geology was spectacular. The nearly vertical west wall rose about a thousand feet. "Your eye doesn't believe it," geologist Tjeerd van Andel of Oregon State said.

There was no single crevasse marking the plate boundary through the center of the rift valley. The topography was much more complicated. Crevasses and faults bisected and ran parallel to the valley. The seafloor had been pulled apart, not forced open by flowing lava. Molten rock had not erupted in any thunderous outpouring; it had oozed out as if from huge tubes.

Nobody had ever seen so many shapes of hardened lava, all needing names. Toothpaste, trapdoor, cousteau, breadcrust, broken egg and phallus basalt would make it into the scientific literature. But there were also peanuts, elephant trunks and swans. And there were the popcorn rocks that cracked, sizzled and jumped across the ship deck with escaping gas. The tiny explosions were apparently triggered by the pressure change from the depths to the surface. The geologists had dredged these before FAMOUS, but they still couldn't explain why other rocks from the same area didn't behave the same way.

When red-hot lava reaches the cold wetness of the ocean it quickly contracts, cools and hardens. The quenching is so rapid that the molten rock has no time to crystallize; instead, it forms a thin crust of pure glass around the black pillows of basalt.

The geologists inferred that the pillows at the Mid-Atlantic Ridge were geologically very young or "fresh" because they were not covered by sediment. "It was an incredible experience," Moore said. "Every outcrop was different, and in an area of such extremely fresh rock. It was a geologist's dream."

Moore and Bryan decided that perhaps the reason nobody found an above-normal temperature was because the hot water cooled as soon as it came in contact with the normal frigid water at the seafloor. A fissure, where the water was confined, seemed like a good place to look for it. About midway through the expedition, on *Alvin*'s 526th dive, they discovered one.

"Look at that!"

"Oh...this is one we can go down into it's so big."

"Let's do that."

"*Look* at that mother!"

"The time is 1409 and we are in the fissure. The width of the fissure..."

"Oh, my lord, the *size!* Look at that thing!"

"We've sunk down, oh maybe six feet into this fissure..."

"Depth is 2552 [8373 feet]. We seem to be touching both walls. The width is, ah, about twelve feet. The width of *Alvin*. Well ...that's funny."

"*Alvin*, this is *Lulu*," an impatient Ballard said from the surface. "Are you still at station four? Better get under way. Mission time is running out."

"We're trying." Jack Donnelly replied. "We don't seem to be able to rise."

Bryan later recalled:

When we crossed large fissures, we thought, gee, maybe we could actually get down inside one of these things. We circled this fissure that looked like it had plenty of room, more than enough room for us. So we swung around and settled down into it, as far as we could at the widest end.

We sat there for a few minutes to see if the thermal sensor picked up a temperature anomaly. It didn't. So we sampled some rocks and then decided to move on up the fissure to see if there was anything there. Jack tried to lift the sub. It wouldn't go anywhere.

It was a really spooky feeling. We would go up maybe half a meter and feel the sub bump against something. Jack tried everything, up, forward, back, and we hit something each time, not knowing what it was. It was as if somebody put a big lid over us.

When *Alvin* circled the fissure before descending, Bryan and Moore had taken detailed notes; their recorded observations saved their lives. The fissure widened to the north. From the drifting marine snow, they deduced that the current ran from north to south. As *Alvin* descended into the crevasse, the current must have pushed it toward the narrow end and to the side because the submarine's nose pointed northwest. Donnelly reproduced each movement in reverse, taking more than two hours to inch *Alvin* up out of the crack.

"We're clear and under way again and proceeding to our next station," Donnelly said into the underwater telephone.

Those at the surface, especially *Alvin*'s other pilots, who had struggled to keep up a constant stream of calm words to those in the trapped submarine, could hardly believe the three men wanted to continue the dive.

In a nearby fault French scientists sampled basalt with peculiar red, yellow and green streaks. The chemical composition of the colorful rock was quite different from that of the surrounding basalt. It had far too much iron and manganese which was deposited extremely slowly, especially at a young seafloor like the FAMOUS site; and there was much less nickel, cobalt and copper in this rock than there was in all the other samples. The French theorized that this basalt had been altered by heat or hydrothermalism, the result of seawater percolating down through the cracks and fissures, being heated by magma chambers—the same roiling ovens producing all that magma—and returning to the frigid ocean floor with dissolved minerals. It was just a theory.

From the boundary of creation the scientists had collected a total of 3000 pounds of rocks, sediment cores, water samples, and more than 100,000 photographs. *Alvin* had made eighteen dives; *Archimede* and *Cyana* had made a combined total of twenty-seven dives. The bathyscaph would never dive again; it was too big, too unmaneuverable, too expensive.

Xavier Le Pichon, who was so skeptical about submersibles, wrote on a postcard mailed to an American colleague: "I was wrong, now I believe."

19

After FAMOUS it looked like the end of the world. Everybody was gloomy as hell. No money, no money, no money.
   Larry Shumaker, Alvin Group manager

It would take years to understand fully the lessons of FAMOUS, but one thing was immediately clear from what the geologists had seen with their own eyes. The Mid-Atlantic Ridge did not fit into their preconceived notions of an orderly topography of a mammoth crevasse with volcanoes spewing lava on either side. Long before any papers were published in the scientific journals, FAMOUS was declared a great success. Why and how, without the only record that was supposed to matter in science? It was the first large scientific project with submersibles and was well covered by the international press. Plate tectonics was one of the hottest topics in oceanography, and the international diplomacy and technology, of both the underwater boats and the navigation, worked. *Alvin* made a record eight straight dives in as many days; its average was six.

So, it came as a shock when, after FAMOUS, Larry Shumaker called in his team again to announce funding trouble and to tell them to look for new jobs.

The annual grants from the Office of Naval Research had decreased steadily from a high of $1,250,000 in 1966 to $460,000 in 1969. Since 1968 other sources, including the Defense Department, National Science Foundation, and National Oceanic and Atmospheric Administration, had made up the difference, but just. *Alvin*'s total monies for 1971 were a meager $608,000. The second year of paying customers brought it to $904,000.

ONR suggested making WHOI a "no fund equipment loan" of *Alvin* and *Lulu*. In mid-July when the indigent duo was at the Mid-Atlantic Ridge, the Chief of Naval Research made it official by informing WHOI that "there will be no Navy funds available after this present extended contract expires . . . [on Dec. 31, 1974]."

On September 4, Paul Fye and the executive committee of

the WHOI trustees decided to make one last effort, and if within two weeks there was still no hope of funding, the *Alvin* project would be shut down. The next day, Fye and WHOI provost Art Maxwell, armed with booklets of press clippings of FAMOUS, went to Washington, D.C., and talked to every friend and ally they could think of.

The director of the Defense Department agency said he could only offer funding to develop new instrumentation. But he tried to be supportive. "[He] made the point," Fye wrote after the visit, "that we should not be so overwhelmed with our present crisis that we fail to look at the long-term possibilities. I take this as an optimistic point of view that he believes that somehow we'll muddle through the crisis and continue to have an ongoing program. I hope he is right."

At the Atomic Energy Commission, Dixie Lee Ray said she wanted to help but her budget wouldn't allow it. NSF's director wondered if *Alvin* could be put on the shelf until another FAMOUS-like project came up. The glimmer of hope Fye had felt after leaving the Pentagon shriveled. But NSF promised to meet with NOAA and the navy to try to work out some kind of joint funding arrangement. NOAA said it would do likewise. The navy said it would meet with NSF and the Defense Department to try to come up with a solution.

Fye wrote:

> I am encouraged... but not unduly so. Clearly they have heard about the problem, they had been studying the problem before we got here, and there was a great deal of sympathy and understanding. Money is short, projects do have to be cut and even though we have touched their heart strings, I am not sure we have touched their purse. I believe we will get a response soon and I think we must immediately plan without waiting for their response how we would begin to wind down this project.

Fye and Maxwell went back home to wait. The two-week deadline passed, followed by another two weeks and finally, a solution. NSF, ONR and NOAA agreed to jointly fund *Alvin* as a "national facility."

It was an old concept with a new twist. American oceanographic labs and universities could not afford to run their ships without substantial federal assistance, so the government paid most of the operating costs for most of the research ships in the country. However, the lab directors still retained full control of their ships.

It could be argued that *Alvin* already was a national facility;

although it was owned by the navy and operated by WHOI, it was built with taxpayer dollars for the national oceanographic community. But when the idea of making *Alvin* a national facility was first suggested, it smacked of a funding gimmick to some federal agencies; to Paul Fye, it meant ceding his control. But WHOI and the government quickly came around to it.

There was no alternative. With the financial and political pressures of Vietnam, navy funding of oceanographic research had dwindled and NSF, which began to shoulder more and more of the ocean science bill, could not afford even to maintain the academic research fleet. The funding climate for oceanography was desperate. Ships in the academic fleet that were not fully utilized had been laid up, some for good. *Alvin*'s NSF program manager, Sandra Toye, said arguments about fairness were heated and frequent.

"There was a lot of tugging between the haves and have-nots," Toye said. "For the institutions, the issues were not just status and control, but in some cases outright survival as centers of marine research and education. Many did not survive. Ironically, the success of FAMOUS reinforced the view of many scientists that only a segment of the marine geology and geophysics community was interested in *Alvin*. There was very little sympathy in the community for keeping it. I think if we had taken a vote then, *Alvin* would have been sent away. In this atmosphere, the obvious and sensible solution was to lay *Alvin* up. It took a lot of vision and courage—or maybe just luck and stubbornness—to do otherwise."

Toye drafted *Alvin*'s life-saving memorandum, which said NSF, NOAA and the navy would chip in $900,000 annually for the next three years to fund *Alvin* as a national oceanographic facility. It was signed on November 7, 1974.

It was not the needed $1.1 million minimum. The agency directors apparently arrived at the $900,000 figure by averaging the *Alvin* project's total past annual funding. The burden of finding the balance would fall to NSF, which soon would become the major patron of oceanographic research.

WHOI sent a single-page flyer with *Alvin*'s picture to the national oceanographic community, announcing the sub's new status and availability. In February 1975, the new twelve-member Alvin Review Committee, comprised of oceanographers from around the country, met in Woods Hole to grade the merit of thirty proposals to use *Alvin*. The Woods Hole Oceanographic Institution, like every other committee member, had a single vote.

20

Alvin's probationary tripartite-funded season began in 1976 with an expedition led by Bob Ballard, now a WHOI scientist and no longer a paid salesman for the Alvin Group. Among his peers, Ballard was considered an expert on submersibles, arguably the most experienced user in the scientific community, and an eloquent champion of getting down inside the deep ocean. The National Science Foundation had rejected his five-year, million-dollar proposal to explore the continental margin along the East Coast. His proposal for 1976 was more modest and practical: to dive in the Cayman Trough, a great gash in the Caribbean that was four times deeper than the Grand Canyon. Two tectonic plates met there and were thought to be sliding past each other.

The team Ballard put together came from WHOI, Oregon State University, Scripps, Wesleyan University, and the State University of New York (SUNY). He also invited a reporter and photographer, and convinced the navy again to use its classified multibeam sonar to map the dive site.

Newcomers were urged to make at least one dive before the expedition. This preparatory "gee-whiz dive," as the scientists called it, was so filled with ooo-ing and ah-ing that serious work was left largely undone; or the tape-recorded observations and written notes taken by the dazzled scientists made little sense.

On FAMOUS dives, Ballard spoke with such exuberance that his voice obliterated van Andel's recordings. After several disappointing playbacks of his tape, only to hear Ballard, van Andel inscribed "BALLARD FILTER" on a cardboard toilet paper roll tube and spoke into it while taping his observations from *Alvin*.

Jeff Fox of SUNY had never dived and until FAMOUS, agreed with the majority of his peers that *Alvin* had little or no scientific potential. Fox accepted Ballard's invitation to join the Cayman Trough team but could not understand the rationale for a gee-whiz dive. He was an experienced geologist, not prone to romanticism or uncontrolled emoting. Ballard persisted and offered Fox a free plane ticket to the palm trees and warmth of the Bahamas where *Alvin* was diving. Fox accepted.

"Ballard was absolutely right," Fox said. "I was blown away by the experience. My dive transcription was filled with 'Wow!' and other great expletives but few geological observations."

Unlike the FAMOUS team, which explored the floor of a rift valley, the Caribbean divers would sample the rift valley walls of the short ridge segments that met at the center of the 23,000-foot-deep Trough.

En route home from FAMOUS in 1974, WHOI agreed to a new depth limit of 12,000 feet for *Alvin*. In the fifteen dives in the Trough, the deepest to 12,015 feet, the geologists saw hardly any basalt. "It was like being in the Alps in the winter with snow covering all the outcrops we wanted to see," Fox said. On small escarpments a few hundred feet high, they sampled plutonic rock, crystalized magma believed to lie 6000 feet below the seafloor. Geologists had seen plutonic rock from drill bits and dredges, and theorized that the material had been thrust from deep within the earth toward the seafloor by faults that were created by moving tectonic plates. But there were no faults in the Trough; it was lined with terraces.

The divers also saw and photographed a strange pink octopus with giant pig's ears in the Trough. Biologist Clyde F. E. Roper of the Smithsonian Institution said it may have been the first time this animal had been seen alive. He wrote:

> Fantastic! Phenomenal photos of the deep-sea octopus!...This is a cirrate octopod...a length of 10–12' makes this by far the largest cirrate known....Let's catch one!

Ballard had brought along ANGUS, the tethered photographic sled built for FAMOUS. At night while *Alvin*'s batteries were recharging, the sled was towed over the Trough. ANGUS' photographs confirmed that the tectonic plates were indeed creeping past each other.

NSF would not pay for the thousands of frames of ANGUS's film. Emory Kristof, the National Geographic Society photographer who met Ballard during FAMOUS, brought the 35 mm film. *National Geographic* had run a cover story with lots of Kristof's color photographs on FAMOUS. Another picture, painted by an artist, depicted *Alvin* stuck in a fissure; Paul Fye had tried to convince the magazine to omit it. The editors agreed only to airbrush parts of the fissure so the entrapment wouldn't look so dramatic.

No more diving into fissures; no diving at all in water deeper than *Alvin*'s 12,000-foot limit. Fye also insisted that an aft-facing

TV camera be mounted on *Alvin* so the pilots could see whatever potential danger lurked behind them. The camera which had filled the spot years before had been removed because of its poor quality, and the big aft prop obscured most of the view anyway. The new camera, hooked to a tiny monitor in the sphere, was far superior; in fact, it was so good that it did not stay long facing rearward. The scientists asked the crew to turn it forward so they could see directly ahead of *Alvin,* the pilot's view.

Kristof had convinced his skeptical editors to produce yet another underwater story by promising to take the world's first photograph of a submarine in the deep ocean. Getting that picture in the Cayman Trough was a far greater challenge than even Kristof imagined. The plan was to mount a camera and strobes on a small frame and take it down in *Alvin*'s basket. At bottom, the pilot would lift the camera package out of the basket, place it on a nice level spot, move the sub back, and flick on the lights to trigger the camera.

The 50-pound camera contraption, the weight limit for *Alvin*'s arm, would not stay put, and once it rolled about 50 feet down slope beyond 12,000 feet, but the pilot slipped down to retrieve it anyway. It took a half dozen tries and as many rolls of film to get *Alvin*'s picture. The pilot who made it work was Dudley Foster, a former jet fighter flyer, who joined the Alvin Group in 1972. Foster propped the camera package on a ledge at 9840 feet, moved *Alvin* back and switched on the strobes. Most of *Alvin* managed to get into the picture. The aft prop was hidden in blackness and an eerie green tint bathed the white submarine and its silver arm. The portrait ran in another *National Geographic* issue with the story of the Cayman Trough dives.

Following the Caribbean expedition, *Alvin* took scientists back to deep ocean stations. The benthic biologists had established new underwater laboratories—near DOS No. 1 off Woods Hole and in the Tongue of the Ocean.

In the Bahamas the navy also used *Alvin* to investigate a Soviet trawler which for days had bobbed silently and mysteriously on the tropical waves. But the only items seen from *Alvin* beneath the trawler were empty Russian sardine tins. Off St. Croix, *Alvin* inspected underwater equipment for the navy and found a classified item that had accidentally dropped into the sea. Closer to home, *Alvin* found a drum of radioactive waste.

When Larry Shumaker first urged his staff to look for other work, Cliff Winget wrote a proposal to the recently formed U.S. Environmental Protection Agency for funds to inspect some of the 76,000 drums of low-level radioactive waste dumped in the Atlantic and Pacific between 1946 and 1962. So EPA oceanographer Bob Dyer suggested exploring an Atlantic dump site about 120 miles off Delaware.

On the first dive in 1975, *Alvin* found many artifacts and containers, including shell casings, ammunition boxes, beer cans and a large plastic bag of trash, but only three corroded drums of nuclear waste. The divers suspected that many of the waste drums never made it to the site. "We started to wonder why we hadn't seen any drums on several dives and then we realized that people had short-hauled them," Winget said.

On other dives biologists saw many fish and shrimp. One particular species of fish liked to root in the mud near the waste drums. The animals were not analyzed for radionuclides but the mud was. It held concentrations of cesium that were from three to seventy times higher than background radiation.

In 1976 Dyer returned to raise a drum for closer scrutiny. *Alvin* followed the marks its runners made in the mud the year before and put a harness over a drum at 9131 feet. On the second dive, the harness was attached to a line and winched from the surface onto the University of Delaware ship *Cape Henlopen*.

It was leaking. Winget, who was splashed with the dribbling fluid, had to strip on deck and get hosed down. The drum was only slightly radioactive. In the six years since it had been dropped, most of the radioisotopes had decayed or seeped out.

Having visited the slow-moving tectonic plates at the FAMOUS site, several geologists were anxious to explore a faster-spreading center. What causes some plates to move faster than others is not well understood; but the scientists know that the spreading centers in the Pacific move much faster, at about the rate a thumbnail grows, than those in the Atlantic. And this greater activity, they reason, would be accompanied by much more heat.

Tjeerd van Andel of Oregon State and Dick Von Herzen of WHOI hatched the idea of a Pacific expedition right after FAMOUS. They asked the National Science Foundation to fund *Alvin* dives at a spreading center near the Galapagos where they planned to look for warmth, the elusive evidence that cold water percolated down into cracks in the seafloor along the boundaries of tectonic plates and returned to the surface with the heat and metallic deposits of volcanism.

There was much indirect evidence for hydrothermalism in the deep ocean, particularly near the Galapagos, where the first and most dramatic clues were found in 1972. Sonar had picked up mysterious mounds as high as 50 feet. Some speculated that the mounds, which on the chart paper looked like little fuzzy black balls, were cauldrons of super-hot water. Also, towed thermometers had recorded temperature anomalies: a few tenths of a degree above normal in the mud and a tenth of a degree higher in bottom water—seemingly trivial, but significant in an ocean so deep, massive and cold that it cooled hot rock instantaneously. That was the theory, anyway.

Back in the 1960s, warm and extremely salty water had been found in the Red Sea with unique sediments rich in metallic deposits. But these "hot holes" were considered possible only in a newly forming ocean basin—a spreading center—like the Red Sea.

Despite the evidence, many scientists remained skeptical about hydrothermalism, and many more distrusted the tiny temperature anomalies. It was extremely difficult to measure anything accurately in the deep ocean on such fine scales, especially when the recording instrument was dangled from the surface. And deep-sea thermometers could be complicated. Those used off the Galapagos measured the voltage change (caused by temperature) with a quartz crystal, that number was calibrated with another thermometer of platinum, and both sets of numbers went into a computer.

The logistics of a Pacific expedition seemed like a nightmare. WHOI wasn't sure *Lulu* could make such a long journey, and the Alvin Review Committee questioned the fiscal soundness of going so far for only one project and a handful of scientists.

Van Andel tried to convince his colleagues to organize their own Pacific expeditions with *Alvin*. In 1975 he wrote Scripps biologist Bob Hessler, who made the first deep science dive in *Alvin*.

> *Dear Bob*
>
> ...we hope to prove not only that ALVIN is more than a WHOI piece of private gear but truly a national facility, but also that high quality, sophisticated science can be done with it that is not possible otherwise, and in proportion to its cost. The program in the Pacific is designed to make a point on both issues.
>
> However... the logistics are still quite unfavorable. I am therefore in the process of alerting other possible users to the opportunity. Additional programs would greatly increase the cost-benefit ratio and much improve our chances of actually securing ALVIN time and laying the groundwork for the long-term support for the little sub....

"It is one of those things," Hessler replied, "where one is bound to be able to do something interesting, simply because so little has been done so far. Nevertheless, I think it would be better if I did not join in. I have irons in so many fires...."

If he didn't want his own diving project, van Andel offered, would he like to join their geochemistry cruise?

Naw, what *for?*

Hessler's words would come back to haunt him.

Try as he did, van Andel could not drum up more Pacific business for *Alvin*. He took another tack in a letter to WHOI:

> I still firmly believe that it is this program, as well as the one by Ballard in the Cayman Trough, that are going to be influential, if not critical in determining whether at the end of the current three years there is going to be further support. The other uses proposed so far to the [Alvin Review Committee] are of the same kind that has formed ALVIN'S main diet in the past. If they could not assure support then, they will not do so now. I therefore believe we have a common interest in solving this problem. Am I right?

Yes, he was right. At the Review Committee's urging, WHOI held an open workshop in May 1976 to discuss the future of *Alvin*-conducted science to show the federal funders that there was ample interest in the submarine. Only nineteen scientists (nine from WHOI) attended.

But besides whatever public relations needs it might fill, the Galapagos expedition was the most exciting science proposed to the Review Committee, and it supported the project.

NSF's response to the proposal was measured. "Do the mounds really represent some manifestation of hydrothermal activity and is there any way to further establish this before putting to sea with a diving program?" NSF's program manager wrote. "Could this... be better substantiated with further field work... before diving on the ridge?"

Hoping for more and better evidence of hydrothermalism, NSF agreed to fund an expedition in 1976 to the Galapagos site, but only for dredging and lowering thermometers and other instruments from a ship. At the same time, the agency funded another group of scientists, primarily from Scripps, who wanted to tow at the same site the Scripps' sled Deep-Tow, which carried cameras, thermometers, magnetometers and sonar. Pending the results of both 1976 Galapagos missions, NSF said, it might fund the original proposal to take *Alvin*.

As it turned out, both expeditions gathered more evidence, although it was more of the same and still indirect, for hydrothermalism. But NSF agreed to fund *Alvin* dives at the site in 1977. The existence of hydrothermal vents in the deep sea sounded like a foregone conclusion in NSF's press release announcing the project:

SCIENTISTS TO EXPLORE HOT SPRINGS AT BOTTOM OF PACIFIC TO LEARN HOW METAL-RICH SEDIMENTS ARE FORMED

Scientists will dive to the bottom of the Pacific Ocean in a 22-foot, three-man submersible in February to explore for the first time vents in the sea floor through which metal-bearing hot waters flow.

The findings are expected to help scientists understand the formation of metal-rich deep sea sediments, the history of the chemistry of sea water and the transfer of heat from the earth's interior into the oceans.

Dr. John B. Corliss, assistant professor of oceanography at Oregon State University, the project coordinator, said the expedition's findings could determine whether hot springs could some day have economic value....

"We'll have to proceed with caution to avoid exposing the submarine to waters that are very hot," Corliss said. "We'll be working with unknowns. No one has ever seen these things up close...."

In fact, no one had ever seen those things, but Corliss' comments clearly reflected his confidence in what he and several of his colleagues expected to find.

In February 1977, some fifty scientists and technicians from Oregon, Massachusetts, California and Texas boarded *Lulu* and her escort *Knorr* and headed to the site 200 miles off the islands that had inspired Charles Darwin.

The search for hydrothermal vents was not among Bob Ballard's research interests. He was asked to join the team for his skills in navigating and using ANGUS as a scout for dive targets. Ballard reasoned that a typical eight-hour dive was too precious to waste on aimless wandering. ANGUS could be towed all night and cover far more area than *Alvin*. If anything interesting showed up in the photographs, the divers could go to that very spot to investigate.

The Navy had again provided a bathymetric map made with classified sonar. But the key to the navigation was held by Kathy Crane, a Scripps graduate student, who had been on the 1976 Galapagos cruises. As part of her Ph.D. thesis work, Crane had prepared a detailed geological and thermal map of the area by matching up the ten spots of temperature anomalies with photographs taken at the Galapagos site.

Also on the Galapagos expedition were three *National Geographic* photographers who for the second time brought ANGUS's film, this time color. And van Andel invited a reporter from a California newspaper.

Buffeted by the trade winds, *Lulu* cruised at "supersonic speed," according to the crew, across the Caribbean Sea through the Panama Canal and into the Pacific for the first time. *Alvin's* white sail had been painted bright red, a safety-inspired measure designed to make the sub more visible at the surface.

On site the Scripps "black box" was used to interrogate the two transponders left behind the year before, and the pinging tubes obediently responded. Triangular navigation nets were established at each. Transponders Faith, Hope and Charity were dropped by the mysterious mounds. Sleepy, Dopey and Bashful went down in the area of the highest temperature anomaly, which the scientists called the "Clambake" because Deep-Tow's photographs captured so many clamshells there. One picture depicted a cluster of shells and a beer can.

The two-ton ANGUS with its chirping pinger and temperature sensor was then lowered on a tether and flown a few yards above the sea floor, snapping a picture every ten seconds. The run was an all-night vigil for the ANGUS crew, a group of WHOI technicians that Ballard had assembled. He made them popcorn, massaged their shoulders and cheered them on like a football team as they endured the drudgery of their lot: watching the telemetered signal which would indicate a temperature change and the sweeping dark lines of the echo sounder pinging softly in the laboratory. The echo sounder chart told the underwater story of steel careening into rock. The ANGUS team would paint on the sled: "Takes a Lickin' But Keeps On Clickin'." The *Alvin* crew called it the "Dope on a Rope."

During its first nightly flyover, ANGUS snapped some 3000 pictures of monotonous naked black rocky sea floor, including thirteen frames, about three minutes' worth, of white clamshells. The time, imaged on each frame, indicated that the thirteen shots were taken when the temperature changed slightly at about midnight. ANGUS could not determine temperature, only a temperature change. What the clamshells had to do with a temperature anomaly was anybody's guess.

Crane could not help but wonder if there was a connection. She had seen too many photographs from the earlier expeditions of clamshells—big, long, empty shells, in piles. And here they were again from ANGUS. A senior Scripps scientist had assured her that a passing ship had a clambake and threw the garbage over the side, just as they all did on their research ships. There was another mystery in Deep-Tow's photographs—little white spots; a senior WHOI scientist insisted that they were just spilled chemicals from the film processing.

What interested the geologists was the tiny temperature spike, so van Andel and Corliss headed for the Clambake on the first dive.

Water Baby

Pilot Jack Donnelly steered *Alvin* over black pillow lava. A long thin white plastic pole with a heat sensor on the tip swung out from *Alvin*'s side, and in the sampling basket were two bright pink plastic milk crates. The digital temperature readout in the passenger sphere held steady at a normal 2°C or 36°F. The counter was rigged to beep with a thousandth of a degree temperature rise.

Within about fifteen minutes of touchdown at about 8000 feet, the sensor beeped and the flashing red numbers indicated a hundredth of a degree rise in temperature. Suddenly *Alvin* was surrounded by life.

There were huge clamshells, stark white against the black elephant-skin basalt; brown mussels; a big bright red shrimp; a couple of white crabs scampering over the basalt; white squat lobsters; a brittlestar; a large pale anemone. Donnelly poked the anemone with the temperature probe and it spread its tentacles. A many-legged creature pumped itself out of sight. Something sticking out of the bottom. Coral? A pretty little pale orange ball that looked like a dandelion gone to seed. Rocks covered in white streaks, what looked like pigeon droppings. Worms? A white crab climbed onto *Alvin*'s pink milk crate and fell off. The water got foggy. Six of those white crabs. A whole cluster of those little peach-colored puffballs. And for as far as they could see, more clams and mussels—or were they oysters?—tucked in among the bulbous, black basalt. The hard earth, the basalt they had come for, was stained with the brilliant red and orange and ochre of iron sulfide—clearly the result of hydrothermal fluids. The basalt had been altered by heat, and for such young rocks, there was an extraordinary amount of alteration.

"We are sampling a hydrothermal vent," the shaggy-haired Corliss announced into the underwater telephone.

His graduate student Debbie Stakes, in the navigation van on *Lulu*, was surprised to hear of success so soon.

"Debra," Corliss said, "isn't the deep ocean supposed to be like a desert?"

"Yes," said Stakes, remembering her only high school biology course and an ecology course from graduate school.

"Well, there's all these animals down here."

Alvin came upon another small clambake area with a tiny temperature anomaly and more life. After four hours on the bottom, the submarine headed home with two pieces of multicolored basalt, bits of clamshells and five live mussels. The water samplers had not triggered properly; they held only about a pint instead of gallons of the water that was a fraction of a degree above normal.

Corliss and van Andel were apparently unaware that most of the animals and the unlifelike things they passed on their dive had never been seen before. Corliss made no connection between the animals and the above-normal temperature. He was ecstatic with the two tiny temperature anomalies and the stained rocks. Van Andel later joked that Corliss rarely looked out the window, he was too busy watching the temperature readout.

Van Andel was perplexed at the abundance and size of the clams. He knew it took a lot of energy to make a shell the size of his shoe. But where was all that food? In the photosynthetic by-products of the tiny plant and animal remains wafted to the deep sea in the eternal flux of marine snow? Surely not.

The next morning *Alvin* took van Andel and Jack Dymond to the mounds area where they hoped—expected—to find hot water geysers. There were none. The so-called mounds were more like enormous stalagmites, and the geologists renamed them "the Alps." Although the mounds were in the general area of the Clambake and the terrain resembled that of the first dive site, the divers saw no clams or "significant increase in life." The water samplers worked, but the collected fluid was at the usual temperature. But the Alps were streaked with brilliant yellows, purples and greens—the colors of minerals, and the divers suspected that the mounds were not basalt but sulfide. For seawater to precipitate sulfide particles, it had to be hot.

The camera recorded something the geologists did not report: delicate stringlike things with little pale orange balls and what looked like heaps of spaghetti draped over rocks.

On the third dive, Corliss and John Edmond, an MIT geochemist, headed back to the Clambake. It was Edmond's first dive ever.

In an hour and a half, *Alvin* was flying over glassy pillow basalt that sparkled in the darkness at 8184 feet. The temperature sensor beeped. Clusters of huge white clams. Brown mussels. Those strange dandelions. Dozens of white crabs.

Corliss instructed the pilot to stop and sample water. Edmond noticed a large purple anemone as he gazed out the porthole. The water seemed to be shimmering like the air above a sidewalk in summer. He blinked, looked again. It *was* shimmering. If Corliss really hadn't paid much attention to the abundant weird life outside the porthole during his first dive, he did now. His face was mashed against the clear plastic.

Alvin passed through a foggy haze. Fish, those dandelions, crabs, something was trying to climb into the pink milk crate. Fish, more crabs, dozens of crabs, and more brown mussels and white clams, everywhere. The temperature was an astounding 8°C or 46°F. *Holy shit!*

Edmond and Corliss were shouting. They did not retrieve any

animals but surfaced with gallons of water for the chemist, rocks for the geologist, and several clamshells. They didn't know that three squat lobsters had climbed onto the sub and hung on for the ride up.

It was already dark when *Alvin* surfaced after one of its longest dives, nine hours and fifteen minutes. The lobsters fell out onto *Lulu*'s deck, dead. In *Knorr*'s laboratory Edmond opened a water bottle and jumped back from the stench of rotten eggs; the repulsive odor was carried throughout the ship by *Knorr*'s excellent central air conditioning. It was difficult to breathe in the lab.

Hydrogen sulfide. The water they collected was full of it, and a hundred times more radon than it should have had. Seawater all over the world was supposed to have the same basic composition. The water making them gag was not merely warmed; its composition had been fundamentally changed—by heat. Seawater must have circulated down into the cracks and fissures to the underground heat which caused the sulfate in the seawater to react with the iron in the basalt to make hydrogen sulfide. The heated fluid also picked up radon from rocks. It could not have done so without heat.

Hydrogen sulfide. Swamps and sewage smelled just like the *Knorr* laboratory. A certain kind of bacteria live in swamps. In processing the sulfur compounds in decaying bogs and wastes, they give off that rotten-egg smell. These chemosynthetic bacteria were discovered almost a century ago. They can use various inorganic materials instead of sunlight to reduce carbon dioxide to organic carbon. But the inorganic material is produced by photosynthesis, and the microbes have been considered little more than a curiosity. Little photosynthetically produced fixed carbon makes it to the deep sea. Two miles down off the Galapagos, did the inorganic material, hydrogen sulfide, come from altered seawater? Is that what nourished the vent animals? Was dinner in the hydrothermal vent a bacterial soup? In the total absence of sunlight was chemosynthesis—not photosynthesis—responsible for life? Could life get its energy from volcanism, not the sun? But how?

"A whole lot of things sort of fell into place," Edmond said. "About halfway into the cruise, we realized that regular seawater was mixing with something. It was a unique solution I had never seen before. We all started jumping up and down, we were dancing off the walls. It was chaos. It was so completely new and unexpected that everyone was fighting to dive. There was so much to learn. It was a discovery cruise. It was like Columbus."

On the next dive, 10°C (50°F) was recorded. "Hot water," the scientists wrote in their log.

Alvin returned to the Alps on its fifth dive and accidentally bumped into an edifice which shook like a gigantic mold of Jell-O. The scientists asked pilot Dudley Foster to repeat the maneuver. The mound trembled again. Foster jammed the temperature probe into the heap and broke through the orange, yellow and green crust which he likened to an overripe watermelon rind. Water oozed from the small hole. Inside the temperature was 15°C (59°F).

Besides water and rocks, the divers retrieved six mussels and two clams. The clams reeked, like the water, of hydrogen sulfide, and their flesh and blood were bright red—humanlike.

Every night ANGUS flew over new territory in search of targets for the morning dive. More navigation triangles were established. Snow White was dropped to replace Grumpy which refused to ping on command. Bashful and Dopey lived up to their names and were replaced by Prince Charming and the Wicked Witch. For more precision, the divers tied their empty beer bottles with nylon line to mark the vents.

On *Alvin*'s eleventh dive Edmond and Corliss saw long slender white tubes. Worms? Snakes? These things had bright red tips. And there were scores of limpets, a snail-like creature, and something they called "segmented worms."

Every day, with every treasured haul and discovery, the explorers hoped against their good sense for another dive, knowing they could anticipate losing a third of their dives to mechanical problems or rough seas. But the Pacific was flat, the *Alvin/Lulu* crews were giving their all and *Alvin* made eleven full dives in as many days, a record broken on the second leg with the thirteen straight dives remaining in the expedition.

On the second leg, *Alvin* followed white crabs to a new vent that Corliss named the "Garden of Eden."

"The crabs increased in number," he wrote, "the water grew milkier and we passed . . . first through a band of 'dandelions,' and then into the vent area with numerous tube worms growing in crevices between broad flat pillows with warm water streaming up past them." The temperature was 17°C (63°F).

In the Garden of Eden they were surrounded by thickets of the red-tipped snakelike creatures standing upright in tubes that swayed gently in the shimmering water. Blue-eyed pink fish swam languidly in the cloudy water, fluid so filled with hydrogen sulfide that it would poison other life. Some of the white tubes were ripped like the bark of birch trees. The pilot poked one with the sub's pincers and the red tip disappeared into its long tube. A crab pinched another tube and it recoiled. White crabs crawled onto *Alvin*.

One sampled rock was covered with a black material that looked like pieces of steel wool. Nobody knew what it was. They later learned that it was sulfide. There was no doubt about the flakes of iron sulfide, also called pyrite, coating another rock. Some of the basalt was bleached white, which reminded Debbie Stakes of the rocks she had seen on a recent trip to Geysir, Iceland. "Geysir, which is the place we get our word geyser from, is an area of hot springs and all the rocks there were bleached white," she said.

The Galapagos team wasn't prepared to handle the hundreds of retrieved animals. They didn't know how to pickle them anyway; on a geology-chemistry cruise, there was no reason to bring formaldehyde along. But *Alvin* escort swimmer John Porteous kept a small jar of formaldehyde on hand in case he found an unusual animal. He gave it to Stakes, whose job on the expedition was to curate the rock samples.

Neither were there any jars to hold the animals. Stakes put them in the plastic Tupperware boxes intended for the most fragile lava samples. Plastic wrap from the galley made handy sleeves for the eighteen-inch-long tube worms. When the Tupperware ran out, Stakes made little baggies and worried about what to use when she ran out of cellophane. "At this point, for all we knew there were Martians down there living in hydrogen sulfide," she said. The animals took all the plastic wrap and every drop of formaldehyde.

Biologists radioed the team, reminding the geologists and chemists to take bacterial swipes, to handle the specimens carefully, to freeze one of everything, and to put the most delicate animals in buffered formaldehyde because they might disintegrate in alcohol. Stakes was cautioned to dilute the formaldehyde with distilled water or risk dissolving the shells. But there was no distilled water; she used seawater.

One unusually long radio message came from Paul Fye, declaring that since there was no biologist on the expedition and both vessels were WHOI ships, all biological specimens were the property of WHOI. However, the expedition leader, Jack Corliss from Oregon State, refused to abide by this claim. Fye was not the only one making demands.

"The biologists assumed we didn't understand the full significance of these organisms," Jack Dymond said. "They were telling us to abandon our program and undertake biological studies. We didn't take to this very openly."

After much discussion with his peers at sea, Corliss announced that he would ask NSF and the Smithsonian to recommend who should get the animals. NSF was paying for the expedition. The Smithsonian employed a national pool of taxonomists who identified and classified new life forms.

John Porteous wrote almost daily to his girlfriend, a WHOI biologist:

Feb 77

Now, are you ready—The water coming out of the vent is...loaded with hydrogen sulfide...there are white clams and brown clams. ...The largest one I measured was a white one...10¾ inches long and 4⅝ inches wide....

I have specimens for you....I do not have a white clam as they have found only one alive....I have also collected some lobster like critters that get into the sub while it's on the bottom....

We could use a biologist down here. Jack Donnelly and Jerry van Andel both described a gastropod which is also down there. It's white and looks like a whelk about three inches long. Seems to be feeding on a brown slime that is on the rocks near the vents. Aneorobic slime?

22 Feb 77

...Today I saw some of the hand-held pictures from the sub, and everywhere there is a vent down there, there are great stringlike networks of worm tubes or something like that. It looks like someone dumped spaghetti all over the lava rocks. I really think the biology story here is as big if not bigger than the geology story....Boy, I wish you were here.

24 Feb 77

The water samples keep coming in and the chemists keep going crazy trying to analyze it. Plus the water from one site is not the same as water from any of the other sites....There is more than a thousand times more dissolved hydrogen in the water of one site and the radon count is almost a hundred times higher.

26 Feb 77

Today the last dive of this leg was made on a new site. So, guess what? New clams, of course and two specimens of limpets, and giant tube worms. The tube feels like nylon pipe and is about 15 inches long....The gill plumes extend out of the tube for about a foot and are bright red. Also the area is called the Dandelion Patch because of an animal which looks exactly like a dandelion gone to seed. There are hundreds of them in this one area....

In all of these oases are the crabs...hopping up onto and into the sub. Now the fish of the area come to these warm vents and sort of lounge around in the warm water. They actually get into the warm stream and roll over and bask....

All of the life is clustered around these vents. Away from the vents, life is as it usually is at that depth. Kind of boring. The warm areas are very small, 4–5 m [13–16 feet], but the warming effect seems to reach out to maybe a 10–20 m [33–66 feet] radius. There seem to be a lot of these vents, but we have only really looked at four of them. . . .

All those round grey spots in Deep-Tow's photographs were not from sloppy developing; they were the pretty dandelions.

From five hydrothermal vents, including a so-called extinct vent which was littered with clamshells, the scientists had amassed a rich booty: two clams, several crabs and squat lobsters, seventeen tube worms and parts of them, other worms, an eel-like fish, more than a hundred limpets, a coral, and shell fragments. They also had more than a hundred pieces of rock, three trays of sediment, and 88 water samples. ANGUS had been towed over about 85 miles of seafloor and snapped 57,000 pictures; 20,000 more photographs were taken from *Alvin*.

After the last dive on March 21, *Knorr* and *Lulu* steamed to Panama where the U.S. naval base donated dry ice and three pressurized aluminum packing cases for the pickled and frozen vent animals. Corliss, Dymond and Edmond carried them aboard as personal baggage on the plane to Miami. U.S. Customs officials warily eyed the crates and the three fully bearded men in open-toed sandals, T-shirts and jeans. They were at one of the world's busiest entryways for drugs.

"What's in those boxes?" a customs man demanded.

"Scientific specimens," replied Corliss, who wore a gold pierced earring.

Suddenly a pressure relief valve on one of the shiny aluminum cases popped and began to hiss.

"Open them."

Wonderland

21

Miami customs agents had a look inside the hissing aluminum box, and finding chitlins instead of marijuana, were not interested in seeing more. The hippy scientists were allowed to check in their weird animals as personal baggage, and the precious biological booty finally reached those whose expertise was life, not rocks.

Almost all the specimens were unknown. Some of the mussels were really clams, and the Oyster Bed held mussels. The dandelions which came back only in photographs stumped everyone, although the biologists knew they were animal, not plant. The dandelions seemed suspended in the water, but up close, the scientists saw that each ball was actually anchored to rock with thin transparent strands. The "flower" was a full blossom of many "petals" which had to be individual animals. The dandelion was probably not one, but a whole colony of—well, something.

Even before the geologists and chemists had returned home in March 1977, a dozen biologists from around the country, led by Fred Grassle of WHOI, started planning a return trip to the Galapagos rift. The biologists' research proposal did not fully pass the conservative peer review. That any of it passed in a year when most funds were already committed, Grassle believed, was due largely to the efforts of his former student and *Alvin* user Tracy McLellan, who was then a program manager at the National Science Foundation. She managed to funnel the support from a special fund, and in January 1979, *Alvin* returned to that fantastic patch of water west of Ecuador where it would spend more than half the year.

Alvin, a lefty for fifteen years, became ambidextrous for the return trip to the Galapagos. Emory Kristof, by now a seasoned marine photographer, promised his editors the first footage in the deep sea taken with a solid-state (no tubes) color movie camera. For him to make good on his promise, *Alvin* needed a second arm. The Alvin Group, which had wanted another limb for a long time, scraped the bottom of three different funding pots and bought a hydraulic arm that could lift 200 pounds. The miniaturized movie camera, developed and lent by RCA, was mounted on it. And this time, NSF paid for the ANGUS film, and Kristof's colleague Pete Petrone developed a method of using seawater to process it.

Titanium had replaced more pieces of corroded aluminum on *Alvin*, including the stern propeller shroud and the frame, now two feet longer to someday accommodate more batteries. A third fiberglass ballast tank was added for six more inches of freeboard to keep water from splashing in at the surface. The crew had hung a shower curtain from the hatch to protect the electronic gear from the salt spray. Rumor had it that the curtain was meant as privacy for the rare urinating female diver. Now the curtain was replaced with the "bathtub," a moat with drainage valves ringing the base of the sail.

And *Alvin* could dive deeper than ever before, thanks to the Environmental Protection Agency, which wanted to return to the Atlantic dump site in 1978 to retrieve a second drum of low-level nuclear wastes almost a thousand feet beyond *Alvin*'s 12,000-foot limit. No argument came from WHOI's front office, which was in transition; Paul Fye was retiring. The navy obliged, approving the Alvin Group's request for certification of 4000 meters or 13,120 feet. With the extra depth, *Alvin* took EPA divers back to the dump site in 1978 and the same year, on *Alvin*'s deepest dive, biologists established a deep ocean station off St. Croix at 13,179 feet.

At the Galapagos site in 1979, navigation nets were laid with transponders Faith, Hope, and Charity, and Sloth, Pride, and Gluttony, and the first divers plunged gently through the foaming surface waves, bound for Clambake I. But the Clambake was nowhere to be found.

On the second dive, January 19, biologists Jim Childress and David Karl saw a white bush. No, Karl corrected himself as *Alvin* neared the object, it wasn't a bush. It was... *spaghetti?* Yes, those weird worms, lots of them.

"Spaghetti on the rocks, fissure on the right," he announced. "Spaghetti is sticky, a mucuslike network. The strands are about several millimeters, length is variable up to a meter.... Spaghetti. The entire area is outrageous.... More spaghetti. The outside cameras should be recording this. It appears to be on the tops of rocks, hanging down over the edges.... This is fantastic! Spaghetti is draping the rocks of this whole region...."

The spaghetti was sampled with Cliff Winget's slurper, a sophisticated version of his ten-year-old sketch of a vacuum cleaner for fragile animals. The slurper also carried insulated chambers for animals and other containers for water. It was a six-martini job, Winget said.

At the surface, the biologists identified the spaghetti as an enteropneust, a common worm-like creature known to burrow in sandy mud. Why these enteropneusts lived entangled in piles on rocks was anyone's guess. A half dozen years later, the creature was classified as a new genus and species of acorn worm. But its behavior remained a mystery. Acorn worms were deposit feeders, but these creatures did not nibble food on the rocks. Their heads floated up and out, as if in greeting, or more realistically, feeding.

Chief scientist Grassle put Bob Ballard on the third dive, hoping his memory and knowledge of geology would steer them to Clambake I. It didn't, but they reached a new vent—the Mussel Bed, dominated by curvy rows of black and brown bearded mussels.

By now the pilots knew enough to follow the "crab gradient," as they called it—the increasing numbers of crabs at the perimeter of every vent which led *Alvin* to shimmering, milky water and bizarre life.

On their dive, Howard Sanders and Bob Hessler followed the crabs to the Garden of Eden, where tall red-tipped tube worms swayed gently. When *Alvin*'s claw squeezed a red head, the worm quickly disappeared down into its long white tube. The creature also ducked when pinched by a crab. Of all the vent animals, the crabs and squat lobsters were the cheekiest. They fought over scraps, chased each other and climbed over *Alvin*. The biologists watched a crab sneak up on an open mussel and try to poke its claw inside before the mussel clapped shut. Another crab picked up a mussel that was at least as big as it was and tried to carry it away.

At 12:30 Hessler announced their depth of 8159 feet and suddenly broke into song. He was a happy scientist. "The thing that gets to my soul the most is to go down there and finally be able to see the bottom and wander over it by moonlight," he said later. "I felt like a person given an opportunity to visit an uninhabited world."

And so he was. He and Sanders decided to stray off their set course and leave the Garden of Eden to see another fraction of this alien world that stretched across two thirds of the planet's surface. But a short distance from the vent, *Alvin*'s gyro blinked out; the pilot didn't know north from west, and surface control lost track of the sub. Chattering porpoises had masked *Alvin*'s chirping transponder, or so those on *Lulu* thought.

Hessler calmly announced into his tape recorder: "We are lost. Decided to go west, not really knowing where we are. It doesn't matter what direction we go in...." The biologists snapped to attention at what they beheld. It was not a vent.

"We saw these huge columns rising up and these Grand Canyons below us," Sanders said. "And then we came to a gigantic overhang. We went a little ways. This huge ledge was on top of us. The whole dimension of the world had shifted. Suddenly we realized that we had shrunk down to the point where we had become the microorganisms and *Alvin* was a ladybug. It was literally like Alice in Wonderland. Our whole perspective of dimension had changed. It altered my view of the world profoundly. I shall never forget it."

It took about a year for geologists to figure out what this Wonderland was. Hessler and Sanders had seen what scientists have since called a collapsed lava lake. When lava fills a depression on a rocky seafloor, the frigid water bubbles up though the lava and hardens it into columns and a top crust. The rest of the lava drains out of the depression, occasionally leaving a huge cavernous pit with the columns still holding parts of the roof.

On another dive, Kathy Crane and Emory Kristof discovered the Rose Garden, which had the highest temperature, 23°C (73°F) recorded at a vent. There the black basalt was covered with thousands of translucent anemones and, at twelve feet, the tallest tube worms ever seen.

Grassle, for one, didn't believe the reports about tube worm length. The longest one collected in 1977 was eighteen inches, which seemed incredible enough. But being properly skeptical on a dive was more difficult. On one of his *Alvin* dives, Grassle saw a fist-sized hairy creature, pumping itself through the water. "That's like nothing I've ever seen," Grassle said. "Oh boy, it's incredible. I have no idea what that organism is. It looks like an undulating hairbrush! Well, the hardest part is going to be figuring out how these animals got here in the first place."

The unloading of specimens was like a child's Christmas morning. The tube worm was as different from other worms, one biologist proclaimed, as horses were different from cows. What in the world was this unearthly thing that had neither mouth, nor gut, nor anus? How did it eat? Neither had it legs or any other means of locomotion. It stood in one spot all its life. The tube itself was extremely tough and difficult to cut, even with a razor. The longest tube of all the tube worms collected was eight feet and the worm inside was five feet.

The dandelions reached the surface as unrecognizable mush. Crew chief George Broderson, hot on the trail of a solution, pried a grungy magnet from a cabinet door in the galley and affixed it to a circle of plastic hinged to a long clear plastic tube. All *Alvin* had to do was place the cylinder over a dandelion and release the stretch cord; the circular lid would snap shut and the magnet would keep it closed. It worked.

Everyone excitedly huddled over the dandelion catcher's first prey. It was a new species of siphonophore, a gelatinous animal related to the Portuguese man-of-war, which to the unskilled eye it resembled not a whit. In a glass dish on deck, it disintegrated in an hour before the scientists' incredulous eyes.

The baited fish traps caught no fish but many crabs. The new species of brachyuran crab resembled its shallow-water cousin but was blind. Many survived the ninety-minute ascent in insulated chambers but didn't last long at ambient pressure.

Hundreds of other animals were collected, including a tiny crustacean with teeth at the tips of its eyestalks where eyes normally would be. Hessler reckoned the animal scraped its food (bacteria) off rocks and mussel shells with these bizarre eyestalks. He knew of no other animal that gathered lunch with its eyes, but it was the only explanation he could think of.

The giant clams were given the species name *magnifica*. A new species of mussel needed a new genus. Pearls were found in some of the molluscs. There were many different types of worms, at least two unknown species of limpets, three unknown kinds of anemones, and new whelks, leeches, featherduster worms, snails, squat lobsters, brittle stars, copepods and bacteria. Virtually all the life was strange, and the biologists were confounded that there could be so much flesh in one place in the so-called desert of the deep sea.

At the end of the nineteen-dive expedition, on the eighteenth dive, Clambake I was finally found. One of *Alvin*'s square drop weights from 1977 led to it. It was only a few hundred yards away from the Mussel Bed. Although navigation had greatly improved since the early *Alvin* dives, searching in darkness with no benefit of landmarks was frustrating. The sub's discarded drop weights were such helpful landmarks at frequently visited sites that the pilots began painting the dive number and date on the steel squares.

The solid-state movie camera had worked on the 1979 expedition, but the *National Geographic* photographers were not entirely happy. They had lost another camera, which dropped to the seafloor on a tripod. After obediently shooting *Alvin*, the tripod was supposed to rise to the surface, but it didn't.

Lulu's radio was heavy with the traffic of the expedition's ups and downs. WHOI required all its ships to call in twice daily, at 9:30 A.M. and 3:30 P.M. Eastern Standard Time. *Lulu* had to call in a third time, after every retrieval, to report that *Alvin* was safe on its cradle. Often the ships had no more to say other than all is well, but from radio WLSX there was usually plenty issuing forth.

Good morning, WHOI [HOO-EE], this is Lulu ...

17 January
URGENTLY NEED SPARE PARTS FOR [NEW] MANIPULA-

TOR, ESPECIALLY SOLENOID VALVES, RELIEF VALVES, FLOW CONTROL VALVES AND STANLEY HYDRAULIC MOTORS

                                                            FOSTER

26 January
PLEASE HAVE MAIL FORWARDED TO PANAMA.... CLIFF WINGET, PUMPING SYSTEM WORKING WELL. EVERYONE AGREES DESIGN EXCELLENT

27 January
WILL MOVIES ARRIVE PANAMA FOR NEXT LEG?

                                                            MASTER

30 January
MOVIES AND BRAY OIL DID *NOT* ARRIVE

                                                            FOSTER

14 February
NO ONIONS
NO EGGPLANT
PINEAPPLE SCENT BAD
TOMATOES GOING BAD

    Two months after leaving the Galapagos, Kristof received a strange message: the lost camera and tripod were on the stern of a Starkist tuna fishing boat in Panama. The fishermen had caught it in their nets 400 miles north of the Galapagos. The film was intact.

    In November, biologists met *Lulu* in Costa Rica for a return visit to the Galapagos vents, primarily to retrieve the instruments and traps they had left on the bottom at the beginning of the year. The afternoon before their departure, Bob Hessler was beside himself—he had forgotten the formaldehyde. A visit to the only two pharmacies in Punta Arenas produced only about a half gallon of the pickling fluid, not nearly enough. The closest city, San Jose, was a six-hour drive one way, providing a car could be found. It would take days, maybe weeks, to have the material shipped from the states and passed through customs and still it might not get through customs in his lifetime. Someone suggested Jose, the town undertaker.

    Jose sold the scientists his entire supply of embalming fluid, about two gallons. "I just hope nobody dies," the undertaker said. "You know, it's the hot season. If somebody dies I'm going to be in big trouble."

    That November the dependably calm equatorial Pacific threatened to behave like the North Atlantic. The first dive was canceled because of high seas, but dawn brought good weather and luck.

Surface control vectored *Alvin* directly to the Rose Garden for the first of eleven dives in as many days.

Approaching a vent with pilot George Ellis and Hessler, ichthyologist Dan Cohen spotted his prey—a blue-eyed pink fish, the only fish seen at the vents, hovering in a cluster of tube worms, which he referred to as "pogos," for *Pogonophora*, the presumed phylum.

"Looking out of the forward port at a fantastic jumble of pogos, about half with red tips protruding," Cohen said. "The longer I look, the more crabs I see working through the bases. As the crabs climb over the tips of the pogos, the pogos retract into their tubes with fairly rapid motion. A jungle scene."

Cohen tried everything he could think of to catch a vent fish. First he laid down hooks with various baits—salmon eggs, bacon and supper leftovers. No nibble. Next he turned the hooks upside down and left them unbaited. That didn't work either. Then he put bait in enclosed containers. These caught only crabs.

He tried a fish narcotic, hoping to sedate his prey. When that didn't work, he tried poison, a last resort. He would have to kill them.

On December 10, *Alvin* took plastic bags of poison, and Cohen and Hessler, back to the Garden of Eden.

"Two vent fish looking directly down on a large area of shimmering water," Cohen said to his tape recorder. "Pogo comes out of tube end... several broken clamshells. What has broken them? Predators?... the film in the first camera is not winding properly. How was the first roll? Time will tell. A short, dark curled holothurian. Looks like dog shit. Now I see seven vent fish. Are we drawing them in? Three vent fish. The one nearest to me has his head in a small hole and is swimming vigorously downwards. Normal behavior?"

The pilot set down the remaining bags and burst the plastic to release the poison.

"The bags on the bottom are oozing solution which has turned milky," reported Cohen. "The fish appear interested. Fish swim around one bag, nosing it. No effect. Fish swim through and up the plume [of milky poison]. No effect. Six fish are nose down several feet beyond the bags in and out of the plume. No effect...."

Cohen and Hessler took pictures, waiting for a fish to expire. But after a half hour, they realized the poison was not working. Not only were the vent fish entirely unaffected by it, they seemed to like it. Four fish swam blissfully in and out of the empty plastic bags, like felines savoring catnip vapors.

They did not snare, live or dead, a single vent fish. The other work from *Alvin* was more productive. Ken Smith of Scripps dropped mussels into plastic cylinders attached to electrodes that measured respiration. It was the kind of *in situ* experiment impossible to perform except in person in the deep sea. The vent animals' rates of respiration, metabolism and growth were comparable to those of shallow-water animals.

Bacteria had moved onto the slates left behind in February. Microbiologists were amazed at the profusion of bacteria in the vents.

The scientists presumed that the many areas of empty clamshells were extinguished vents. Therefore, vents could not last long, perhaps only a few decades; otherwise the shells would have been dissolved. But while individual vents are short-lived, the vent ecosystem has been around for a long time. The closest relatives to some of the limpets dated to the Paleozoic Era, 250 million years ago.

Little else was clear about this ecosystem, which seemed unique in the world. The question was: just how dependent was it on chemosynthetic bacteria which drew energy from the earth—from *rocks*, not the sun. Presumably, the vent animals ate the bacteria. The clams and mussels took the microbes in through their filtering mouthpieces. Perhaps the spaghetti worms absorbed the nourishment through their skin. The tube worms? They were being frantically scrutinized for a hole, some opening.

The incredible science story even made the *National Enquirer*; the tabloid ran an entirely factual story under the headline WEIRD NEW CREATURES FOUND 1½ MILES DEEP IN PACIFIC.

Friends of Grassle's in Australia sent him a clipping from the magazine *Weekend*:

GIANT WORM TURNS UP

WHEN Dr. Frederick Grassle, of the American Oceanographic Institution, was lowered 1½ miles into the dark depths of the Pacific Ocean he found a strange new world. Where cracks in the seabed let warmth from the earth's core heat the water he saw sea spiders 1 ft. across, worms 8 ft. long which stood upright on their tails and shrimps with eyes on stalks.

22

Women—as members of the minority aboard ship (in most cases) you are likely to receive more than the usual amount of male attention. Enjoy it, certainly, but deal with it graciously. . . .
     A Manual for Science Personnel Aboard Research Vessels of the
     Woods Hole Oceanographic Institution, 1975

Scripps told us to wear big bulky sweaters.
     Kathy Crane, marine geologist

Dive 647 will be a survey dive for the Navy. . . . They are short-handed for people to handle ropes so there'll be one empty seat. Sure wish I was a guy strong enough to do this. Really feel in the way sometimes. . . . Actually I'm not allowed to do some things I could, which is very frustrating.
     From biologist Ruth Turner's journal, April 19, 1976

This is ridiculous.
     Tanya Atwater, marine geophysicist

Tanya Atwater joined the 1977 Galapagos cruise midway to get a training dive for an upcoming *Alvin* expedition in the Atlantic. Baggage in tow, the thirty-four-year-old oceanographer followed the only other female aboard *Lulu* to her "room."

As chief scientist, Jack Corliss was responsible for assigning berths; this he did with consensus. Those diving slept the night before their dive on *Lulu* and after the dive returned to the comforts of *Knorr*, where most of the science party lived. Jerry van Andel was assigned permanently to *Lulu* because he said he preferred the company of the mother ship to that of his peers.

With one exception, the women slept on *Knorr*. Debbie Stakes wanted to stay on *Lulu* because, she said, that was where the samples were unloaded and she was curator. *Lulu*'s skipper said it was strictly against WHOI rules. But Corliss ended the argument by threatening to take the skipper to court if his graduate student could not sleep on *Lulu*.

As all four berths in the bunk van were filled and no female was allowed in the Tube of Doom, Stakes slept on a moldy mattress on the narrow grimy floor of the electronics van. The spot was available from midnight to 6 A.M., when the electronics technician banged on the door to awaken Stakes, who would get out of his workshop, mattress in tow.

That's where Stakes took Atwater.

"Are you *crazy?*" Atwater said. "This is ridiculous."

Atwater walked the few steps to the bunk van and placed her luggage on a bed. "This is where we're sleeping," she told Stakes.

Nobody argued. "It was easy to get away with it because people were moving around so much," Atwater said. "It was really like a camp-out."

A lot of people followed the day's samples to *Knorr*, which had the lab space *Lulu* lacked, and by midnight or so it was too late to be ferried back to *Lulu*. Sleeping assignments were so boggled up that few paid attention to them. People on *Knorr* were sleeping on the floors in the laboratories and the chief scientist's cabin, which was also against the rules.

When Atwater and Stakes moved into the bunk van, *Lulu*'s skipper asked them to keep it a secret.

Woods Hole had no written policy governing women at sea, only an understood rule of at least two women on a ship or none at all. WHOI's seagoing senior female scientist Betty Bunce was an exception. In 1959 Bunce was kept off ships of other institutions because of her sex, but she demanded and was allowed to go to sea that year on a WHOI ship as chief scientist, "much to the captain's horror," she said.

Other unwritten rules about women at sea confused the issue. Some oceanographers refused to let women on their cruises. Women with degrees in science fumed when the senior scientist they worked for hired a male with the same or fewer qualifications to fulfill the seagoing days of a research project.

In March 1973 a group of women established a women's committee to represent the 188 females at WHOI—12 on the science staff and 176 in support positions. The committee immediately called for an investigation into the injustices of salary, job titles and promotions for women, and recommended the enunciation of a policy for women at sea and a manual written by women for women going to sea.

The women did write a booklet, but when it was finally printed in 1975, the 39-page *Manual for Science Personnel Aboard Research Vessels of the Woods Hole Oceanographic Institution* said nothing about the minimum two females. The booklet was an excellent primer for anyone going to sea for the first time, but it addressed women only three times. Under the "Miscellany" of "Shipboard Life," it read:

> On some ships women are prohibited from the crew's quarters. Ask a ship's officer.
>
> When someone of the opposite sex is visiting your stateroom, gossip can be minimized by keeping the door open. (Required by some Captains—find out.)
>
> Women—as members of the minority aboard ship (in most cases) you are likely to receive more than the usual amount of male attention. Enjoy it, certainly, but deal with it graciously, avoid undue favoritism, and remember that small incidents, both pleasant and unpleasant, can take on exaggerated importance in such close quarters and with the sense of isolation from real life that accompanies shipboard living. It is of primary importance that you handle this attention with persistent good judgment and good taste. Failure to do so can result in difficult situations, as well as jeopardize the credibility of other women with legitimate needs to participate in future cruises.

The topics Bunce had suggested including were deemed too delicate. She said women should be told not to put sanitary napkins or applicators into toilets; not to pay more attention—or even appear to—to one man than to others; and not to carry tales home from sea. "What a man does in a foreign port is his business, even if he's married," she said. "You belong to a club at sea. If you think anybody can have a private life on the same ship after sixty days even with three port stops, you're nuts. The only private thing you have is your mail."

The all-male camaraderie of research cruises had a long tradition. Seagoing oceanography was intrinsically macho. It was strenuous and often dangerous—launching heavy equipment (to say nothing of TNT) over the side of a bobbing platform; operating machinery; getting filthy dirty, bruised and wet; and tossing in a narrow bunk, too miserable with nausea to be scared out of your wits by the fury of a storm.

The old superstition that women were bad luck at sea persisted into the 1970s. And those who rejected such myths rationalized that a ship was no place for a female. Of course not everyone shared such sensibilities, but a decade after the United States Congress made sex discrimination illegal, female oceanographers across the country were still fighting for the right to go to sea and do their science. They dealt with the same indignities and outrages that women in any profession faced; and they met these in the classroom, in the laboratory, and in that quintessential male world of the ship.

As a summer student at WHOI in 1965, Atwater joined research cruises with the requisite second woman. "They told me that women are a big nuisance to take to sea," Atwater recalled. "They said we do this only for special women. It's a big privilege."

When Atwater went to the West Coast for graduate school in 1967, she understood what Paul Fye meant when he said WHOI was a "pioneer" in women's rights. At least WHOI routinely allowed pairs of women to go to sea. Scripps rarely allowed women on it's ships. Atwater's thesis adviser fought for her, reminding his administrators that sex discrimination was against the law.

"The first cruises at Scripps I went on, I seemed to be the number one subject in the planning meetings—what to do with me," Atwater said. "Women were poison. For my first ten years at sea I always wore the biggest, baggiest clothes I could find."

"This weirdness about sleeping put women in bad situations, especially on the smaller ships," she continued. "On one ship they made me a special bunk under the chart table in the lab. There was only one room with six bunks, and Scripps would not let me in that room. It took a long time to get them to realize that the safest place for a lone woman to sleep is in a room *full* of men."

In the early 1970s on a University of Miami ship, the winch operators ignored the instructions of geologist Kathy Crane. "The first mate supported me," Crane said, "but a few men on the bridge absolutely refused to accept any orders from a woman."

Research vessels are microcosms of society; every sailor and scientist adjusts his or her behavior accordingly. The deck beneath their feet is finite; if they are unhappy during a cruise, they can't get off. If they want privacy, there are few places to find it, although the newer ships have more two-person cabins and small libraries. At any time, a sailor may be sleeping; loud noise is inconsiderate. A scientist's very presence in many areas on the ship is unwelcome because he or she is a guest in someone else's home. The men who made *Lulu* home were a thoroughly masculine community until the first women scientists jogged their sensibility about the crew's rights—about what could and could not be done to their home and who could and could not come into their living room.

Ruth Turner and Tracy McLellan had slept on the escort *Gosnold* in 1971 and 1972. On their first trip, *Gosnold*'s skipper Harry Siebert asked the older scientist how she wanted to handle the bathroom facilities. "What do you want to do with the two heads?" he asked. "Should we label one Men and the other Women?"

"Hell, no, Harry," Turner replied wearily.

Both women were at turns frustrated and angry at the forced segregation. McLellan almost missed a critical cruise because until the last minute she could not find the token second female to accompany her. The twenty-two-year-old biologist sensed a group phobia; she heard members of the crew say it was bad luck to have women aboard ship. Turner's natural curiosity made her eager to witness one of the pre- or post-dives at a deep ocean station, but she was not allowed to remain on *Lulu* even that long. "I always felt we missed an awful lot," Turner said. "I was just terribly annoyed."

A formal procedure was established in 1971 in case Turner or McLellan needed to use the head on *Lulu*. (They did.) Locks were put on both doors to the head and before a woman could enter, a man had to inspect the room to ensure it was vacant. After each dive, the women were whisked in the whaler to the escort ship.

Debbie Stakes, the first woman to live alone on *Lulu*, was nearly raped. Late one night off the Galapagos, an *Alvin* technician entered the electronics van and got on top of her. They struggled but he easily held her down. She screamed and he finally left. Stakes reported the incident to the chief pilot who told her attacker that if he did that again, he would be fired. He didn't try it again.

After the incident Stakes slept fitfully, worried about the next man or men who might drop by. The van had no lock. She didn't know if her screams had been heard by others or if she was being talked about by the crew. She was in that discomfiting limbo of feeling outrage, fear, and irrational guilt and shame. "It was really hard being out there," she said. "I worried about being ambushed. There was all this pressure on women to not make trouble, to not tattle. Everyone was scrutinizing me to see if I was going to pay attention to an individual. It became a favorite pastime to see how far they could get with me. I was given a half hour a day for a shower and every day everybody would know when I was taking a shower. The pressure was constant."

Bill Page's electrician van, which doubled as his bedroom, became a forum, especially for a hard-core third of the crew strongly opposed to having women aboard. "Everyone came to cry the blues in my van," Page said. "I preached that women were here to stay. They listened, but I don't know if it did a damned bit of good. One *Alvin* crew member eventually quit because of the women. He was really distressed by it. Another crewman was totally destroyed about having Debbie and Tanya on board. He felt the girls had broken up the relations between the men."

"But the male scientists were the ones who put the women down the most," Page continued. "I watched them. The male graduate students would suddenly have their turn in the sub on the third or fourth day out, but if they were female graduate students, they might not get a turn at all."

The women scientists who lived on *Lulu* in the months following the Galapagos expedition found a generally unfriendly crew. "The men were really in a stir," said New York geologist Janet Stroup, who boarded *Lulu* in April 1977. "They didn't want to see us, didn't want us on board at all. We got very real antagonistic vibes from the crew. In fact some of the men told me they never wanted women on board ever again." Biologist Barbara Hecker knew something was wrong when she sat with a group of crewmen on deck and every man got up and walked away.

Stakes was not the last woman to have to fight off an intruder on *Lulu*. On another expedition a female scientist was jumped at night walking across the deck. A crewman on the bridge heard a scream and switched on the lights to illuminate a tall woman giving a sailor hell from the safe perch of an engine box.

Some skippers were ill at ease with women, especially when they were in charge. A female chief scientist said she once was informed by a mate: "Captain told me to tell you to tell your girls to please keep your underwear out of the head." The scientist went to the head, which also held the washer and dryer, and found the offending green underwear, obviously a man's. She dangled them under the skipper's nose.

Those locks on the doors to the head were arguably the most unpopular renovation ever made to the mother ship. Someone—that is, some one woman—would forget to unlock both doors. Going through the head to reach the deck or the pontoon was a standard shortcut; the crew didn't like meeting up with a locked door or having to go up the stairs to get to the combined bathroom, laundry room and passageway.

Many women did not bother to find an escort to the head and many ignored their shower time slot, set by the skipper. They refused to obey on principle, but also because the rules were a nuisance that made life at sea more difficult than it already was. Nobody, man or woman, wanted to take a shower at an assigned slot of between 6 and 6:30 P.M. when she or he would be elbow-deep in mud curating samples until after midnight.

Life on *Lulu* was hard on everyone. In *Alvin* the constraints of a seven-foot ball assured magnanimous equality for all—everyone used the Human Element Range Extender or HERE, the same used by pilots of light aircraft. It was a short plastic cylinder from which a funnel flared at the side or top. Nobody enjoyed urinating in the submarine; the urine receptacles had lids, but there could still be a stench and any strong odor was intensified in the damp seven-foot sphere. The odor of tuna sandwiches, the divers learned by experience, was overpowering. And of course, there was the embarrassment. A woman had to undress to urinate; a man could perform the function with minimal effort and display, but some men just could not pee in the sphere.

Many women and men heeded Ruth Turner's advice to "just live like a bedouin goat; you don't drink anything." Some experimented with various formulas for keeping a bladder quiescent for eight to ten hours. Fred Grassle recommended scotch the night before a dive. Ken Smith took red wine. Craig Smith had a bottle of beer. For some, nothing worked; for others the dehydration was so effective it made them sick. After a successful three-day dehydration, one scientist got the shakes.

Scripps Biologist Sue Garner remembered her first challenge with the HERE. She was five feet eleven, the pilot was six feet four, and the other passenger was at least her height. "There wasn't a whole lot of room with all those legs in there," Garner said. "They supplied this little canister with a tiny funnel. It was very, very awkward to pull down your pants while squatting and try to keep your balance with your legs all tangled together and somehow balance over this tiny hole."

She missed.

Finding nothing but the standard HERE on a French minisub, an American female biologist used a wine goblet, the widest-mouthed item in the sub. She said she filled it ten times. Perhaps this was the implicit risk of the standard French submersible gourmet lunch that started with paté and Beaujolais.

The public ritual of unloading *Alvin* after a dive could include full HEREs held aloft with great exclamations of accomplishment by the *Alvin* technicians who eagerly looked for the urinators. Each diver was responsible for emptying his or her HERE and washing it. For the recalcitrant, the chief pilot might bang a full HERE on a table in the mess while supper was under way, announcing that this was so-and-so's property and there would be no dive in the morning unless so-and-so took care of it.

Another receptacle was the Piddle Pack, a small zip-lock plastic bag with a sponge. Unlike the HERE, it was disposable. But it was impossible for women. Pilot Dudley Foster remembered watching a woman look at a Piddle Pack for the first time. She thought he was kidding. He thought she needed directions, since the words on the package were in Japanese. The Piddle Pack was short-lived.

A few women made special clothing. Barbara Hecker made a kaftan to cover her "from neck to toes" while crossing from the head to the bunk van, and she added snaps along the inseam in a pair of pants, but never had to unsnap them. Atwater made a jumpsuit with a zippered inseam which she did use.

On September 16, 1977, *Lulu* got her first female crewmember who was hired as a temporary "messman" for five weeks. At least one other female, also filling in for a vacationing messman, was hired. Both were liked. The crew endured the transition to a more egalitarian mother ship; they came to like having women aboard.

Some pilots took advantage of the increasingly greater numbers of females in the science parties and asked them if they'd like to join the "Mile Down Club." Like the earliest pilots, the newer ones bragged about the amorous records they allegedly set inside *Alvin* under water. But the Mile Down Club is certainly as mythical as females being bad luck at sea.

23

The geophysicists are the poets of oceanography. Their lot is not to pry off chunks of earth but to measure the invisible and infer characteristics, eventually to arrive at some elusive truth of what is happening beneath the seafloor. If they can learn what happened a year or a million years before, they can better predict what will happen in the future. Their measurements of magnetism, gravity, electrical conductivity and the propagation of sound waves through rock speak the language of squiggles, blotches, and stripes, which they translate into characteristics of the planet—how elastic, porous and dense it is.

Such measurements were usually made by equipment towed from a ship. More deeply towed instruments on sleds such as Deep-Tow yielded finer scale recordings. If they took down their gear in person with *Alvin*, some geophysicists wondered, could they get even more precise measurements?

Tanya Atwater, Ken Macdonald and a few other young Scripps geophysicists hauled their sleeping bags and several different kinds of borrowed magnetometers to Mount Shasta in the Cascade Range, hoping that one of the instruments would give a true recording of the magnetic reversals recorded in the hardened lavas of the ancient volcano—the same kind of magnetic field they expected to find under water.

"All of our friends who work on the magnetism of rocks thought we were crazy," Atwater said. "They knew that if you take a compass into a lava field it goes crazy. We knew where to go for the reversed magnetic rocks on Mount Shasta. Of course there weren't any pillows there, but we looked for big rocks with reversed magnetics and pretended that we were in the submarine flying over seafloor, thinking about how we were going to perform the experiments, how we were going to use the instrument from *Alvin*—mounted in the basket or would we have to wave it like a magic wand over the rocks?"

The standard towed magnetometer did not work on the volcano, but another kind did, and in the summer of 1978, a group of earth poets, led by Atwater, dived on *Alvin*'s first geophysical expedition. Their study site was at the Mid-Atlantic Ridge, partially contiguous to the area visited in 1974 on the FAMOUS dives.

The magnetometer was mounted on *Alvin* on a nonmagnetic ledge to which Atwater sewed bright orange fish netting, and the wiring was piped through a penetrator to a box of moving paper inside the sphere where four tiny pens drew squiggles and straight lines, registering magnetism on an unprecedented scale. The magnetometer towed from a ship recorded a very general average field from the many layers of magnetized rock beneath the seafloor; on *Alvin* it recorded the magnetic polarity of individual pillows and individual flows.

To the scientists' surprise, they found an unFAMOUS-like place. Instead of a narrow, V-shaped valley, this was wide and U-shaped. Instead of bulbous pillow basalts, the area was covered mostly in wide flat sheets of basalt, a common sight around volcanoes on land. Sheet flows were seen for the first time under water at the Galapagos ridge in 1977. Seeing them again confirmed that major underwater eruptions did not always form pillow basalt. And the valley, despite its nearness to the young FAMOUS valley, was old. The magnetic signal was unclear, which hinted at what the FAMOUS divers had suspected—that volcanic eruptions under water were episodic.

On one of their dives together, Atwater and Debbie Stakes wore homemade T-shirts with a silhouette of *Alvin* and the words: "A Woman's Place Is on the Bottom." As they descended 8338 feet, they knew Arlene Blum was leading to the top of Annapurna in Nepal an all-women team who had raised most of the money for their expedition by selling 15,000 T-shirts which said: "A Women's Place Is on Top."

The 1978 geophysical cruise met everyone's expectations, success which, Atwater knew, was due in large part to the *Alvin* and *Lulu* crews. The pilots had competed with one another to fetch the most samples. "Reachin' Ralph" Hollis, who collected fifty-nine rocks, was declared "(Official!) GRAND CHAMPION."

Atwater later wrote Hollis, the chief pilot:

> I realize that these long trips to the centers of the oceans cause many very real hardships for everyone aboard Lulu and for their families at home. It is only with the special care and effort of each individual that such trips can be accomplished in reasonable harmony and that good work can be done.
>
> Both from the point of view of pure science and from that of exploration of these last frontiers on our planet, the special effort was well worth it. The mid-ocean ridges are an unearthly land of delicate, glassy volcanic castles, towering cliffs, great fissures, and

chasms. It is a wonderful world that land-locked geologists could never begin to comprehend without this firsthand experience that you have helped us gain. The Lulu/Alvin tool is truely unique in the world, not a little because of the pride and craftmanship that all of you bring to your work.

On behalf of our group, and personally as well, I thank you.

A second geophysical expedition with *Alvin* was planned for the following year at 21° north latitude on the East Pacific Rise, a section of midocean ridge in the Gulf of California that joins the San Andreas fault system. Oceanographers had been coming to this site since the 1950s, primarily because it was close to Scripps in San Diego. By now it was well surveyed by surface-deployed instruments, including towed magnetometers. The ridge was about 5 million years old, time enough for the earth's poles to have reversed 19 times.

Lulu and the Scripps ship *Melville*, carrying about two dozen geologists and geophysicists from institutions in California, Massachusetts, France, New York, and Mexico, rendezvoused in April 1979 at 21° north. Scripps' Deep-Tow mapped the unsurveyed half of the study area while a navigation net was established. ANGUS began its runs and almost immediately picked up temperature anomalies. Soon Deep-Tow began to detect tiny temperature spikes. Was it a vent 1800 miles from the Galapagos? Some of the oceanographers thought so.

In 1978, French and American scientists diving here in *Cyana* saw large clamshells, mostly dissolved, and peculiar mounds and cylindrical stacks several feet tall. "We didn't know what they were then," French geologist Jean Francheteau said. "One wears blinkers in the ocean." A retrieved chunk of a mound was given a cursory description, packaged and carried home to Brest, where it sat for six months until a visiting American geologist saw it and realized what it was. It was sulfide.

But some of them were not so sure that a vent was nearby, and the chief scientists, Fred Spiess and Ken Macdonald of Scripps, who had been planning this cruise for two years, were not about to abandon their plans, not yet anyway. Besides, the funding was for a geophysical expedition.

On the first dives, they took gravity measurements which were more than a hundred times more detailed than those taken from the surface; the data suggested that subterranean magma chambers were not so large or so deep as expected. *Alvin* also banged a big hammer on the seafloor to measure the transmission of acoustic energy through the crust, and the magnetometer again drew straight and wiggly stripes in the deep sea.

After recording gravity on the fourth dive, Bruce Luyendyk and Jean Francheteau passed several tall columns which ANGUS had photographed the night before. This time the geologists were certain that they were seeing sulfide. The night before the ninth dive, ANGUS photographed jumbo clamshells and sensed more temperature changes. Bill Normark of the U.S. Geological Survey and French volcanologist Thierry Juteau, who were scheduled for the next dive, decided to investigate.

Normark was relieved not to have to immediately take another tedious gravity reading. If the submarine wasn't absolutely still, the gravimeter wouldn't work, and it could take a half hour to get *Alvin* perfectly still. During the last gravity measurement Normark's glasses slid down his nose; by pushing them back up with a finger he threw off the instrument and they had to start all over.

Surface control navigated *Alvin* to the first of three possible vents. Normark and Juteau were excited to see basalt stained with the colors of sulfide. However, it did not impress the pilot, Dudley Foster, who had been to the Garden of Eden. He moved on to the second target 300 feet away.

At about noon, Foster reported to the surface, "Coming into a dead clam area." He jotted down the coordinates triangulated from transponders Moe, Larry and Curly Joe. And then Foster saw it, what he likened to "a locomotive blasting out this stuff."

The "stuff" resembled thick black smoke, but of course it wasn't; there is no smoke in the ocean. It was spewing from a slender stack about six feet tall. On either side of the stack were two pyramid-like edifices, about four times as wide, with rounded tops.

"There were these funny little fish lying over one mound," Normark said. "They were very strange, very lumpy. They looked like pieces of intestine."

The observers were extremely curious, especially about the black stuff rushing from the chimney as rapidly as water from a fire hydrant. *Alvin* nudged closer. Water shimmered. The temperature was obviously above normal. If they had had any idea how much above normal, they would not have moved in.

Perhaps something was inside the stack belching the black smoke. They wanted to see. Foster knocked down the top half of the chimney with *Alvin*'s sample basket.

"It made a much wider opening but the black smoke came out even stronger," Normark said. "We really didn't know what to think."

Foster grabbed the temperature probe with *Alvin*'s pincers and stuck it in the smoke. The digital readout shot to 32.767°C (91°F).

He urged the scientists to report the temperature to the surface; he knew that the highest temperature recorded at the Galapagos was 23°C (73°F).

Alvin engineer Jim Akens, who built the temperature probe, knew something was not right. He had calibrated the probe based on 32.767°. "It was a number I remember," Akens said. "How could I forget it? Two to the 15th power minus one divided by a thousand and you get 32.767. That's what I used to set the scale to be accurate between 0° and 25°C, which was already almost twice as high as anybody expected to get, except for Jack Corliss, who wanted it to reach 100°C. And we all thought he was crazy."

Akens told Foster that 32° wasn't an accurate temperature. Try again.

Foster did, this time holding the probe in the smoke for a good long while. But the same thing happened, the numbers shot to 32.767°.

Alvin climbed above the fuming stack of rock and hovered directly over it. But it was impossible to see through all that smoke, so Foster gave up and headed for the third target, which reminded him of the Galapagos oases. There were no smoking stacks but much life—giant clams, tube worms, crabs. Thinking the thermometer was broken, he didn't take a temperature reading. Someone at the surface reminded them to do a gravity station, but it was too late for that.

As soon as *Alvin* surfaced, Akens looked at the fickle thermometer. The probe consisted of a thermister, a tiny glass bead attached to two wires. The wires snaked through a penetrator to a box of electronics inside the hull that converted the thermister's electrical resistance (which varied with temperature change) to frequency signals. The thermisters were taped to a long grey PVC rod. Now the plastic rod was charred black and the tip was melted.

The engineers searched through their manuals until they found one that gave the melting point of PVC—356°F or 180°C. In fact, the "smoke" was much hotter. On subsequent dives, the recalibrated thermometer recorded 662°F or 350°C. It was hot enough to melt lead.

"A lot of people still didn't believe it because they didn't think the thermisters were that accurate," Akens said. "But after every other dive, I would recalibrate the thermister again to make sure it was not damaged."

Hard to believe, but true. On the eleven remaining dives, *Alvin* discovered five other vents of belching chimneys—dubbed "black smokers" and "white smokers." Some were as tall as thirty feet. The black smoke consisted of particles of iron sulfide or pyrrhotite and zinc sulfide or sphalerite. The white smokers emitted slow-moving streams of clear or milky fluid containing particulates of silica and barium sulfate or barite.

Among the 300 pounds of sulfates and sulfides collected from the stacks was a mineral that had never been identified before in nature. It was magnesium-hydroxysulphate-hydrate, subsequently called caminite, from the Latin *caminus* for chimney.

"These samples were extraordinary," said Scripps geochemist Rachel Haymon. "We learned that the mineral anhydrite, which forms from sulfate and calcium, had the primary role of making the chimneys. Anhydrite forms when seawater is heated but it dissolves at low temperature. So, at the vents, the anhydrite dissolves after the hydrothermal activity ceases. We didn't know caminite existed in nature, although it was turning up in the lab when seawater was heated. Pretty science-fiction-like stuff."

In 1983, Haymon and her colleague Randy Koski would find vent fragments—pieces of sulfide covered in fossilized worm tubes—in the Bayda sulfide copper mine in the southern Arabian peninsula. In the Oman mine, the sulfides were embedded in ophiolites, rocks that were part of seafloor created 95 million years ago on a midocean ridge.

"Before the discovery of the black and white smokers, we had no idea about the morphology of sulfide deposits at the seafloor," Haymon said. "The deposits in the geologic record have no intact chimneys; they don't stand up to preservation. And we had no idea that there would be a biological component."

Now, they knew; they had *seen* it. Hydrothermal fluids were producing fresh sulfide polymetallic ore deposits. Minerals precipitating from these fluids stained the basalt with brilliant colors or bleached it white, or formed columns and mounds—underwater ore mines. Geologists usually reserve the word "ore" for minerals that are economic to mine; deep-sea mining is still an unrealized business. However, the amount of sulfides at the vents, especially in mounds, is as massive as some of the biggest ore deposits on land.

Like the Galapagos vent water, the 350°C (662°F) fluid that gushed from the columns had high concentrations of hydrogen sulfide, methane and primordial helium. The isotopes of the latter two clearly indicated a mantle source, that is, they could spring from nowhere else except deep within the earth's interior. Unlike any other seawater known, the black smoker fluid was truly hot—more than three times the boiling point at sea level. Chemists said the high pressure in the deep sea kept the water from boiling. The fluid also was rich in three sulfide ore-making elements, containing

100 parts per million of iron and a few ppm of zinc and copper, elements usually measured in seawater in parts per trillion.

When hot seawater passes through and by rock, the composition of both rock and water are profoundly altered. Cold seawater seeps down through the cracks and fissures in the ocean floor at spreading centers. As the water circulates underground near a magma chamber, it is heated and the chemical transformation of water and rock begins. Oxygen, magnesium and sulfates are removed from the water. Virtually all of the oxygen and some sulfate react with the iron in the rock to form dissolved hydrogen sulfide; the balance of the sulfate combines with calcium to precipitate anhydrite. The magnesium combines with hydroxyl ions and silicates to create hydroxysilicate, greatly increasing the acidity of the circulating fluid. The water is now so acidic that it dissolves sulfide and metallic elements in the basaltic crust.

When it reaches the seafloor, the acidic metallic water mixes with cold, alkaline sulfate-rich seawater. During this mixing, calcium sulfate precipitates from the heated seawater and metal sulfide minerals precipitate from the cooled, diluted fluid. These minerals accumulate around jets of gushing "smoke," making structures like the chimneys. The iron sulfide precipitates as pyrite and pyrrhotite; sulfides of copper and iron make chalcopyrite; and zinc deposits as sphalerite and wartzite.

Some so-called chimneys, called "snowballs," were more rotund than slender; they were covered by six-inch-long white tubes made by an unknown creature named the "Pompeii worm." There were also jumbo clams, tube worms and spaghetti at 21° north, but no mussels, anemones or pink vent fish. The lumpy fish Normark likened to intestines was an unknown species of eelpout.

There were no biologists aboard, but thanks to Haymon there was formaldehyde. Haymon had wanted to do her Ph.D. thesis on vent sulfides, and based on her innate optimism and the piece of sulfide *Cyana* had retrieved in 1978, she anticipated finding vents off Baja. She was so confident that she asked biologists how to curate animals; they told her to take sealed containers, formaldehyde and borax to raise the pH in the formaldehyde. Haymon didn't know until the first vent life came to the surface that the Scripps stockroom had given her Boraxo instead of borax. The animals were soapy but well preserved.

Neither were the geophysicists prepared to sample water, but they made do with the bottles on Deep-Tow. They couldn't place them directly in the vents so the fluid was diluted, but even diluted, it told an astonishing story.

The fluid was also loaded with chemosynthetic bacteria nourished from the hydrogen sulfide. More and more scientists now believed that the vent communities were nourished by these chemosynthetic bacteria which did not need sunlight to reduce carbon dioxide to organic carbon. Life instead existed in total darkness, with energy that ultimately came from the heat-generating radioactive decay within the planet.

Many scientists remained skeptical, however, that volcanism was the main source of nourishment. They argued that photosynthetic products in the plant and animal parts that rained from the sunlit surface water to the midnight depths played a significant role in nutrition. While this might be the situation in bogs and swamps, others argued, only about 1 percent of the photosynthetically produced organic carbon reached the deep sea—not enough to sustain the thriving oases of life at the warm Galapagos and hot 21° north vents.

While that debate raged, and rage it did, there was no doubt about how sulfide ore at the seafloor was formed and what played a major role in regulating the global ocean's chemistry.

According to the traditional model of the ocean's chemical budget, the constituents of seawater come from river runoff and weathering. But for decades chemists had known that the model was flawed. It did not account for all the ocean's minerals. And there was too much manganese and iron which accumulated in the deep sea in nodules and on basalt rocks. Where did the extra iron and manganese come from? And there wasn't enough magnesium. Where did it go? If all the salt came from rivers, why wasn't the ocean getting saltier? And what controlled seawater's pH? Why was it always between 7.5 and 8? Something was missing—some source and some sink. And it was a big something because it had a global effect—seawater around the world was basically the same. Now, suddenly, there were answers to questions about the ocean's chemistry that nobody ever had been able to resolve.

Collectively, the vents play a major role in regulating the uniformity of the ocean's chemistry for certain elements, including iron, manganese, lithium and rubidium. In comparison, the contributions of these elements from river runoff, weathering, and dehydration are minuscule.

Chemists calculated that a volume of water equivalent to the entire world ocean circulates down (as deep as four miles) through the ridges at plate boundaries every 10 million years. In geologic time, that is extremely brief. If the chemists are right, the ocean has made this journey only six or seven times since the dinosaurs died. And oceanographers reasoned that many more hydrothermal vents must exist at midocean ridges, but probably only at the fastest spreading ridges. (They soon would change their mind about the latter.)

"*Alvin* found what we had thought would never be found," MIT chemists John Edmond and Karen Von Damm wrote. "... [T]he discovery of hydrothermal activity at the submarine ridge axes demonstrates that the new and the unique are not the exclusive province of spacecraft missions to other planets. It is said that the earth is the water planet. Now it can be said that the chemistry of the water and of ocean sediment bears the prominent imprint of volcanic processes."

And they were just getting started.

24

> *East Pacific Rise 21°N*
> *RV* Melville *12:25 A.M.*
>
> *Hello Shoshannah—*
> *Floating around in the middle of this big ocean thinking heavy thoughts, as I always do out here. What an amazing cruise. We have all these superstar scientists out here this leg, studying the amazing part of the seafloor. . . . We've found incredible stuff out here—hot springs gushing out onto the deep seafloor, flocked with benthic communities of foot long clams, 3 ft. high tube worms, & other critters thriving in the warm, bacteria-rich water, like a deep-sea health spa. We've also found 30' high mounds of sulfide deposits (pyrite, sphalerite, chalcopyrite, native sulfur, etc.)—deep sea ores. . . .*
>
> *For all the great science that's happening it hasn't been such a happy cruise. . . . I look around at all the prima donna scientist assholes . . . and wonder if I'm going to end up like them. Their values stink. . . .*
>> Personal letter from a member of the 1979 expedition
>> to the East Pacific Rise

> *The clams hit the fan.*
>> Bob Corell, chairman of the Alvin Review Committee
>> from 1977 to 1987

> *If you can imagine war at sea; yes, I really mean it. There was so much pressure to succeed, to find more, to be the first. It was as if people became Neanderthals out there.*
>> Kathy Crane, marine geologist

> *Welcome to the Hydrothermal Vent Olympics. The first event of the day will be the clam toss.*
>> Announcement made aboard ship

The clam toss was the only event, too. Most people just want to forget it.

Biologists had set up an assembly line for curating the clams and mussels on a return cruise in 1982 to the Galapagos vents. The physiologist measured their metabolism; then passed them on to the geneticist who excised a tiny shred of tissue; the next person froze certain portions and pickled others, while the shells went to another. Every bit of mollusc was numbered, described, and logged. Speed was crucial because the molluscs had to be alive for the first few steps.

But at one station along this assembly line, a dead clam turned up and a few biologists complained: someone was not moving quickly enough. In response, the physiologist at the front of the line screamed: "You want a *clam?* I'll *give* you a clam!" *Whack!* One jumbo *Calyptogena magnifica* splattered its human-colored blood across the deck.

No one said a word. In the morning, the start of the Vent Olympics was broadcast anonymously on the ship's intercom, and at the expedition's end, the hot-tempered physiologist was awarded a plaque to commemorate his solitary medalist standing in the "clam toss." On another *Alvin* cruise he slammed his fist into a bulkhead and broke his hand.

What happened? "There were too many people, too many demands and a lot of pressure to get my work done," said the gold medalist, a successful, respected, and well-funded scientist.

Discovery got the best of others, too. It was hard enough competing for too few federal research dollars. You had to scramble to be first, to get the credit, especially for the scoop of a lifetime. The vent discovery transcended oceanography; scientists with no interest in the ocean were interested now.

Vent samples were stolen and bitterly fought over, as were dive seats and the authorship of papers. After the flap over the declaration that all biological samples from the 1977 Galapagos cruise belonged to WHOI, *Alvin* manager Larry Shumaker gave notice on that cruise that any animal which dropped off the sub onto *Lulu*'s deck was WHOI property. (The only animals that did were the freeloading white crabs.)

Requests for the animals came fast and furious, especially furious because there were not enough of the first vent animals to go around. Eight scientists from WHOI, Harvard and the small private Marine Biological Laboratory in Woods Hole jointly requested *all* the vent biology samples. They were turned down by Jack Corliss, who as expedition leader was responsible for the samples.

Corliss went for advice to the Smithsonian which, he said, was the one place in the country that probably had legal rights to specimens obtained with federal funding. Based on the Smithsonian's recommendations, the tube worms, molluscs, limpets, and other gastropods went to Meredith Jones at the Smithsonian, Ken Boss at Harvard, Vida Kenk at San Jose State University and James McLean at the Los Angeles County Museum. Since the Smithsonian had no bacteria specialists, Corliss gave the frozen slime and some water samples to John Baross and a few crabs to a graduate student, both at Oregon State. Not famous Harvard or Stanford or the major marine labs Scripps and WHOI. Some WHOI scientists were indignant; a few went behind Corliss' back to the Alvin Review Committee, hoping to get samples.

The Review Committee chairman, Bob Corell, winced at the memory. "Boy, there were some Woods Hole people who were really really angry with me," he said. "The question of who would have those samples became very contentious because we had something unbelievable. People saw careers built around that. My phone just flew off the hook for days. Some of those guys were on my doorstep every hour. When you have something that dramatic—this is a once in a century finding, this, here on the earth. There were people who saw Nobel prizes sitting out there. It was that sort of mentality."

Victor Vidal, a Scripps graduate student, could have cried. Vidal had been studying chemosynthetic bacteria from shallow water off Baja California, and expecting similar microbes from the deep Galapagos site, had asked Kathy Crane to bring him a water sample from the 1977 cruise. But Crane left the expedition after the first leg, and the cultures she had faithfully collected for Vidal and entrusted to a colleague were lost—at least, they never reached the farsighted graduate student.

A besieged Corliss in August 1977 wrote a four-page, single-spaced letter to thirty-eight people and the National Science Foundation to explain again the disposition of the samples and suggest holding a symposium at Oregon State for anyone interested in the vent biology:

> We sent the samples to the taxonomists whom the best independent biological group we could come up with recommended.
>
> I can only say to them, these WHOI-Harvard biologists, with respect for their high stature as scientists, that I hope they can, through this memo...see my perspective on this whole process. It is my hope that they will participate in our discussions of the hot springs at our November meeting, and that we can explore this fascinating phenomenon together. I would like to point out the rationale behind my actions in this matter. I certainly have no self-interest in spending my time this way. I have been accused in statements made to our program manager at NSF of "wasting hundreds

of thousands of dollars" because I didn't give samples to the right people, I have spent hours on the telephone....

I have acted first under the premise that the decisions regarding distribution of such samples are the responsibility of the Principal Investigators, the clearly stated policy of our funding agency....

Secondly, I have acted under the premise which I believe the Alvin Review Committee should clearly support, that *Alvin* is a national facility, and that we should make an effort to make it available to an increasing body of scientists.... I believe it is important to avoid the appearance that *Alvin* is a facility which is passed around among a closed group of insiders....

I'm sure this is more than you ever wanted to know about this subject. It is certainly more than I ever wanted to write about it. I hope it helps to serve to put it all behind us....

Scientists and *Alvin/Lulu* crew members displayed vent rocks, clamshells and tube worm skins on their fireplace mantels and office walls. The tough-as-Teflon tube worm skins, which dried to look like parchment, were proudly autographed and notated with dive numbers and dates. A scientist suggested selling the giant clamshells and using the proceeds to buy *Lulu* an icemaker, which scientist and crewman alike coveted in the tropical Pacific.

Thefts on the expedition to the East Pacific Rise in 1979 were more covert. Nobody admitted taking water samples and sulfides or unwrapping the sulfides and carelessly rewrapping them so that the labels were mismatched. Ken Macdonald noticed some of the missing samples months later at institutions other than Scripps, the home lab of the expedition's chief scientists. "It was incredible," Macdonald said. "Some things went on that I never saw before or have ever seen since. There were a number of large egos who wanted to carve out as much of this discovery for themselves as they could. It got pretty hot and heavy."

Of course everyone wanted a piece of chimney—the French, who had missed the black smokers on the 1978 cruise by only 160 feet; the U.S. Geological Survey, whose geologist was on the dive that discovered the black smokers; and the Mexicans, who were invited on the expedition in the interests of diplomacy—the Mexicans who gave the foreigners clearance to dive five days *after* the science party was scheduled to sail.

The cruise participants weren't home a week when three geologists from WHOI and Harvard wrote for sulfide samples. Not getting a response, they wrote again ten days later. Macdonald and Fred Spiess, who led the 1979 black smoker expedition, decided that the curator, Rachel Haymon, would catalogue the sulfides and send thin slices to those interested on a first come, first served basis.

It was a relief to Haymon, who worried that her Ph.D. research would be threatened by the thefts at sea and the clamor for samples. In the preface to her thesis she mentioned the challenge of overcoming "the resentment I felt toward those who attempted by less-than-ethical methods to 'scoop' me; the word is a journalistic term for an act which has no legitimate place in science. The disillusionment I felt initially over the avaricious behavior of highly reputable researchers severely shadowed my attitude toward my thesis for many months...."

Everyone who wanted a piece of chimney eventually got one, and Haymon earned her Ph.D. in geology and geochemistry. Likewise, the vent biology—everything from mites and leeches to the giant clams and the little Pompeii worms that lived in a microcosmic Vesuvius—was soon shared by a worldwide network of about fifty scientists.

With the hindsight of great discovery made with a national facility, the Alvin Review Committee established guidelines to prevent samples from migrating, unrecorded, onto mantels. All samples, "without exception and regardless of whether collected intentionally, incidentally, or accidentally" were to be curated aboard ship, and copies of the curation reports would be sent to WHOI, which would also store the original photographs and film from every dive. A year after the cruise, geological samples would be stored in one of seven national repositories; and after three years, biology samples would be sent to a national repository, such as the Smithsonian. In practice, however, the disposition of samples was ultimately up to the chief scientist, who could ignore the new policy with impunity.

It was an imperfect policy. It did not address every thorny issue, such as a chief scientist who does not make the discovery but insists on the credit for it. The scientists had to share an imperfect pie. There were not enough funds for all the desired dives; not enough time on a dive for all the desired tasks; and not enough luck to make every day free of problems with the submarine, the ships, the weather and the foreign governments that held absolute authority over their presence at some dive sites.

Even easy-going Fred Grassle, who never raised his voice, got angry once. He didn't get upset when almost everyone on the second expedition to the Galapagos vents complained about being short-changed for one reason or another. He wasn't too upset when four dives were aborted or shortened because of problems with *Alvin*'s hydraulic system, or when *Lulu*'s cradle broke and they had to travel 1100 nautical miles to the U.S. naval base at

Panama for repairs because Ecuador, the nearest port, demanded copies of all their scientific data translated into Spanish on arrival. Grassle didn't get mad until one of his peers accused him of trying to secretly make off with a vent anemone. It was a preposterous charge. Grassle had sacrificed his own work to try to satisfy everyone else. There were so many senior scientists and "guests" that there hadn't been room for the technicians and graduate students who usually curated samples. When Grassle realized that the senior scientists had forgotten how to perform this dirty and tedious chore and, in fact, were destroying the samples, he took over the curation, spending time that he otherwise would have devoted to more senior-level pursuits.

Fifteen of the thirty-three people on the 1979 Galapagos expedition had nothing to do with science. They were from the National Geographic Society and WQED, Pittsburgh public television, on board to film *Alvin* at the vents for a TV special called "Dive to the Edge of Creation". They complained about not getting all the dives they had paid for, and the scientists complained that the media were taking dives that were rightly theirs. The filmmakers got nine seats of the total twenty-nine dives, and nine seats was three more than any scientist got. Tjeerd van Andel and Ruth Turner got only one. Turner gave her second dive away to the photographers after Bob Ballard called a meeting in search of a volunteer to sacrifice a dive. The bottom line was hard to take: the media were richer than the scientists.

Figuring it would be good publicity for *Alvin*, Ballard had asked ichthyologist Dan Cohen to give one of his dives to John Denver, a favorite singer of Ballard's. Cohen told Ballard to give Denver one of *his* dives. Denver never dived.

The demands of dealing with the press brought new pressure and frustration. Ballard was an eloquent spokesman for ocean science and always happy to talk to reporters. He was eminently quotable and still one of the most experienced deep-sea divers in the world. *National Geographic* had sent him to Scotland to narrate a film on the elusive Loch Ness monster and he was the star of the television film "Dive to the Edge of Creation," filmed during the 1979 Galapagos expedition. The first version of the production included only one dive which featured geologist Ballard wearing a Cousteau-like watch cap and singing with a John Denver tape, *Take me home, country roads*... as the camera told a story about biology. A second version was more accurate, less corny, and included a dive with biologist Grassle.

The East Pacific Rise team had agreed not to talk to the press until the end of the cruise, when the chief scientists would call a

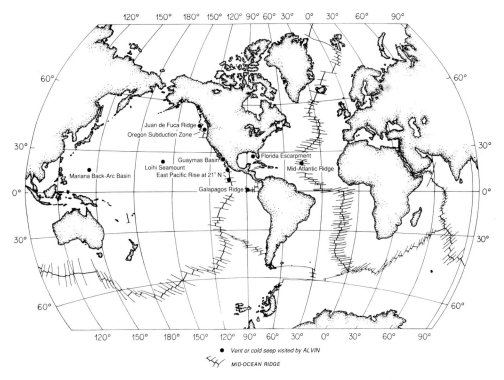

More than fifty hydrothermal vents and cold seeps have been discovered but fewer than half have been visited. (E. Paul Oberlander)

Allyn Vine, at right, explains the new submarine to visitors. (WHOI)

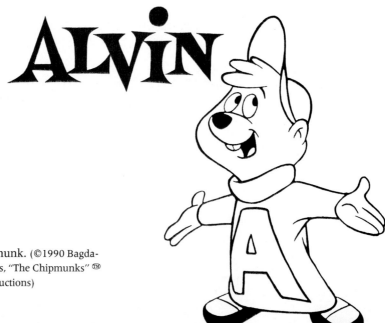

Alvin the Chipmunk. (©1990 Bagdasarian Productions, "The Chipmunks" ™ Bagdasarian Productions)

The bathyscaph *Trieste*. (from Auguste Piccard, *Earth, Sky, and Sea* Oxford University Press, N.Y., 1956)

General Mills's *Seapup*. (courtesy Bud Froehlich)

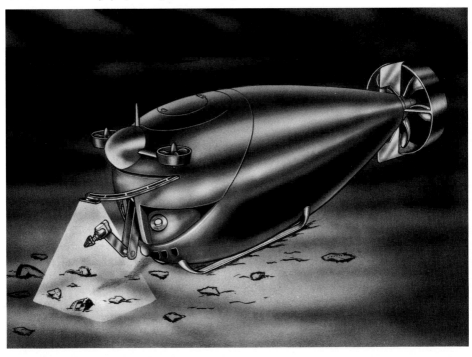

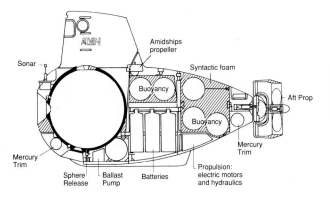

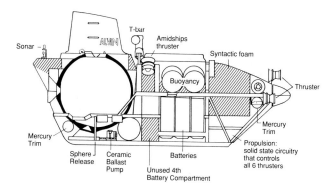

Schematics of the earliest *Alvin* (left) and the modern *Alvin*, still basically the same craft built around a seven-foot passenger sphere. (E. Paul Oberlander)

Welding an early *Alvin* passenger sphere at Hahn and Clay in Houston. "She sure looks purdy," the welding supervisor, Larry Megow, wrote on a photograph of the naked steel ball. (courtesy Bud Froehlich)

Top: The "bird cage," photographed in 1989, a frame of aluminum tubing in the passenger sphere which supports virtually everything inside, even the floorboards. Exposed portions of tubing, at just above the heads of sitting passengers, are handy grips. (Rodney Catanach) *Bottom, left:* The new steel passenger sphere being lowered into the sub in 1964; note electrical wiring in penetrators around the center porthole. (courtesy Bud Froehlich) *Bottom, right: Alvin*'s upper section, with sail attached, being lowered onto the passenger sphere after an annual overhaul circa 1968. (WHOI)

Alvin in Miami in 1967. A brow of syntactic foam was added above the center viewport. (WHOI)

The earliest *Lulu*. (WHOI)

The picture of a young Lulu Vine, Allyn Vine's mother, that hung in *Alvin*'s first mother ship. (Rodney Catanach; courtesy Allyn Vine)

Alvin on its cradle aboard *Lulu* circa 1968. The small shop vans with equipment for servicing *Alvin* are at the front and sides. The bridge, flanked by outboards, is at the stern. (WHOI)

Aluminaut.

Spanish and American officials with the hydrogen bomb that was retrieved from the Mediterranean in 1966. Rear Admiral William Guest is second from right. (U.S. Navy)

This cartoon ran in the Houston *Post* March 21, 1966, with an editorial. "Until the last it was a story of high drama right out of James Bond," the editors wrote. "The end though, when it came, was pure Walt Disney. It was, as the children's stories might title it, a Stubby Little Sub That Could which finally solved the whole problem by finding that bomb in 2500 feet of water. It had the unlikely name of Alvin. Never give up on the little people of the world."

'I WOULDN'T BE SO EMBARRASSED IF THE ANDREW JACKSON OR THE GEORGE WASHINGTON OR THE NATHAN HALE OR THE SAM HOUSTON OR THE WOODROW WILSON FOUND IT ... BUT ALVIN!'

(Cartoon, Tom Darcy)

A hapless swordfish rammed its sword between *Alvin*'s upper and lower sections in 1967, narrowly missing a viewport. (WHOI)

A traumatized *Alvin* — crippled and disfigured but safe on land following its rescue in 1969. The flag Adelaide Vine made is at left. (WHOI)

Cliff Winget, at left, hands Skip Marquet one of three baloney sandwiches that were untouched by decay after spending ten months at the bottom of the ocean inside *Alvin*. The bouillon in the imploded thermos and the apples were in the same amazing state of freshness. (WHOI)

Getting *Alvin* back onto its cradle after a dive. Pilot Mac McCamis, in the sail with the steering control box of toggle switches, shouts directions to the linehandlers on *Lulu*'s pontoons. (WHOI)

Retrieving *Alvin* with the A-frame system on the *Atlantis II*—escort swimmers ready to loop the main line around *Alvin*'s T-bar. (Rodney Catanach)

Alvin freshly hoisted from the water at the *AII*'s stern. (Rodney Catanach)

On *Lulu*'s deck, pilot Mac McCamis at the board listing the day's dive assignments. (WHOI)

One of many cartoons by geologist Conrad Neumann depicting life aboard *Alvin* and *Lulu*.

#52

Dive # 1160

24 NOV 81

Position: 20° 51' N, 109° 04' W

Location: "21-North" (New site on offsite rift to SW, 8 km)

Obs: K. Kim (SCRIPPS)
Obs: S. Mercado (MEXICO GOV'MT)

Depth: 2572 m

Time: 8 hr 19 min

Description: Found vent area fm ANGUS coordinates. Lots of tube worms, young clams, one black/white smoker w/ temps up to 350°C. Very rough terrain. Plume makes it difficult coming down in area. Spaghetti + crabs.

pillow basalt hill

tube worm garden

~5 m

white chimney + organic part of chimney w/ small black smoker in middle

A page from pilot Bob Brown's log describing a dive to a hydrothermal vent on the East Pacific Rise at 20° north.

Geochemist Harmon Craig about to savor whopper escargot from a Mariana vent. (Andy Campbell)

Spoke Too Soon?

Courtesy *News-Times*, Portland, Oregon.

press conference. But those who left the ship after the last leg were greeted at the dock with the latest *Newsweek*, which ran three color photographs across a page with the caption: "Oceanographer Ballard and big red worms: An underwater Disneyland of extraordinary animals and minerals." Ballard had made the vents come alive, and while he took no credit that wasn't due him, his colleagues were furious. Some of the press were agitated, too. A television network had demanded *National Geographic*'s underwater footage of the Galapagos vents, which the magazine refused to cede. The network sent a lawyer to NSF. But NSF could honestly say that the photographers had paid for their dives, so the film did not belong to the taxpayers.

To many of his peers, Ballard was too interested in high-tech toys like *Alvin*, and too frequently featured as author or subject in the popular press, which many scientists distrusted. And Ballard kept talking about a cockamamy idea of finding the *Titanic*, not a lofty pursuit of basic science. Since the early 1970s, Ballard had been thinking about how to find the *Titanic* and investigating funding and publicity possibilities with *National Geographic*, Walt Disney Studios, and television stations in Japan, Sweden, and Britain. In 1977, an opportunity came along.

Chrome George, chairman of the board of Alcoa, offered his company's drill ship *Seaprobe* to WHOI free of charge. *Seaprobe*'s "moon pool," through which the drill pipe was lowered, was especially attractive to oceanographers; it was easier and safer to launch equipment through a hole in the center of a ship than over the side. During about a year of use, however, WHOI scientists also learned the vessel's weaknesses. It rolled like crazy and suffered engine trouble. Once *Lulu*, which had never passed anything except concrete piers, zipped by the drill ship. But *Seaprobe* still appealed to Ballard. The big claw at the end of the drill string could lift more than 200 tons. It had already recovered an airplane. Besides, it was free.

Ballard and his friend Emory Kristof planned a cruise with *Seaprobe* to determine if high-quality photographs were possible in the deep North Atlantic where the massive Gulf Stream, a "river" of many strong currents, flowed and stirred up bottom sediment. They wanted to know if the muddied waters of this powerful current could hide a ship as massive as the *Titanic*. Their plans made the *Washington Star* under the headline "Ghoulish Fishing in the North Atlantic."

In the fall of 1977, Ballard and a few WHOI technicians, Kristof, and other *National Geographic* photographers, left Woods Hole on *Seaprobe*. The goal was to try to find and photograph a "target," a clump of scrap metal, which they would push off the ship. The high-tech searching instruments and cameras would be attached to the end of the drill string.

On their first night at sea, the *Seaprobe* crew laid off at 11:30 P.M. after a day of lowering sixty-foot lengths of piping through the moon pool. As they slept, Ballard and company began looking for the "target" with their sophisticated gear, almost all of it borrowed. The sonar came from Westinghouse, a positioning system from Honeywell, special cameras and strobes from the navy and Benthos, a Cape Cod electronics firm.

At 1:41 A.M. the monstrous counterweight at the top of the derrick plunged to the deck in a thunderous crash and splintering of wood. The lights went out; arcing wires sprayed sparks and whipped across the deck. Ballard's technician Earl Young frantically looked for the main power supply. He saw a lever, pushed; the arcing stopped. The Woods Hole and *National Geographic* people stared silently into the black well of the moon pool and passed a bottle of bourbon in wonderment at the might that could have killed them had they been on the deck. The drill pipe, weakened from the ship's acute rolling, had snapped at 3125 feet, causing the attached cables to rip apart and drop the counterweights. It was not the first accident on the *Seaprobe,* but it was the last under WHOI's operation. Woods Hole sent it back to Alcoa.

Insurance paid for the half million dollars of lost equipment but did not assuage Ballard's humiliation. He was up for tenure at WHOI and doubted that he would get it. Unwilling to bear the shame of the rejection he anticipated, Ballard moved his family to California in 1979 for a sabbatical at Stanford University. His hope of finding the *Titanic* didn't fade. The following year he faced more disappointment when someone else tried to beat him to it.

Lamont Geologist Bill Ryan had no special desire to find the *Titanic,* but he needed big money to develop a multibeam sonar system as good as the navy's classified version. "The federal funding agencies were reluctant to take a risk," Ryan said, "so we found this Texas oil millionaire Jack Grimm, who was willing to take a chance as long as the sonar was used to look for the *Titanic.*"

No sweat.

SeaMarc, the sonar Ryan and his team developed with Grimm's money, turned out to be a superb tool; some thought it was even better than the navy's sonar. In 1980, more Grimm money financed an expedition to use it. Also aboard was Deep-Tow and its developer, Fred Spiess. Deep-Tow, which carried a side-looking sonar, magnetometers, a TV camera, 35 mm camera and lights, could transmit data live via electric cable to monitors at the surface. With Deep-Tow and SeaMarc, Grimm had the best remote search tools in the world.

The remote sleds searched at 41° 46′ north, 50° 14′ west, the coordinates tapped out on wireless by a brave sailor on April 14, 1912, as the *Titanic* was sinking. After twenty days of looking, nothing was found. The group returned the following year. Grimm was certain that a large curved image captured by a color video camera was a propeller from the *Titanic*. The scientists didn't think so. In 1983, the team returned again, but the weather was so bad they couldn't even get their gear over the side.

25

The Alvin *Group were not normals. It was because of their mentality; there were a lot of characters. I was telling my wife this, and she said, "Yes, but stop and think, George, you're no different from the rest of them. Normal people live with their families and mow the lawn and go to church on Sundays. You can't say that normal people spend most of their life on the ocean eating in galleys."*
So maybe that's what it is. Maybe we're not normal to start with.
Alvin *crew chief George "Brody" Broderson*

From sixteen to nineteen men made *Lulu* home; they were about evenly divided between the sailors who crewed the mother ship and the mechanics, electricians, and electronics technicians who tended and piloted *Alvin*. Some were after adventure. Others were escaping something or someone—the law, a wife, society. Or it was just a job.

Everyone had a nickname. There was Chicken Man, the Weasel, Bubble Butt, and Saniflush. The Little Pisser was arrested for relieving himself in a parking lot. Charlie Tuna, El Kabong, and Captain Crunch were skippers.

Strawberry Fields walked naked across the deck one day in front of reporters. Rib Roast was a cook. There was Foo-Foo, Horny, Turtle Man, Flubadub, The Bird, Spider Man, and Meatball. Soggy was an escort swimmer. The Murderer had a bullet hole in his side, which he acquired while trying to kill himself after he killed his wife and the man in bed with her.

Baby Duck was pigeon-toed. He was also six foot four, had a schooner etched in a front tooth, an anchor tattooed on one cheek, a pony tail, and a long beard that separated into two strands. And there were Pickles, Uncle Hard Ass, Buffalo Bob, and Joe the Plumber—who was a real plumber until he was hired as *Lulu*'s cook. Ready Kilowatt had a potentially dangerous habit of attaching the plus wires to the minus terminals in the submarine. Bare Bear slept with a teddy bear and his girlfriend's underwear over his head. The Fat Tourist ate in his bunk.

Some nicknames stuck so well that the crew would forget a man's real name. Some of them hated their nicknames. Chicken Man complained that he was losing the crew's respect.

When the retired naval officer Jack Donnelly first came aboard as a member of the team, crew chief George Broderson said, "Well, weeell, if it isn't Immaculate Jack." The name stuck.

Alvin was the Little Pig and the Fat Devil and worse. But these were for only the *Alvin/Lulu* family members who would not let an outsider get away with disparaging their sub.

"When I first came on board," Tjeerd van Andel said, "*Alvin* was sitting in the center deck like a little white goddess and I got the feeling that we were supposed to curtsy. If you leaned against *Alvin* and left a smudge mark someone would rush up and wipe it off and stare hard at you."

"If you got near *Alvin*," Bob Ballard said, "Brody would spot you, tap you on the shoulder, and say 'What do you want?' And *Lulu* was a Mad Hatter's tea party. Those guys were so eccentric. When you got on *Lulu* you had the sense that you were an intruder."

The crews from WHOI's other ships made fun of them. During FAMOUS the *Knorr* crew pelted them with eggs and potatoes. An egg cracked on the current meter atop *Alvin*'s sail and baked in the sun. It wasn't until the next dive that the crew realized a fried egg had made the meter inoperable.

The mother ship community was bound in a kind of wartime camaraderie by both their mission to support *Alvin* and the physical constraints and hardships of their home. They were tough; they had to be. During launch and recovery, every man used all his strength, because it took everything he had to hold the lines and keep *Alvin* from bashing into *Lulu*. They were frequently up to their knees or waists in water. In rough seas the lines snapped and they worried about losing a finger. They fell overboard and worried about losing an arm or a leg to a shark or to one of *Lulu*'s huge outboards.

On a few dives Bill "Soggy" Page felt sharks rub against him. On seeing a couple of sharks circle the mother ship, a recently hired technician told Page: "This is the way I see it. It doesn't have to be both of us. I have a college education. You dive." Page dived alone and lived to tell about it; the college man didn't remain with the group long.

Launch and recovery took virtually all hands, including the galley staff and members of the scientific party. While the skipper steered *Lulu,* at least ten people handled the lines from the pontoons; two were in the whaler to fetch the escort swimmers and the second pilot who motored *Alvin* out; an engineer lowered and raised the cradle while someone stood with a pipe at the hoist to poke apart the chains when they overlapped.

Steve "Mongoose" Allsopp turned twenty-one in 1973 when he first sailed on *Lulu* as a messman, the lowliest job in salary and status. "As we left port," he said, "I felt this hand on my shoulder and this voice said 'The last messman we had here went out for a fourteen-day trip and he was sick for thirteen days.'"

The voice came from Page who was seasick most of his first few years on *Lulu*. The mother ship didn't move like any other vessel; it kind of sashayed a figure eight, which they called "Lulu's dance."

Another challenge Allsopp faced was the cook. Big Ed Brodrick had bid a sad adieu to the galley in 1969 on doctor's orders to lose seventy pounds. The new cook ran, cleaver held high, after scientist and messman alike, chasing them out of the galley, the mess and, when his temper was really toasted, across the deck.

He got angry at Allsopp for refusing to put out the butter for supper. The messman said it was moldy. "I keel you when you sleep," the cook threatened. He was almost fatherly about it. "I keel you," he'd say and nod. "I keel you."

Allsopp thought he was eminently capable of the deed. At least he said he had killed a man before, and one sultry night in the Bahamas, the night someone tried to steal *Lulu*, he volunteered to do it again.

Lulu was anchored about a hundred yards off shore. At about 3 A.M., Allsopp was startled at what sounded like the *whoosh* of *Lulu*'s air compressors. He heard the noise again. *Lulu*'s two engines were running! At the bridge he found a tall bare-chested man in wet shorts.

"What do you think you're doing?" Allsopp asked.

The man replied in a clipped British accent: "I am starting the engines."

Allsopp called to the mate on watch in the mess. "Could you get up here; we have a problem."

When the mate reached the bridge, he said, "Who's *he?*"

"How the hell do I know!" said Allsopp.

The Bahamian offered that he boarded with another man. As the mate guarded the Bahamian, Allsopp crept through the ship, looking for the accomplice and awakening everyone with whispered words of caution.

The cook began sharpening his knives. "We keel him *now?*" he said, *clink, clink.* "No, not now," Allsopp said. In ten minutes, the cook checked the status of the situation. *"Now?"*

By the time the police arrived, the Bahamian was trembling. The crew learned later that he had escaped from prison and robbed a gas station at knifepoint before swimming to *Lulu* with the intention of sailing to freedom.

Joe "Rib Roast" Ribeiro, who stayed for more than four years, a long tenure for a *Lulu* cook, kept a coffee can at his side as a portable urinal. Laziness wasn't his only motivation. Getting to the head above the starboard pontoon meant descending a ladder and crossing the deck; it was a short journey, but one that could be fraught with surprise, especially during a storm in the darkness of early morning or night when deck planks might be missing. Some of the plywood was eventually replaced with steel grating, but these, too, ripped off in heavy seas.

Within the confines of *Lulu*'s ninety-by-fifty-foot deck were three eight-foot-long shop vans and *Alvin;* the space left over could fit in about half a tennis court.

It was a damp home. The watertight doors leaked. In only moderate seas water rushed up through the gratings at mid-deck.

Except for a dim red light, it was always dark and dank in the starboard pontoon, with the bunks stacked three high. Chuck Porembski, an electronics technician, had a bottom bunk. "I could put my fist on my stomach with my thumb up and my thumb would touch the mattress above," he said. "To this day, I still sleep with one knee pulled up high and a leg out to keep from falling. If Kenny [Costa] was not in his rack and we took a roll, I would end up on the floor."

Kenny Costa, a *Lulu* chief engineer, was known to sleep through the cacophony of seismics. His tubemates packed inflatable life vests beneath his mattress, and as he snoozed pulled the cords to release the carbon dioxide in the small cartridges on the vests. The idea was to rapidly—instantaneously would have been ideal—elevate Costa up to the ceiling of the pontoon. But the vests slowly inflated and Costa, still snoozing, was enveloped by his mattress like a giant taco.

An arguable advantage to *Lulu*'s smallness was that everyone had to get along, and if you had an occasional bad day, that was OK, your family understood.

Broderson was the resident papa-san and patriarch. He was at turns loved and disliked. His impatience could erupt into a foot-stomping tantrum. He had no patience for a seaman who couldn't splice line or a scientist who put grimy hands on his clean white submarine. He also could be kind and compassionate. He bought ice cream for the curious children at the docks, and he was quick to know a sailor with a broken heart needed attention.

Complaints found their way to him and he invariably kept the peace, sometimes behind his own hunting knife, but usually through common sense. That his orders were obeyed, even for the least popular jobs, like stacking the filthy drop weights, was a testament to the respect he commanded.

"All right, you foofniks!" he would yell. "Haul iron!" And if someone objected, Broderson had ready another line. "Do you eat? Yes? Ah! You *do* eat. Then you work. We all eat, we all work. All right, you spaghetti-eaters! Haul iron!"

Everyone was a foofnik in Broderson's nonsense vocabulary, which featured prominently in his sudden loud emotings or sotto-voce soliloquies. Your first witness to these displays might make you laugh, unless of course it was you whom Broderson was imitating, your tics and funny mannerisms. You didn't have to do anything special to fall victim to Broderson's irreverence. He so enjoyed deflating the swollen egos of Ph.D.-holders.

"Hey, fatboy, you ready to dive?"

To the geologist cracking rocks on deck: "HEY, YOU CRAZY SON OF A BITCH! WHAT THE HELL DO YOU THINK YOU'RE DOING! THOSE PIECES OF ROCK ARE FLYING INTO THE HYDRAULIC SYSTEM OF MY SUBMARINE!"

Naturally, this crew of aliases nicknamed their customers. Ruth Turner was Lady Wormwood. Ken Smith, who measured respiration and oxygen [O_2] rates at DOS No. 1, was the Egg Man. The Bug Man collected bacteria. Beethoven's coiffure resembled that of the composer. Goat Man had bad body odor. And there were The Good, The Bad and The Ugly. The Mad Russian.

Most of the scientists understood that they needed this family. But the mother ship and submarine crews were not sure that WHOI knew. In 1976 they aired their grievances in a memo:

> LULU's only potentially favorable accommodation, the galley, has become an area of shame and neglect. The walk-in reefer, host to unsavory growths of molds, has been kept locked against inspection and only cleaned when the characteristic odor of decay became unbearable. The fact that food associated with this condition was often served as part of the remedy was hard to keep secret....
>
> Repeated mishaps and a man overboard due to oil-slick pontoons coupled with the insecure and often elusive decking on LULU, compound the possibility of injury.... The lack of air circulation in the berthing area and the subsequent morning headaches are another area of concern....
>
> In conclusion, it is evident that LULU by nature is substandard and often must make do with the hand-me-downs of the fleet. We take pride in our ability to perform as a group with the professionalism for which we are famous and ask nothing more than fair consideration in the aforementioned areas.

They really did care. During *Alvin*'s funding crises they came up with solutions. They suggested using less expensive steel or concrete instead of lead for *Alvin*'s drop weights. Rags could be washed and recycled rather than thrown away. If a scientist didn't use all the film on a roll, the film could stay in the camera for the next diver. The administrators were surprised but unimpressed with the recommendations that might save hundreds but not thousands of dollars. In another decade, however, the crew's suggestions would become standard operating procedure.

Following the crews' memo, the paint locker across from the head was made into a communal room called the Lounge, about as big as the back of a camper and with the same kind of furnishings, a small table hugged by two benches. A few shelves (the Library) were supplied with books by WHOI's librarians, and another new air conditioner went into the Tube of Doom.

WHOI knew that life aboard RV *Lulu* was difficult, but there were no funds for a new mother ship; there were hardly enough funds for *Alvin*. The front office had appointed special panels to investigate *Lulu*. One internal report from the late 1960s said: "The users feel perfectly safe in *Alvin* once it is free of *Lulu*; most feel safer in *Alvin* than on *Lulu*." A 1974 report read:

> Many people think that LULU is an uncomfortable miserable ship and that it presents more safety hazards than *Alvin*.... Relatively speaking, LULU remains the most hazardous part of the operation; there are at most 3 people exposed to danger on ALVIN and up to 35 on LULU to say nothing of ALVIN's complete dependence on LULU. There is no question that LULU is miserable as a "home" at sea; this does not mean she is not safe.

Neither of course did it mean *Lulu* was safe. Some crewmen were outright scared in heavy weather. They feared that *Lulu* would be swamped if anything happened to either pontoon, the only means of keeping the ship afloat. During a severe storm, one technician slept in his skin-diving gear with an air tank in his bunk. Bill Page refused to sleep in the Tube of Doom; he built himself a bunk that folded down on hinges in the electrician shop van. Pilot Ralph Hollis did the same in another shop van.

According to the engineers' calculations, if *Lulu* was rammed or a pontoon flooded, she would float sideways and half up in the air on the good pontoon. Small comfort. What if the bad pontoon was the Tube of Doom?

26

Alvin gained generally high marks on its first report card as a national facility from 1975 to 1977. With the traditional evaluation of peer review, a half dozen scientists from around the country anonymously assessed the sub's performance, concluding that funding should be extended another three years. A reviewer wrote:

> *Alvin* has become an extraordinary tool in investigations of life processes in the deep sea. Experiments and quantitative observations of the fauna of the seafloor have become possible only with *Alvin*, and have advanced greatly our knowledge of this otherwise inaccessible environment.... The continued support of *Alvin* is necessary for the expansion of our understanding of the deep sea. I strongly urge funding at the requested level.

Despite the agreed-on allotment of $300,000 each from the National Science Foundation, the navy and the National Oceanic and Atmospheric Administration, it had not been an equal split, and the ante was never big enough; it had fallen on NSF to find the extra dollars. Money was so tight that the Alvin Review Committee considered a request to let the country's only national oceanographic facility become a movie star. "We got a letter from Walt Disney Studios with Mickey Mouse on the letterhead," Sandra Toye of NSF said. "It made everyone at the committee laugh but the request was considered seriously, although it never panned out."

Operating costs for *Alvin* were about the same as for any of the big oceanographic ships, not including the escort ship which was necessary on major expeditions not so much to chaperone *Lulu* but to berth scientists and provide laboratory space. Larry Shumaker said he had to "lie, cheat, and steal" for more money every year. "WHOI would not put any of its own money in; that was an absolute," he said. "Come November, we usually would have spent all our budget, and Skip [Marquet] would start padding next year's budget. Skip is basically a genius. We'd take our budget into NSF, and Sandra Toye would say, 'I know you've got padding in here but damned if I can find it.'"

Each year the budget grew by about $100,000; by 1980 it was $1.9 million. The time between overhauls got longer and *Alvin* made a record 117 dives in 1979. But the idea of it being a WHOI boat persisted. A peer reviewer complained in 1977 that *Alvin* needed "new blood" because "the preponderance of work done is still allotted to WHOI scientists or their 'friends.'" But that was only natural; it was WHOI—primarily its biologists and geologists—who had had faith in this tool when few others did.

Any scientist who wanted to dive was required to send a research proposal to a federal agency, usually NSF, whose peer reviewers determined the quality of the science. Having passed this muster, the proposal endured a second perusal by the Review Committee, which looked at the feasibility of using *Alvin* for the proposed work. Diving for the sake of simply exploring was unacceptable. If the work could be done remotely from a ship, the committee rejected the proposal. The committee said no scientist, no matter how lofty his or her reputation, was guaranteed a dive.

While the committee may have been true to its ideals, the reality of political pressure sometimes overruled its decisions. Some scientists with funding from NOAA or the navy tried to avoid NSF's tougher peer review standards. "There was always tension trying to accommodate the navy and NOAA users without eroding the quality or distorting the schedule," said Toye. "It was never an entirely ideal case of peer-review quality for *Alvin* users. The navy and NOAA, in effect, pounded the table, and said so-and-so has got to use this for so many days or you don't get our $300,000."

Having to endure two peer reviews was a kind of double jeopardy, especially when the committee rejected a proposal approved by peer review. Or vice versa. To be clear about their intended use of *Alvin*, the investigators had to know beforehand what they wanted to accomplish under water. That could be a real head-stretcher for those who had never dived. And rejecting the look-see dives disavowed the special ingredient of serendipity in some of *Alvin*'s major discoveries.

There was disagreement about whether *Alvin* was essential for certain tasks, but with experience the committee grew more astute at making that call. More and more scientists became adept at using the extraordinary tool, and after the vent discoveries there were twice as many excellent proposals for *Alvin* as there were dive openings. Some would have to wait two and three years for a dive.

With enough customers to fill two subs, what about the other deep-diving subs, *Turtle* and *Sea Cliff*, a.k.a. *Columbus?* They were not outfitted for science or *Alvin*'s customers. "Those operations were totally chaotic," said biologist Jim Childress, who dived in *Sea Cliff* more than twenty times. "One year we'd go out and have a great pilot, and the next year the guy couldn't even drive the sub. The Alvin Group has always been a lot more professional."

Alvin was the most experienced deep-diving craft in the world. Nothing else came close. By 1980, the end of the second three-year funding period, there was no serious threat of losing funding, and for the first time the Alvin Group began to relax. By then there were thirty-three papers devoted solely to the Galapagos vents. The only record that was supposed to matter in science—technical journal papers—finally existed for *Alvin*, and what a stunning record it was.

Most of the divers were biologists and geologists, followed in frequency by chemists and geophysicists. The sphere had acquired new comforts. The boat cushions were replaced with fireproof foam crescents above the windows for foreheads. On the long rides down and up, scientists and pilots brought cassette tapes of music. George Ellis brought his Green Hornet tapes from the 1950s radio show. Some brought videotapes. *The Sound of Music*, a favorite of Ralph Hollis's, featured on several ascents. "Ralph had a whole drawer full," WHOI biologist Rose Petrecca said. "You took two videotapes, one for up and one for down." With only one palm-sized monitor, *Alvin* was hardly ideal for movie viewing. The passenger on the starboard side had to watch the movie reflected off a mirror.

Alvin and *Lulu* stayed in the Pacific exploring mostly vents from October 1978 until mid-February 1980, when *Alvin* made its thousandth dive. That year, WHOI celebrated its fiftieth anniversary with a buffet dinner for three hundred at a Boston hotel. The centerpiece was an ice sculpture of *Alvin*. And later that year in California, Shirley Temple Black and her husband Charlie held a "tube worm party" in Redwood City as a fund-raiser for WHOI. The Blacks, who were WHOI trustees, asked for the loan of a big vent worm. Hoyt Watson from Woods Hole's development office borrowed the huge tube worm kept at WHOI's public affairs office. It was too big to fit into anything, so he severed it and packed the halves in two clear plastic tubes, the kind the geologists used for mud cores. At the small Cape Cod airport, the pilot saw Watson's unusual baggage and declared: "You're not getting on *my* plane with *that*." It was the volatile pickling fluid he objected to. Watson quickly poured the liquid down a toilet, and tubes in tow, was allowed to board the plane. The pickled tube worm was given the place of honor at the Menlo Country Club atop a bed of exotic flowers flown in for the occasion from Hawaii.

On return expeditions in the early 1980s to the East Pacific

Rise, *Alvin* and *Cyana* discovered more vents. Towed instruments picked up temperature anomalies as far north as the Juan de Fuca Ridge off British Columbia and as far south as Easter Island. The ANGUS team found so many vents that they kept score by stenciling little red smokers on the camera sled. By 1982, ANGUS sported more than a dozen stencils.

One of the new vent fields was in the Guaymas Basin about 400 miles north of 21° north in the Gulf of California. Like the other vents, it was at a spreading center on a portion of midocean ridge, the East Pacific Rise, and was populated by tube worms. But unlike the other oases, the Guaymas site was covered in sediment saturated with a brown ooze and carpeted with green, yellow, and white mats. "We didn't know what it was," said Fred Grassle. "We called them the furry mats and the styrofoam mats." Besides hydrogen sulfide, the water was rich in ammonia and hydrocarbons. At the surface it smelled like diesel fuel. So did the waxy globs and patches of sediment which melted to a liquid that looked like gasoline. *Alvin* had dived to a natural oil refinery.

Because it ascended through sediment, not basaltic rock, the hydrothermal fluid in the Guaymas Basin was quite different from that at the other vents; it was four times more alkaline than seawater and nearly depleted in iron and other ore-forming elements. Water heated by underground magma chambers passed through the organic-rich mud and cooked it to produce petroleum. As the water cooled from the surrounding frigid water, the oil fractionated, leaching heavy components such as tar.

There were also huge mounds, columns, spires, pagodas, and mushrooms of sulfide more than 30 feet high and 160 feet across. "Perhaps in an enlightened future," Scripps geologist Peter Lonsdale wrote, "the region will form a 2000-meter [6560-feet] underwater park where *Alvin*'s successors will treat tourists to sights that rival Yellowstone."

The green and yellow mats, it turned out, were formed by a new species of *Beggiatoa* which the divers called "jumbo bacteria" because they could see the microbes without a microscope. The "styrofoam" mats were made of sulfur crystals, perhaps cooked *Beggiatoa*. The ocean floor was hot enough to charcoal a piece of wood which held a thermometer that was stuck into the sediment. Besides oxidizing hydrogen sulfide, the Guaymas microbes produced methane.

Much progress had been made keeping the vent animals alive and in one piece. Having learned that old clam buckets didn't do, biologists experimented with insulated containers. One of the best was Childress's plastic box in an aluminum frame. Its success was soon celebrated with T-shirts depicting the coffin-shaped container and the inscription: "Happiness Is a Full Coffin."

In 1982 *Alvin* dived on two seamounts, both active volcanoes not far from 21° north, but not on the East Pacific Rise. The basalt on the seamounts was fresher than the volcanic rock at the nearby spreading center. While little life and no hydrogen sulfide were found, extensive sulfide deposits and warm water were unmistakable evidence of hydrothermal activity. One seamount, called Red, was covered in red oxide and "all sorts of weirdo chimneys and spires," Lonsdale said. "It was discharging water that was 10° to 20°C [50°–68°F] warm." The other seamount was named Green for its cloak of bright green copper minerals.

The mud on Red and Green was a strange gelatinous mix of bacterial filaments and iron-rich material, which one geologist said looked exactly like chocolate pudding.

Both seamounts were draped in mounds of yellow bacterial fluff which the divers tried vainly to collect. "We finally decided to fill up *Alvin*'s basket with this fluffy yellow stuff and see what survived," Rodey Batiza, a University of Hawaii geologist, recalled. "Ralph [Hollis] drove *Alvin* right through one of these things on Red, about a couple of meters high." As it turned out, the kamikaze ride was not an effective way to sample material the consistency of pudding or fluff.

A later expedition took *Alvin* to five seamounts at 9° north on the East Pacific Rise. These volcanoes were given names like MIB and DTD, after the repeated inflictions the scientists subjected themselves to on ship, where a Jane Fonda exercise tape commanded: "Make it burn!" and "Don't touch down!"

On this seamount chain the divers found no pudding-like mud, no microbial fluff, no mineral deposits, no warm temperatures. But there was young basalt. Why the fresh rock but no other evidence of recent volcanism? Why was the chemistry of the seamount rocks so different from those at the ridge? And how could there be so many different kinds of lava?

"No one had ever seen lava like the lava we got from the seamounts," Lamont geologist Dan Fornari said. "We were looking for one basic kind of lava, but we found many different types, and the more places we sampled, the more different kinds of lava we saw. Even on one seamount we got three different magma compositions, and there was not one big magma chamber but several."

"We don't understand it all," Jeff Fox said. "But the chemistry of the lavas indicates that the magma has a different source from that of the midocean ridge. That means there is some very intricate plumbing at depth which isolates and channels the magma. We're talking about a property of the upper mantle that was not known to exist. Seamounts are not just pimples on the face of the seafloor. They're telling us something different about the dynamics of the upper mantle."

The debate about whether the vents were entirely chemosynthetic gained momentum in 1981, when symbiotic bacteria were found inside vent animals. While examining tube worm tissue, the Smithsonian zoologist Meredith Jones saw curious white flecks on a section of trophosome, the lumpy mass of tissue at the base of the trunk. He happened to mention this during a lecture at Harvard, and a graduate student, Colleen Cavanaugh, jumped out of her seat and shouted: "I'll bet those are bacteria! They've got to be symbiotic!" She was right.

Symbiosis explains how the tube worm is nourished, but not how the bacteria get into the worm. It took six more years to explain that one. In 1985 Jones found a microscopic opening in a juvenile tube worm, an animal about a sixteenth of an inch long. Jones described the opening as "a single snoutlike structure that appears to sort of snuffle up bacteria." As the animal matures, the slit seals up. How? "I don't know," he said. "It's now you see it, now you don't."

The bacteria in the tube worms oxidize hydrogen sulfide and with this chemical energy fix carbon dioxide into organic carbon (proteins, lipids and carbohydrates)—just as in photosynthesis. The ultimate energy source at the vents is geothermal, energy that derives from the radioactive decay in the global nuclear reactor that is Planet Earth.

This energy needs no sun. But chemosynthetic bacteria need oxygen; seawater is full of oxygen, which is a byproduct of photosynthesis. The blood-rich plume of the tube worms takes in both oxygen and hydrogen sulfide, needed by their symbiotic bacteria.

But in 1983 chemosynthetic bacteria that do *not* use oxygen were discovered at a vent at 21° north. *Methanococcus jannaschii* produce methane, another food source for vent life. *Methanococcus* needs no sun. But these microbes don't seem to provide most of the nourishment at the deep-sea oases. At least, the hydrogen-sulfide–oxidizing bacteria appear to be the key providers.

The concentration of microbes at vents is three to four times higher than that in the ocean's sunlit surface waters, the photosynthetically productive portion of ocean, the part once believed to be the only food-producing area. There is more biomass in a cubic foot at a vent than exists in the same volume of a rain forest. "No cornfield on land, no mussel bed in shallow water is as productive, by area, as the vents," WHOI microbiologist Holger Jannasch said. "And a cornfield has one season a year; the vents have constant growth. It's so fundamental, it's hard to believe." Why, Jannasch wondered, didn't a science fiction writer think of this before? Jannasch wrote:

> Most astounding is the fundamental fact that our planet is indeed able to provide energy for the sustenance of large animal populations at locations where this terrestrial energy is in no competition with solar energy. Has this ever been considered before? There is only one indication in the literature. The eminent ecologist G. E. Hutchinson of Yale University stated more than a decade before the deep-sea vents were discovered (in his book *The Ecological Theater and the Evolutionary Play*): "The internal heat of the planet, mostly of radioactive origin, in theory would provide an alternative to incoming radiation though we have little precedent as to how an organism could use it." Now we know that this is indeed happening, that there are whole ecosystems living in the deep sea and that they are run by microorganisms.
>
> Few discoveries in science come entirely unexpected. This is one of them.

In his search of the literature, Jones learned that the kind of tube worms found at vents were not entirely new; the first species of *Pogonophoran* had been described in 1914. "The person who worked up the worms, which were dredged on an expedition to the Dutch East Indies, came across this thing that didn't fit that well so he set it aside as an aberrant worm," Jones said. "He probably noted it didn't have a gut. In 1933, a Russian described another aberrant polychaete, this time from the northwest Pacific. Again, it didn't fit. In 1944 someone noticed similarities between the two aberrant worms, and that was the beginning in 1944 of the phylum *Pogonophora*. Another Russian in the 1950s made all sorts of deep-sea dredgings and got a tremendous amount of *Pogonophorans*."

By 1960, some 100 specimens, all small and skinny, had been identified, but they still remained curiosities. In 1966, an American scientist diving in the minisub *Deepstar* off California, found pencil-thick *Pogonophorans*. But no one paid much attention to these creatures until the discovery of the Galapagos vents in 1977 and the unknown jumbo tube worms.

One of the most intriguing facets of vent life is the ability of the creatures to thrive in the noxious broth of hydrogen sulfides and other compounds that would poison other life, including human. The iron-rich hemoglobin that carries oxygen in human blood moves both oxygen and hydrogen sulfide through the tube worm, but the vent creature actually detoxifies the extremely high concentrations of sulfide. The other animals, less well understood, seem to handle the poison differently. While the giant clams are hemoglobin-rich, the mussels are not. These molluscs break down hydrogen sulfide into thiosulfate, which is nontoxic.

Unlike humans, these animals are adept at regulating the iron content in their bodies. From her laboratory at the University of California at Santa Barbara, Tracy McLellan is looking for that regulator, hoping it will explain not just more about vent life, but human iron-deficiency diseases, some of them life-threatening.

"It certainly has given us a whole new perspective on environments with hydrogen sulfide which we had pretty much ignored before," Childress said. "Once we saw what was going on at the vents and we realized we had sulfur-oxidizing symbionts, there was a tremendous rush to find more of these habitats where sulfide and oxygen are available to organisms—in mangrove swamps, eel grass beds, sewer outfalls. It's opened up a whole new physiology of how animals deal with hydrogen sulfide."

Scripps biochemist George Somero, for one, went to sewage outfalls in Los Angeles to collect *Solemya*. The "sewage clam," as it has come to be known, is gutless, loaded with symbiotic sulfur-oxidizing bacteria, and like the vent mussels reduces hydrogen sulfide to the nontoxic compound thiosulfate. "It's one of the nice things sewage has done for us lately," Somero said. "There must be forty or fifty species of clams discovered in sewage outfall areas and swamps subsequent to finding the earliest vent species. These animals use hydrogen sulfide and symbiosis. It's actually very widespread. The irony is no one paid attention before." Somero and his colleagues have found higher concentrations of sulfide in the sewage clam than have been measured in any of the vent animals.

When *Alvin* wasn't at vents in the early 1980s, it went to canyons, faults and deep ocean stations in the North Atlantic and the Caribbean; and it took down biologists who wanted to observe and try to capture plankton, the free-floating creatures of the ocean that drift in mid-water. Rich Harbison and Larry Madin were most interested in the large gelatinous zooplankton or "jelly animals," as they call them. Perhaps the most familiar are the jellyfish or medusae. Others include siphonophores, which consist of long chains or colonies of separate animals dependent on one another. Among the most fragile and beautiful are the ctenophores or comb animals, named for the cilia aligned on their bodies like teeth in a comb.

Collecting jelly animals is difficult, often impossible. Nets towed from a ship rip the delicate creatures; some explode and others disintegrate at the surface—like the vent dandelion, which is a kind of siphonophore. To gather these animals more gently and observe them in their own environment, a handful of oceanographers began scuba diving in the early 1970s. In 1976 Harbison and Madin jumped at their first opportunity to go far beyond scuba depth with *Alvin* off the New York coast.

"We were about a quarter of a mile below the surface in one of the most studied areas in the whole world ocean," Harbison said. "And bango! Here was this new animal no one had ever seen before." It was a mostly transparent mass, about four inches long, gracefully pumping its ice-tong–like self through the dark water. Like a tiny Roman galley with all the oars sticking out the sides, Harbison said. Only these "oars" were comb-like plates. It was a ctenophore, glittering like a prism. The color display was caused by the ctenophore's combs diffracting the light from *Alvin*.

Harbison and Madin counted ninety-three of the same species on their first dive. *Alvin* pilot Dudley Foster succeeded in catching one, which the scientists named *Bathocyroe fosteri*.

Harbison and Madin have since caught more than two dozen unknown species of ctenophores and debunked some long-accepted wisdom about this hermaphroditic animal. Some species, including *B. fosteri*, have separate sexes. Not all ctenophores are small; the biologists saw one that was about five feet across. The surprisingly high diversity and abundance of ctenophores in the deep sea suggests that they play a far more important role in the deep ocean's ecosystem than anyone supposed.

And there were, and are, always those elusive "things" that refuse to be captured and often don't tarry long enough to be identified. In dives off San Diego, Harbison and Madin glimpsed a creature they dubbed "the mystery monster." "We don't know what this thing was," Madin said. "It looked like a blue lampshade, about two feet high and three feet in diameter, and it was hollow. It appeared to be able to swim but we couldn't tell how."

On a dive off Santa Barbara, they tried to catch a "gigantic" siphonophore. "It was about sixty feet long and it looked like a one-inch diameter rope," Harbison recalled. "We saw large numbers of these and sucked one of them into the suction collector. We were really pleased, thinking that we finally got one. There was nothing in the can at the surface. It had fallen apart and turned to slime. But we have it on video."

Much has been learned about life in the deep sea since Deep Ocean Station No. 1 was established off Cape Cod in 1971. *Alvin* has taken scientists to other sites of flat muddy ocean bottom off the Bahamas, southern California, Panama, Puerto Rico, St. Croix and Mexico. From their repeated visits to these seafloors of red and milk-chocolate clays and sands, biologists have learned that life in the deep sea is vastly more complex and richer than they were taught and than what they themselves had taught their students.

They have learned firsthand the mighty importance of the small deep-sea creatures that live in, on, beneath and just above the mud in cycling nutrients and chemically and geologically transforming the seafloor. The mud dwellers are savvy creatures that make sophisticated choices about their nourishment and environment. They build burrows, tubes and mounds, and irrigate their homes by moving water: the worms undulate, the clams pump, and the shrimp wave their arms, constantly drawing in fresh water, flushing out the stale. Like earthworms, they churn up the floor of the sea by eating the organic tidbits on the sediment.

At the Panama Deep Ocean Station, staked out in 1981, scientists scattered tiny plain and protein-coated glass beads. The kinds of beads found in the animals' guts contradicted the popular belief that deep-sea creatures were like vacuum cleaners and would eat anything. These mud dwellers were finicky; they chose smaller beads over bigger ones but never turned down a protein-coated glass bead no matter what its size. Some preferred smooth beads to the coarse ones and vice versa. The motivation behind the various preferences is still a mystery.

"After the experience in the Panama Basin you realize the deep sea is composed of a number of different habitats," said biologist Bob Whitlatch of the University of Connecticut, who with Fred Grassle designed the glass-bead experiment. "While they all may be muddy and have species that feed on sediment, each area is different. I was struck by this. You have as much complexity in the deep sea as you have in shallow water. It is not a simple homogenous mud bottom. About half of the things we brought up from the Panama Basin had never been described. We called them Species A, Species B, and then we went through the alphabet and started over again, so some were AA and MM."

Before oceanographers got inside the deep ocean, they knew little about the kinds of animals that lived in canyons or on seamounts because it was virtually impossible to dredge these geologic features. Diving on seamounts off Acapulco, Lisa Levin of North Carolina State University found giant protozoa, one-celled microscopic creatures common everywhere there is life. The specimens Levin found were one-celled but the cell was the size of her fist, and for unknown reasons had slightly elevated levels of radioactivity. Levin called them "little hot spots."

The protozoa stick the sediment grains to themselves to make a house and take on the shape of the mud hut—honeycombed, fanned or flowerlike. "The animal is less than 1 percent of the volume of its house, or test, as it's called," Levin said. "They're so big, they're like apartment houses for other animals. The tests are elaborate structures. In the cracks and crevices there are many things—worms, clams, crustaceans, just about all the different taxa of life in the deep sea, and as far as we know, they stay put in one spot all their lives."

This animal, *Xenophyophora*, from the Greek, meaning bearer of foreign bodies, looks like a sponge, which is what biologists initially thought it was when they first saw it in dredges many decades ago. It has been found only in the deep sea.

University of Rhode Island biologist Karen Wishner and her colleagues have focused on zooplankton, the tiny creatures that live a few inches above the seafloor—at DOS No. 1 off Cape Cod, in the Santa Catalina Basin off California and off Mexico. To the biologists' surprise, they discovered that the animals, mostly copepods, the base of the oceanic food chain, were carnivores; and their feeding rates were comparable to their shallow-water counterparts. They ate marine snow.

It turns out that marine snow, that eternal fine shower, is not so fine. To be sure, it is an unending rain, but of many different-sized orbs, globs and strands, ranging from microscopic to several centimeters, of animal corpses, fecal pellets, bits of old meteorites and other cosmic dust, coccoliths, shell fragments, crustacean molts, empty skeletons, pollen from the tulips in your garden, pumice from all the land volcanoes, and more. This snowfall, which exists throughout the world ocean, is home for healthy bacteria, flagellates and protozoans that ride the snowflakes, sometimes in a brilliant bioluminescent light show.

"It was something of a surprise to find that marine snow was so abundant in the deep sea," said University of California biologist Alice Alldredge, who collected the first intact batches of marine snow in the deep ocean from *Alvin* in 1979. She and biologist Mary Silver were the first to describe the concentration and composition of microorganisms on the "snow."

"Mary and I showed that there were actually large viable populations growing on these particles," Alldredge said. "They weren't just debris. They were homes. These particles are really rich in organisms, especially phytoplankton. That was interesting because one would expect that phytoplankton would require sunlight and would not last long in the deep sea. But they were alive." Alldredge thinks the plankton go into a quiescent stage in the dark depths and come alive in the presence of light.

"In surface waters, copepods and krill larvae get on one of these particles and scrape off the goodies," she said. "They kind of short-circuit the food chain by eating marine snow because they can eat bacteria which are usually too small to access otherwise. Off coastal areas, fish eat marine snow, and I'm sure that fish in the deep sea must eat marine snow, too."

Also on marine snow are radionuclides, cadmium, mercury, PCBs, bits of plastic and other pollutants. Can they work their way up the food chain? "There's some evidence that when heavy metals adsorb onto marine snow they are carried quickly to the deep-sea sediments," Alldredge said. "Whether that's good or bad, I don't know. I suppose if you live in the sediments it's not so great. We're just beginning to unravel this food-chain stuff. It stimulates us to think of alternate food chains, but at this point, we're right at the brink of this being an idea whose time has come to be studied."

Karen Wishner agreed. "We don't know enough about the food chain in the deep sea to predict where wastes are going to go," she said. "Plankton is the primary link, but there's an awful lot more to quantify."

"We don't know enough about the basic biology of these environments," Whitlatch said. "We're just scratching the veneer of all this. We can make some good guesses about the impact an insult will have in those environments, but all we can hope to do is try to provide a range of possible responses."

"The seafloor sediments are the only historical record of processes going on in the whole ocean—productivity, food, oxygen and temperature," biogeochemist Bob Aller said. "The sediment in different ways stores all that information, including the ocean's carbon budget in the amount of oxygen and calcium carbonate, the density of animals, and so on. The larger animals control how particles move in the upper mud. Some of them mix it like a blender. In order to know what's on the bottom you need to know its fundamental composition."

From *Alvin* Aller placed tiny probes in the mud off Panama to measure silica, ammonia and other basic nutrients. Peer reviewers in 1977 rejected his request to fund his research because, they said, there was no evidence that animal-sediment interactions in the deep sea were important. But they eventually changed their mind, and in 1981, he got funded.

The notion that deep-sea creatures live longer than their counterparts in shallow water has been laid to rest for good. In 1977 WHOI and Yale scientists discovered in a trawl a tiny clam which they dated to be 100 years old. Nobody has found another clam so old, but it is clear that the old rule—the smaller the animal, the shorter the life span—does not apply to the deep sea, where adult clams as big as jellybeans and worms as long as match sticks live as long as elephants and oak trees; and where worms much taller than the tallest basketball player and clams the size of dinner platters endure only a decade or two.

The first direct evidence of how long deep-sea animals live came from Fred Grassle's mud trays and Ruth Turner's boards at DOS No. 1. Grassle's clams reached maturity in two years; Turner's wood borers were adults in three months. Age has much to do with environment.

Nine unknown species of Xylophaginae, the only known wood borers in the deep sea, moved into the triangular stacks of foot-long pine panels and spruce blocks at deep ocean stations in the Atlantic and Pacific. The tiny clamlike borers would extend a foot against a luscious piece of Turner Tower, and teeth against the wood, rasp away until there was nothing left but the sides of the onion and potato bags Turner kept the boards in.

Turner found 7900 of them on a single twenty-four-inch long board. And that's not all. After the ravenous wood borers' arrival, colonies of worms, hydroids, crabs and other creatures moved in. The worms ate the fecal pellets of the wood borers. The crabs ate the wood borers and the worms. The filter-feeders, including jellyfish and sponges, ate the wood borers' larvae, and big fish ate the small fry. When the wood borers died, other organisms moved into their tubes.

The wood borers lived no longer than six months because by then they had gobbled up their habitat. The creatures in Grassle's mud trays shunned food orgies and lived longer. The vent clam *Calyptogena magnifica* grew 500 times faster than its smaller cousin because it lived amid plenty of food.

Wood, like the rich chemosynthetic bacterial brew at vents, enabled the creation of a whole new ecosystem where before there was none, directly affecting the lifetime of entire communities of different animals. "Wood is doing exactly the same thing the vents are doing—providing nutrients," Turner said. "It doesn't matter what the nutrient is as long as the animal can utilize it. In the case of the vents, bacteria are essential. We haven't proved this yet completely, but we think the medium by which the wood borers utilize the wood for food is by the bacteria in their gills, bacteria that are symbiotic."

After more than twenty-five years of hosting human guests, the deep sea still astonishes. On a dive to the Panama Basin, geologist Vernon Asper brought a country music tape of the Statler Brothers. "I'll never forget it," Asper said. "I'll never forget sitting in the submarine, hearing that old hymn, 'How Great Thou Art.' *Oh, Lord, my God. When I in awesome wonder consider all the worlds Thy hands have made.* . . . Hearing that phrase and seeing the most incredible animals go by the window, animals that you can't even hope to describe, and listening to that song, seeing what you know is a new world. It was a very special experience."

"*Alvin* opened a door for me," Alice Alldredge said. "It allowed me to collect and really observe, really see the material. And that allows me to conceptualize the deep sea. It may not sound like much, but how we conceptualize is directly related to the kinds of questions we ask. The other major thing *Alvin* did was to make me aware of how abundant life was in the deep sea. It really brings home how rich the ocean is, especially the deep sea. It's *teeming* with organisms."

"It's unbelievable just how much can be gained by going to the bottom of the ocean and just thinking about what you're seeing," said Craig Smith, a biologist at the University of Hawaii who since 1984 has been visiting a bottom station off southern California. "Doing deep-sea work without a submersible is like trying to study the ecology of a forest from a zeppelin at night. From *Alvin* you look out the window and see patterns in the sediment and animals, small ones and big ones. An interesting thing about soft sediments is when you get to the bottom it's like reading a book. The activities of animals of the last day or week are recorded on the bottom, and you really need to see it at close range. It changes your impression of how the deep sea works. Regardless of how good your imaging is, there's nothing like being there in person to see it."

"My first dive changed my whole ideas about what the deep sea was like," WHOI biologist Lauren Mullineaux said. "It opened up a whole new world. I think there's no way to understand an environment like the deep sea if you can't physically be there. We could see gelatinous animals lighting up as the sub went down. It was like being in a light show. On the bottom the animals were just so bizarre. Some of them were huge, like nothing I'd ever seen before. It was a complete shock. I went down there thinking I would see an immense mud flat and maybe every once in awhile a fish would swim by. Where ever I looked, I saw animals."

"It's spectacular," Bob Aller said. "It lets you see how alive the ocean really is. The kind of perspective *Alvin* gives, with a human being seeing the lay of the land, that perspective is really incredibly important in understanding how the bottom works. When I first dove it was in the Panama Basin, and I couldn't believe what the bottom looked like. All the activity—the shrimp, the polycheate worms, all that evidence of biological activity. It reminded me of a tidal flat. If I saw a picture of it, I know I wouldn't have gotten that three-dimensional and spatial perspective."

On his first ride to 13,120 feet, Bob Whitlatch was astonished to hear the soft snores of his friend Fred Grassle. Grassle wasn't the only one who snoozed on the long rides up and down. But Whitlatch said he never tired of watching the bioluminescence. "All the way down all through the water column I was struck by the marine snow," he said. "This was deep water, where there is supposed to be very little life and very little nutrients. I was seeing organisms I had only seen before in textbooks. You really realize how thin a veneer of light the ocean has. The light is limited to the upper 200 meters (656 feet). You know that, you tell your students that, but to experience it is something else."

These scientists referred not to the spectacular vents but to the plain old deep sea.

In summing up the Cape Cod bottom station experiments in 1987, Grassle and Linda Morse-Porteous wrote:

> The dynamics of deep-sea communities are more like other highly diverse environments such as rain forests and coral reefs than has previously been thought. Direct observation and experimentation on the seafloor shows the communities in this environment to be equally complex.

A rain forest is the richest, most crowded environment known.

27

I was really quite fond of Lulu. *I thought it rode the seas pretty well. But everybody was always terrified that* Lulu *was going to sink.*
        Fred Grassle, biological oceanographer

Transits could be hell. Sometimes you woke up in the morning to learn that you were thirty miles further away from where you were the night before.
        Dudley Foster, pilot

Lulu *was by far the most hostile unforgiving oceanographic platform I have ever worked on. I have never been so miserable in my life as I was working on* Lulu. *It was very hard to do any science on* Lulu. *You slept, you ate, and you dove. You couldn't even lay out charts on a table. You just essentially hung on for dear life.*
        Jeff Fox, marine geologist

Jon, does Mom know you go to sea on that?
        Alvin technician Jonathan Borden's sister, on seeing *Lulu* for the first time

BEWARE THE TUBE OF DOOM
        Graffito under a pontoon bunk

Everybody got sick on Lulu. Lulu *was sick.* Lulu *was a deathtrap. There was no way to judge which way the ship was going to move. It was impossible to walk straight. Anybody with any sense at all wouldn't have gone on that ship.*
        Ralph Hollis, pilot

Piece of shit. I still have bad memories.
        Ken Smith, biological oceanographer

I kind of liked it. It was so much more intense.
        Jim Akens, *Alvin* engineer

The general condition of Lulu *kind of shakes your confidence in the whole operation. On the other hand, that was kind of made up for by*

the competence and professionalism of the Alvin *support group and the* Lulu *crew. They worked really hard. It seemed that they were working all the time.*
                Bruce Luyendyk, marine geophysicist

Lulu *was a disaster. It's a wonder nobody was killed. We were trying to do serious science on this piece of shit. We'd come back from a cruise and scream our heads off and nobody paid any attention. They thought we were telling old sea tales.*
                John Edmond, marine chemist

I think that Lulu *has outlived her usefulness. . . .*
                WHOI administrator, 1972

The single most important short-term improvement which can be made in the present ALVIN *system is improved surface support.*
                U.S. oceanographers, at Stanford Workshop, 1976

A viable alternative to Lulu must be found. . . .
                Peer review, 1977

In spite of her record of past accomplishments, LULU *is the most serious hindrance to carrying out a worldwide mission of high priority ALVIN research. . . .*
                U.S. oceanographers, in *Submersible Science Study,* 1982

Seas had been choppy and the swells high throughout the daylong ride to Deep Ocean Station No. 2 off Cape Cod in July 1980. It was misery to the inexperienced seagoer Cheryl Ann Butman who spent the time throwing up and nibbling crackers. Once at the dive site, her thesis adviser Fred Grassle convinced her that there would be no nauseating motions in *Alvin.* Butman followed Grassle down through the sail. They played several rounds of hearts during the descent to 12,000 feet and quickly found the bottom station and the five-year-old mud tray, the longest *in situ* record ever of life processes in the deep sea.

After several hours on the bottom, surface control called them back home because of building seas. Pilot Ralph Hollis dropped the side weights and *Alvin* began the two-hour rise.

A hundred feet below the surface, the passengers saw tumbling masses of foaming green water, and soon they were rocking violently. Everything not secure in the sphere came loose; the electronic boxes and spare battery packs fell onto their heads.

Grassle, who feared losing the mud tray in a collision with

Lulu's lower arch, voted to wait out the storm in the sub. Hollis, itching for a cigarette, climbed into the sail, where wet spray extinguished his matches. It soon became apparent that the seas were too wild to get *Alvin* onto the mother ship. And then Grassle's precious mud tray disappeared into the high swells with the secrets of long-term colonization and maturation.

Butman, sitting cross-legged hunched over an airplane barf bag, was throwing up. Seeing tears on her face, Grassle put a hand on her leg. "Do you think it would help if we did something like play cards?"

She shook her head.

"Do you feel sick?" Grassle asked gently.

She nodded.

"Are you scared?"

Nod.

"Do you think something bad's going to happen?"

Nod.

"Do you think we're going to die?"

She sobbed. Grassle put an arm around her. "We're not going to die."

The underwater telephone crackled. A rescue attempt was under way.

"I could see *Lulu* from the sub," Butman recalled, "and the waves plunging. I just couldn't see how we wouldn't get totally mashed. All this time I was crying, I was so scared, but now I really got scared. I know this sounds very unprofessional, but I was so scared."

It took some time to get the Boston whaler, already scarred from repeated bashing against the mother ship, into the water and finally out to *Alvin*. Butman couldn't feel her legs; she thought if she climbed out of the sail just when *Alvin* rolled, she surely would drown. Pilot Bob Brown coaxed her out. "I have to put this on you," he said, holding a life jacket. "You have to climb out for me to put this on you." He reached in and pulled her up, but before he could get the life vest on her, *Alvin* rolled, and a wave covered them. Brown held her against the sail as they dipped into the water.

"I'll tell you when to jump," Brown said, strapping the life vest on her. "You're going to have to jump into the whaler...."

Butman's feet grew tap roots in the whaler's bow and her hands welded to a post. She didn't hear the screams from the others aboard. Grassle made his way to the bow, and holding her tightly, brought her to the safety of the stern.

There was one more jump to go, onto the pontoon, and the whaler had to make many passes for each person to make that leap. Butman was caught by pairs of arms that lifted her up onto the heaving mother ship. Ruth Turner stood her, still fully clothed and still mute, under the shower, undressed her, and got her into bed.

It was impossible to get the whaler on *Lulu*, and when the crew tethered it to the mother ship, the cleats holding the mooring line ripped out.

Whalers had been lost before. A similar rescue had taken place earlier that year in twenty-foot swells off Florida. Because the whaler motor died, the passengers sat for two hours in the wildly bobbing sphere. "I recall that we had chocolate cake for lunch," geologist Bill Dillon said. "Page [Valentine] didn't throw up. I hated him for it."

George Ellis and Bob Brown had remained the night in the whaler tethered to *Alvin*. This time, however, the seas were too rough for that. *Alvin* rode the waves alone, untethered, all night. In the morning Brown swam out to the battered submarine. The screws holding the fiberglass skins had worked themselves loose; one panel was missing, another skin was hardly attached, and there was a crack in another.

Butman slept for twenty hours. Grassle, afraid that she would never dive again, convinced her to make the next dive with him. The day was perfect and so was the dive, but it would be seven years before Butman, an oceanographer, felt comfortable enough to return to sea. It would not be on *Lulu*.

Lulu was always blamed as a source of tension. By early 1977, all three outboards, those junkyard Murray Tregurthas, were in trouble. The steering systems were binding, a shaft broke on one, and they leaked oil. That summer a special panel of scientists and federal bureaucrats spent two days at Woods Hole assessing *Lulu*. To nobody's surprise, the visitors concluded that the 105-foot semi-submerged platform was "only marginally adequate." They wrote:

> RV *Lulu*'s habitability is the poorest of any ship in the academic fleet. Berthing arrangements, availability of only one head, absence of lounges, [there was a lounge; apparently the panel didn't recognize it as such] small messroom, poor ventilation and air conditioning, unattractive appearance, lack of any place for personal privacy are all problems that need attention.

There were also safety problems—the electrical wiring was deteriorating; the shop and bunk vans were extensively corroded and leaked; the emergency escape trunks on both pontoons were more appropriate for a svelte greyhound than a human adult. Nor was *Lulu* in compliance with sewage discharge regulations. The submarine's cradle and handling system, however, appeared to be fine.

Lulu underwent a major overhaul worth $250,000 in improvements the winter of 1977–78. Saloon-type swinging doors replaced the shower curtains in the two toilet stalls, a urinal was installed, and the Tube of Doom was transformed from an open twenty-three-bed dormitory to ten two-person cubicles, each with a folding door. The renovations made accommodations even tighter. Sailor Craig "Mad Dog" Dickson had no empty nook to keep his guitar, so he stowed it on whatever bunk was empty, changing beds with the watches. There wasn't enough room for both bunkmates to get dressed at the same time, and you had to open the door to pull on your pants or fit an arm through a long sleeve. But the new privacy made everyone happy.

The bunk van was so badly rusted that it crumpled when a crane tried to lift it in 1980. It was replaced as a permanent superstructure that included *Lulu*'s second bathroom, a tiny head with an undersized toilet and shower.

Lulu had come a long way from her junkyard roots as the Deep Submergence Research Vehicle Tender No. 1. After nearly two decades, the mother ship had history, personality, even a little class. A old photograph of Lulu Vine in an oval mahogany frame hung outside the mess. And high on the bridge house at the stern was a wooden plaque, whittled by a cook, of *Alvin* straddling two whales. Railings were wound with Turk's heads and other fancy knots of bright yellow nylon line. Paperbacks lined shelves in the Lounge and a small refrigerator held midnight snacks. The crew bought the fridge, TV, a dart board, stereo, exercise weights and the frame for Lulu Vine's picture with a slush fund fueled by their T-shirt business.

There were two T-shirt designs. The first borrowed the motif of the whales and *Alvin*. The other, created by a resident artist, *Lulu* engineer "Buffalo Bob" Barton, depicted a dead rat. This was Bernard.

As far as anyone knew, Bernard came aboard in Puntas Arenas, Costa Rica, and was seen a few weeks later weaving across the deck; then the rat collapsed. "He wasn't quite dead but he looked like he wanted to be," Barton said. "I was overwhelmed by the existential implications of this."

The notion that even a rat could not endure *Lulu* was too delicious to ignore. The crew held a funeral and Barton painted the deceased, eyes shut, holding a wilted flower, beneath the epitaph:

BERNARD DA RAT
Buried at sea 10/11/81 RV *Lulu*

> God bless the little bastard's soul
> where ever it's at.
>
> Both rats and men
> must live with their errors
> so shall we return
> to the Twin Tubes of Terror.

Sometimes Barton painted character sketches of scientists. But he wasn't the only one who rummaged in the paint locker. "Turner Tower" signs with the dates and *Alvin* dive numbers went down with Ruth Turner's stacks of wood blocks mounted in triangular frames. Pilot George Ellis painted a skull and crossbones and "Death to Benthic Animals" on the side of what Grassle called his Bottom Smotherer, designed to unfold like a Ping-Pong table on the bottom and kill a neat patch of tiny mud dwellers.

Lulu had mascots: Fleegle, a stuffed frog; and Mr. Bill, a life-sized headless doll. Both were thrown around, hoisted up the mast and given seats of honor on the bridge or *Alvin*'s sail. The fuzzy green frog rode in the whaler and the submarine and once was towed through the Panama Canal. After repeated abuse, Fleegle eventually lost an eye, and Mr. Bill disappeared altogether.

At the end of some expeditions, the science party and crew put on skits for each other. George Ellis filled his free hours as surface controller by writing poems about the scientists. He limned Alice Alldredge's dives to collect marine snow in 1981, documenting in rhyme some of the frustrations of doing science under water:

> A noted biologist named Alice
> Got her kicks in a coldwater palace
> While spinning like hell
> Her new carousel
> It broke, but she held it no malice.
>
> Up and down did Alice go
> Until her time was spent
> Her pilot suffered all in silence
> But his ear was surely bent.

Compared with WHOI's other ships, discipline was loose aboard *Lulu;* some sailors had no previous ship experience. A skipper once ordered a deckhand to "free the anchor." Taking him literally, the sailor let out the entire length of chain. A relief skipper, Captain Reuben Baker, found a mate on watch steering *Lulu* with his feet. "When you stand your watch with me," Baker told him, "you *stand* your watch."

On a typical research ship which had separate messes and sleeping quarters, the science party and the ship's crew didn't see much of one another. But *Lulu* was too small for anonymity. Scientists, ships and submarine crews lived together.

They swam together during "swim call," which was unique to *Lulu*. When the sub was down and the weather was good, they swung Tarzan-style on a rope into the ocean between the pontoons. Sometimes they lowered the cradle and used it as a raft, taking turns on shark lookout. They drank together at the bar in Georgetown on Grand Cayman, and the Blue Marlin with the dirt floor and the scotch that tasted like kerosene, and at the Chief's Club at Rodman Naval Base in Panama, and the Bridge Cafe on Paradise Island where a big black man played a sweet piano. They got seasick together and nursed each other with soda crackers and tenderness. Ralph Hollis remembered a scientist on his hands and knees. "This guy, the poor bastard, I think he was trying to make the railing," Hollis said. "He got down on all fours and puked through the grating, and all of it came back through the grating and got all over him. We hosed him down." And they worked together—on the pontoons pulling with all their strength, in the whaler, in *Alvin*.

WHOI geologist Geoff Thompson, who got seasick no matter what the weather or craft, hated it. "It was always a battle whether or not I went out," he recalled. "I hated handling the ropes down there and I hated going out in the whaler because I knew I was going to get sick again. I never knew which to do." Rose Petrecca, a WHOI biologist, loved it; she enjoyed Broderson bellowing to pull harder and hold tighter. "He yelled and screamed at everyone," she said, "but I felt kind of honored."

Ruth Turner painted the Lounge, spliced line, and gave the crew haircuts. Lady Wormwood acquired a new nickname when, while dissecting one of the jumbo vent clams, she was heard to exclaim, "Holy mackerel! Look at those gonads!"

When Broderson saw the blisters on her fingers from splicing line, he went after the sailor whose job she was doing. He found him in the mess drinking coffee. "But she *wanted* to do it," the sailor protested. "She was showing you *how* to do it," Broderson said. "As an able-bodied seaman you're supposed to already know how to splice a line. Now, move!"

Broderson was a one-man PR unit that worked the barmaids and the military supply chiefs in the ports of call by bringing them photographs, T-shirts and news of the one and only *Alvin*. If he liked you, he might introduce you to his friends, like the Eagle

Lady in Panama who was reputed to have an eagle tattooed across her chest and a snake on the back of her neck.

When John Steele, the WHOI director who succeeded Paul Fye in 1977, first boarded *Lulu,* he did what most people did: walked up to *Alvin* and touched it. Broderson, ever impartial at meting out his supervisory responsibility, whisked off the foreign paw with a "Hey, hey, hey! You touch that and I'll break your arm!" Steele smiled politely.

While the players changed over the years, the crew was always a can-do team, and many times they were heroes to the oceanographers. When the chain hoist to the cradle broke, Jeff Fox thought it meant the end of the cruise. "Every dive was a question mark right up to the moment you launched," Fox said. "The chain hoist to *Lulu*'s cradle could be very temperamental, or there might be mechanical problems with the submarine, or the weather could be trouble. Now, the cradle was hanging by a thread, and it looked for sure like we'd have to go in to port and abort the whole project in midexpedition. Well, Brody and a few others, out of thin air and ingenuity, found this ratty part in the bilge and worked around the clock to get the cradle working. It didn't work well, but it worked, and we could continue our dives. They did things like that routinely. There were some real bizarre characters you had to accept, but the crew had a tremendous esprit. They were a delight to work with."

Most of the forward underside of the deck ripped away when *Lulu* ran over a buoy in 1979 out of Nassau. "It was bad," Ellis remembered. "The steel beams of the cradle were all twisted. We thought we'd have to find a shipyard. We reported this to Woods Hole and turned toward shore, as all the scientists aboard stood and stared. Brody grabbed hammers, wrenches, torches and chains and got to work. In a few hours he had most of the damage identified and much of it repaired. So the rest of us went to help him. In twenty-four hours the entire cradle was repaired. It was astounding. Brody used the cradle's chain hoist to pull the beams straight. Only Brody would have tried it. Brody looks at wreckage and sees only what is needed to be done to put it working. The rest of us look at wreckage and remember how beautiful it used to be."

At the air force base in Panama, Broderson swapped a sack of blank tape cassettes for the alcohol Grassle needed for samples. One of the military workers told Grassle that his family was involved in all the docks in New Jersey and if he ever needed anything more, just to ask.

American cigarettes were also good wampum in foreign ports, especially at customs. In Mexico it took five cartons, one carton in other places. Items such as razors, jeans, tape recorders and electric shavers regularly disappeared in some foreign ports.

At a Galapagos island where *Lulu* stopped for fuel, her skipper Dick Flegenheimer exchanged cigarettes with the military commanders for permission to let his crew and scientists deboard. "Another problem," Flegenheimer said, "was that the customs people on Galapagos wouldn't let the scientists flying in get off the island until certain things were done. It's called bribery. They had the passport stamp with their hand out, and they said, 'Cannot do, cannot do.' " When Flegenheimer pressed five-dollar bills into their open palms, they could do.

Nothing did any good when Mexico inexplicably impounded *Lulu* for a day and a half at Mazatlan in 1981. Nor did Mexico ever explain why it detained for three months a shipment of *Alvin*'s drop weights shipped from Woods Hole or why it kept the Teflon-coated titanium water samplers for six months and then returned them to Woods Hole. *Alvin* engineer Barrie Walden carted them from WHOI back to Logan airport in Boston. The commercial aircraft would not accept them as excess baggage, so Walden checked them and himself on a cargo plane to Mexico. The three boxes with the eighteen precious titanium samplers did not make it to even the first stop at San Diego. Walden eventually tracked them down on another plane and flew to Mexico to await their arrival. The chemists were desperate to get those samplers to analyze the first uncontaminated hot ocean water. Walden had never bribed anyone before. He wasn't sure how or when to do it or even who to bribe. Unlike WHOI, the *National Geographic* provided expense-account vouchers with a provision of "gifts to natives." As it turned out, Walden didn't have to exchange any money for the samplers. After spending every day for a week at customs, an official suddenly told him he could take his property.

The crew never seemed to get enough movies or be in the right port to receive a shipment of new flicks. Mexico once impounded their movies and sent them back to WHOI as pornographic. They weren't. The only porno on *Lulu* was a videotape (*Debbie Does Dallas*) donated by a scientist.

WHOI's director John Steele, who had come from a small research laboratory in Scotland, said he was most impressed with the crew's "high morale, considering the conditions."

Dave Landry was especially mindful of morale. He had worked on ships for about a dozen years before taking on the job in the summer of 1979 as *Lulu*'s last skipper at Woods Hole. Landry, who had a third mate's license, had never been captain of a ship before. Like most of the mother ship's skippers, he didn't have a captain's license, required by the Coast Guard for bigger ships. When Landry's taxi pulled up to the dock in San Diego and for the first time he laid eyes on his charge, the cabdriver spoke first. "Which is the front and which is the back?"

Landry didn't know either. "*Lulu* was a total wreck," he said. "It was unbelievable, unbe*lieva*ble. I had never seen a ship like that. Cockroaches all over the place. Not a single fire extinguisher worked; the crew used them to keep their beer cold. *Lulu* never would have passed a Coast Guard inspection."

He took a deep breath and for the next four years never looked back. Landry said this was the best job he ever had. Almost immediately he posted rules on the small bulletin board in the mess. On deck, shoes had to be worn at all times and flashlights carried at night. No peeing over the side in the dark. An ice maker, acquired in 1980, stemmed the use of fire extinguishers to chill sixpacks. Everyone got two free beers after *Alvin* was put to bed for the night; the beer bill Landry hid in the ship's grocery accounts.

The eighteen-year-old experiment that was RV *Lulu* was still under way. Despite the added buoyancy, *Lulu* got heavier and sat increasingly lower, until the tops of the pontoons were again almost level with the water surface. Landry admitted fear a few times, like when he tried to outrun a typhoon with only one working engine. "We were running flat-out," he said, "and flat-out was 4 knots, trying like hell to get to Manzanilla in twenty-foot seas. They say God protects sailors and drunks. It's true."

The folks at Rodman Naval Base watched *Lulu*, with about twenty feet to spare on both sides, ram both walls of the Panama Canal. The ships were supposed to honk at each lock. A couple of pilots had stuffed rags in *Lulu*'s horns; instead of BLAH! she went *phhtt*. Backing out of the canal, Landry turned her right at full power and the ship went left, pirouetting 360 degrees into the Pacific.

"This ship did not sail with grace," Barton said. "My heart used to bleed for those engines, the same kind of engines used on a city bus. The fastest I ever saw *Lulu* go was 8 knots in a tail wind."

Despite all the improvements and repairs, it was impossible to keep up with breakdowns. "It was like putting lipstick on a pig," Landry said. In a 1980 report to WHOI he wrote:

> It's incredible to believe that with the exception of the Loran [navigation], all items repaired or purchased outside have either broken down or were delivered not as ordered. The washing machine virtually destroyed itself due to an imbalance in the drum. Vibrations were so bad that the dryer "walked" right off the top of the washer....

When the washer broke, as it frequently did, the crew dunked their jeans and shirts in bilge cleaner and towed them in the ocean from the stern. "It seemed to work," John Porteous said. "At least it got out the grease." Porteous sent his jeans on eight-hour tows.

The constant breakdowns drove *Lulu*'s engineers crazy. Some things they ignored, like the macerating heads that were supposed to mush up the offal from the toilets so it could be treated with chemicals and dumped into the ocean. "The truth is, the sewage went directly over the side," Barton said. "The macerating heads never worked, so I shut them off. Most things I fixed, but I couldn't fix those heads and I couldn't fix the autopilot. I hated that boat and it hated me."

No matter how much space was carved out to store food, fuel and fresh water for the humans and wash-downs for the submarine, the mother ship could stay out at sea for only two weeks at a stretch. And *Lulu* was always a crowded ship.

Almost everyone had had it. The scientists wanted greater range and more days at sea without having to come in to resupply. They hated the psychological conundrum, known as the "Siberia effect," caused by the hotel/escort ship. Moving the center of action from *Lulu* at launch to the second ship after launch split the science party into two communities. The group left out of a planning meeting felt ostracized. No one wanted to be on *Lulu* when a wonderful sample or a dive videotape was being viewed on the other ship.

Communications via radio between ships was rudimentary. It was easy to be misunderstood, difficult to relay detailed research results, and impossible to have a private conversation. The latter was critical when big egos clashed over important discoveries, and there were plenty of both on *Alvin*'s expeditions, which included concentrations of the nation's top scientists, high-powered people accustomed to running the show. The chief scientists had their hands full trying to satisfy the cream of the crop.

A bigger and better mother ship would obviate the need for a second ship. *Lulu* cost about $10,000 more a year than *Alvin* to operate. By 1982 it cost as much to rent hotel ships as it did to operate *Lulu*.

Alvin's funding agencies paid for the oceanographic community to assess the dilemma, and a fifteen-member task force put its consensus in the Submersible Science Report of 1982. It said *Lulu* had to go. Now.

Bob Barton drew a new T-shirt design of the crew's ideal mother ship. There was a merry-go-round for the scientists, a giant slingshot to launch and catch *Alvin*, and slung between the ship's smokestack and radar antenna was a huge volleyball net for stressed-out scientists and crew. And a big balloon for extra ballast was at the bow, or was it the stern?

As RV *Lulu* headed north to Cape Cod in the summer of 1982, after a year and twenty-five days away from home, the crew heaved the washing machine, the last one they would ever have to endure on *Lulu*, overboard.

IV The Submarine Blues, 1982–1989

Chorus: *You sink 18 tons and what do you get*
Another day older and deeper in depth
Ralph Hollis, don't be callin' for me too soon
I was up last night with the submarine blues

I spend my days getting tired and cold
Driving around the ocean floor
Pickin' up rocks and biology
I make my living under the sea

Well here I sit on the ocean floor
This talus slope is really a bore
Do this, do that, make the samples fine
I don't care, it's all overtime

You better drive up that cliff and get to the top
Sample me some mud and get me a rock
Pick up some fauna and some flora too
I can't wait till these folks go back to their zoo

Every single night when the diving is through
Someone's in the ball from the Alvin *crew*
Gettin' it ready for the next dive day
Bitchin' about no raise in pay

>By *Alvin* Pilot Will Sellers, to the tune
>of The Company Store, (You lift sixteen tons,
>and what do you get. . . .)

28

Lulu approached Woods Hole to an unusually large welcoming crowd. A Coast Guard fireboat, hoses at full blast, was outside the harbor. The Woods Hole Chowder and Marching Society Band, made up of scientists and other WHOI employees with musical talent, was at the dock with snare drum, clarinet, accordion and trombone; and a huge banner hung on the back of the brick Bigelow Building:

WELCOME HOME
ALVIN/LULU
390 DAYS   136 DIVES
20,878 MILES
JULY 25, 1981—AUGUST 19, 1982

With binoculars George Broderson recognized the two girls in pink hot pants, black top hats and tails. They were his teenagers, Kye and Jacci, and they held a sign "Old Loveable Fan Club" which matched one of his favorite T-shirts. Tears filled Old Loveable's eyes. "I wanted to die on the damn thing," he said. "They say you should always leave a boat and a wife when you're still in love with them." Broderson was sixty-five years old, and retiring, appropriately with the mother ship that been home to him and *Alvin* for eighteen years.

Lulu's replacement was not obvious. Another twin-hulled ship was rejected as too costly. It was also clear that a between-the-hulls launch system was too limited by rough seas. WHOI's *Knorr* and Scripps' *Melville* had been designed and built in 1969 at extra cost to accommodate a submersible someday. But ships don't always behave as planned. The larger decks could fit *Alvin* but both ships already sat a foot and a half deeper in the water than they should have. And the side crane on *Knorr*, which was supposed to be strong enough to one day lift a sixteen-ton submarine was a "complete turkey," said Robertson Dinsmore, supervisor of WHOI's marine department. And WHOI's other large ship, the

Atlantis II, had much less room than *Knorr* and an even worse weight problem.

As Woods Hole pondered the dilemma, the National Science Foundation warned of funding cutbacks for the research ships. It was almost certain a vessel would be laid up, and WHOI worried it would be the *AII*, the oldest ship in the academic fleet. "We were at a loss for what to do," Dinsmore said. "People were screaming about how bad *Lulu* was and sending us pictures of potential tenders that were worse than what we had. We knew there would be paring back and we were nervous. We made a decision in desperation that the last thing we wanted to do was to bring another ship into the field. So we took another look at the *AII*."

Namesake of the ketch *Atlantis,* WHOI's first ship, the *AII* was delivered to Woods Hole a year before *Alvin*. In 1982 it was ready for a midlife refit, a major overhaul. The timing was perfect. The most promising launch system was a stern-mounted A-frame used primarily by the offshore oil industry; but it was also the same kind of lift that twenty-five years before had handled Jacques Cousteau's shallow-water *Diving Saucer,* and in more modern times the French submersible *Cyana*.

With primarily NSF funds, the *AII* was renovated and fitted with an A-frame made by the Scottish firm Caley. Most of the work on the *AII* was done at Woods Hole; Dan Clark and his gang added the twenty-five-ton A-frame and ballast at a Boston shipyard. WHOI wanted lead ballast in the bow to compensate for the new loads (A-frame and *Alvin*) at the stern, but couldn't afford it because Dinsmore had forgotten to include the costs in the ship renovation proposal to NSF. He was still trying to figure out what to do when he met a navy captain who had worked at the Portsmouth Naval Shipyard and said he was looking for a new job. Dinsmore responded with a laugh, "Well, I'm looking for some lead, actually, a lot of lead, about 200 tons. They don't have any in Portsmouth, do they?"

To Dinsmore's surprise, the captain later contacted him to say he found some lead for the *AII,* and soon a procession of trucks began hauling loads of lead scrap and pigs to the Boston shipyard for *AII*'s bow. But one day the caravan was followed by FBI agents who continued south to the village of Woods Hole to inform WHOI that it was stealing federal property. WHOI returned the lead intended for a new nuclear submarine but was allowed to keep the "other" lead, which presumably had come out of scrapped submarines.

There was no part of *Alvin* for the A-frame's thick main line to hitch on to. To withstand repeated lifting, the hitching post, a titanium T-bar, would have to become an integral part of the sub and ideally go directly over *Alvin*'s center of gravity. The engineers were not confident of their calculations identifying the center of gravity, which had moved over the years with the addition of new and differently shaped blocks of syntactic foam. But wherever the true middle was, the general center lay near the amidships propellers and the five-foot-high sail. One or the other would require modification to accommodate the T-bar.

"Unfortunately for us, the weight and balance calculations for the submarine are marginal," *Alvin*'s manager Barrie Walden said. "Except for how much it weighs in air, these numbers are not easy to find. We knew our center of gravity was kind of iffy so we said what the hell. Strength-wise it made the most sense to put the lift hook slightly behind the sphere."

To make room for the T-bar, *Alvin*'s sail was shortened by about six inches and slenderized by twenty inches. On the first lift with the thirty-foot-high A-frame, *Alvin* drooped tail down. The center of gravity was about three inches behind the T-bar. After several iterations, the engineers settled on adding a small winch to the A-frame for a second but thinner line, which attached to *Alvin*'s tail would keep the sub horizontal in the air.

On the *AII*'s stern, *Alvin* fit in a high-ceilinged hangar, closed on three sides. There was little to compare with the old mother ship. *AII* could stay at sea for a month at a time. At 210 feet, she was twice *Lulu*'s length and carried fifty berths. As *AII* was already staffed, only four *Lulu* crew members joined the new ship's crew. Captain Reuben Baker, who was properly licensed for the 2300-ton mother ship, was in charge.

Alvin spent only a couple of months in the water in 1983, mostly for trials and navy certification, and the following January left Woods Hole for a twenty-one-month journey and more surprises. Five months before its twentieth birthday, *Alvin* would make headlines again.

Diving began in February to inspect the navy's underwater weapons range in the Bahamas, and the following month, geologists and chemists dived off Florida's Atlantic coast to inspect lithoherms, the elongated mounds of rock-hard mud and animals first discovered in 1967. Lithoherms were still something of a mystery. Divers have seen the deep-sea reefs only in the straits of Florida. While the mounds look fragile, they are as solid as coral reefs, and in fact incorporate the same limestone-making process. WHOI chemist Ellen Druffel wanted a core of a lithoherm.

Among the oldest of marine life forms, coral holds a history of subtle changes in the constituents of seawater. Druffel, whose specialty was climate change during the past 2000 years, was studying this record to infer characteristics of ancient climates. Most of her work focused on the symbiotic reef communities of animals and plants in the sunlit surface waters; like tree rings, the shallow-water corals make annual bands. But deep-sea coral is different. In the dark depths where no sunlight penetrates, reefs contain no plants, only animals; and the bands form at longer and irregular intervals, making them more difficult to date. "We had been taking corals from the upper fifty meters of surface waters, but that gives only one dimension of the ocean," Druffel said. "We wanted to see how things change at depth. People have dredged at depth live corals, reefs down there that have no light at all that are actually living. It's really freaky."

Despite repeated efforts, the divers could not get a piece of lithoherm, let alone a core. And navigation was a disaster. The transponder pings bounced off the lithoherms in a confusing mess of echoes, making it impossible to return to the same lithoherm—repetition that was necessary for Druffel's experiment. However, a pilot managed to pry off one lithoherm a three-foot-wide fan coral which Druffel dated at 250 years.

Another limestone mystery, also off Florida, were cliffs, stunning geologic structures that rise almost vertically, straight up for half a mile through the ocean. Geologists were intrigued at how such topography formed in the deep sea. They knew from earlier dives on Florida's Atlantic side that the cliffs were not ancient reefs, as they had thought. After diving on the lithoherms in 1984, *All* took the divers to the edge of the continental margin about 120 miles off Tampa so they could get their first look at the cliffs in the Gulf of Mexico.

On the first dive March 7, geologists sank to the bottom and got to work, taking samples and pictures. A peculiar band of black sediment hemmed the base of the cliffs. The divers also saw a pile of clamshell fragments and a few big whole shells. They thought the shells must have had a long drop from the top of the cliff.

Lamont biologist Barbara Hecker and Navy geologist Ray Freeman-Lynde were next. "Our mission was to sample the base of the escarpment, then go up, taking samples," Hecker recalled. "We started driving to the escarpment and I began seeing all these shells, big shells. 'Ralph,' I said, 'stop; something strange is going on here.' Ralph said: 'You bet there's something strange going on here. This looks like a vent.' "

Before *Alvin* were beds of giant mussels and pencil-thin tube worms about three feet long. Stringy white stuff coated a band of black sediment. The water temperature was about freezing. "I developed a wonderful migraine headache," said Hecker, who instructed Hollis to "grab everything in sight."

The objectives of the expedition immediately changed. The scientists radioed NSF to ask for another dive. NSF said no; *Alvin*'s schedule was too tight. They called MIT chemist John Edmond for permission to use the water samplers he had stored on the *AII* for his upcoming cruise. Edmond said yes and gave them one of his dive days. He was just as intrigued as they and the rest of the scientific community. It didn't make a lot of sense. Hydrothermal vents were known to exist only along the warm and hot boundaries of fast-moving tectonic plates in the Pacific. There was no shimmering water here, but the sediment reeked of hydrogen sulfide and the animals, virtually all unknown species, were clearly related to the vent fauna. The stringy white material had to be bacteria.

Scripps graduate student Charlie Paull, the chief scientist, joined Hecker for the third dive. When they saw a fissure in a cliff wall stuffed with tube worms, they asked pilot Jim Hardiman to nuzzle the basket into the orifice so they could sample water as close as possible to the worms. Hardiman nudged *Alvin* into the opening, and Hecker and Paull excitedly talked into their tape recorders. Suddenly, Hardiman screamed: "Will you shut up so I can think! We're stuck!"

29

Oh, God, please don't let me die at the bottom of the ocean!
   Alvin pilot Jim Hardiman

Virtually every *Alvin* pilot has braved at least one close call, and some passengers have been more frequently frightened than pilots. There have been several instances of burning wires or electrical components. On a 1974 dive, geologist Jim Heirtzler smelled rubber burning. Although the pilot, Jack Donnelly, had a cold and couldn't smell anything, he calmly grabbed a flashlight and wire cutters. "He opened this little electrical box and cut every wire," Heirtzler recalled. "Then he went back and found which wire it was and left that one unconnected and reconnected the others. As it turned out, it was one of the lights on the outside that was overheating."

Years later, an electric short caused a power loss. "We had gotten to the bottom and we were cruising along and suddenly everything went completely black," geologist Jeff Karson recalled. "Everything was gone, no power, no nothing. In the darkness, I heard the pilot say 'Oh, shit!' It was a horrible moment." The pilot took the same course of action that Donnelly had in 1974, and eventually restored the power.

Karson was unlucky enough to be aboard again for a similar experience. "Different things had stopped working during the dive, like the sonar," Karson said. "We knew something was really wrong when the strobes wouldn't work. It was clear that we had unacceptable power losses and we had to head for the surface."

"We were just scared," geologist Geoff Thompson said. "All we could see out the portholes were these spectacular flashes and arcing. There was obviously a major electric short, which scared the heck out of us."

The pilot hit the switches to drop the side weights, but the weights on one side would not let go. Enough power remained to operate a manipulator and remove the rocks in the basket to lighten the load, and with 250 pounds of weights still attached, a listing *Alvin* rose in a big sloppy slow-motion spiral through 13,000 feet. It took about six hours to reach the surface, where the pilot again tried the dump switch and the weights dropped.

When *Alvin* lost power on the slope of a volcano, it slid backward in the direction where there was no visibility. "We could distinctly feel ourselves slipping down the slope," biologist Lisa Levin said. "I distinctly remember Ralph [Hollis] saying, 'Look out the window and tell me if you see any rocks.' "

"We had no lights," Levin continued. "Ralph got a flashlight and started pulling out wires, trying to figure out why we lost all our power. The surface controller talked Ralph through the different-colored wires. Thirty minutes went by and we continued to slide. I was worried. We conceivably could have slid into an overhang. Ralph finally gave up and dropped the weights." The problem had been a blown outside fuse.

None of these incidents was serious enough to use the rebreather face masks or the two fire extinguishers in the sphere. And *Alvin* was never totally without power because it always carries emergency batteries with about four extra hours of juice, more than enough to drop the side weights and run the carbon dioxide scrubber throughout an ascent from 13,000 feet.

Far more serious than an electrical short is an oxygen leak. The vital gas is automatically bled into the sphere at concentrations between 18 percent and 20 percent, normal for room air. Oxygen concentrations as low as 25 percent can start a fire from the simple action of moving a switch, even turning on a flashlight. At that concentration, according to *Alvin* engineer Roger Maloof, an oxygen fire could burn through the insulation, wiring and paint on the control panels within about two seconds; and in less than a minute, the whole sphere would be an inferno.

Alvin has never had an oxygen leak, although pilots once discovered leaking hoses in oxygen tanks that were about to be loaded into the sphere. "It was just plain luck that they started leaking outside the sphere," Maloof said. "Most oxygen leaks you can hear, but not all of them." In the sphere, an alarm rings when the oxygen level is too high or too low; and pilots monitor an aircraft altimeter that measures the barometric pressure which, if the oxygen and carbon dioxide are balanced, should be at one atmosphere. The three men who smoked cigarettes at the bottom of the Mediterranean in 1966 were very lucky.

To remove carbon dioxide, the air in the sphere is pushed by a small fan through a canister of lithium hydroxide or sodium hydroxide (soda lime). Because the scrubber is so noisy, it is run intermittently; thus, carbon dioxide levels fluctuate. On a few dives pilots have forgotten about the scrubber until a passenger with a severe headache, a symptom of carbon dioxide poisoning, missed it; thus, "scrubber on" was added to the message every pilot gives at the surface before being granted permission to dive. Even with the scrubber on, some passengers develop headaches because what's too much CO_2 for one person may be too little for another. But given a choice between too much oxygen or too much CO_2, the pilots have no doubt about which side to err on.

Hypothermia was only recently recognized as a potential problem, and vacuum-packed down sleeping bags were added to the sphere. "We realized that in an emergency, the guys would die before the life support died," Maloof said. "Within twenty-four hours in the near-freezing temperature of the ocean, people don't make the right decisions. They become irrational and that's the worst thing because that's the time when you need all your wits to figure out how to get off the bottom." The sleeping bags, like the emergency rations, have never been used.

Outside the sphere, the environment can be full of surprises. Rocks have tumbled onto *Alvin* and currents have slammed the sub against cliffs and propelled it into pirouettes. *Alvin* has even been forced downward by what one pilot called a "waterfall," currents cascading over a cliff. All it takes to move the sub is a weak (2 knot) current. One trouble spot is off the East Coast from Miami to Cape Hatteras where the Gulf Stream flows.

Geologist Bill Dillon remembered being pushed backward in the area. "We were near the bottom and this current of dense water moving full speed ahead pushed us back alongside a cliff," he said. "I could see that we were going to hit, so I told the pilot, 'Ralph, we're going to hit!' He said 'I know. I know.' This happened on several dives, so we got used to it. The current smashed us into the cliff several times, damaged lights, and *Alvin* was gouged by the face of a cliff, which also took our emergency transponder."

Currents and sliding rocks never caused any more than the kind of exterior scars Dillon mentioned, and the pilots became adept at extricating the sub. In the waterfall, for instance, the pilot pushed a manipulator against the cliff wall to pop *Alvin* out of the cascading current.

But at vents, with temperatures as high as 400°C or 752°F, the risks are different. Although the temperature gradient is so steep that only six inches from a smoker the water is cool, *Alvin*'s skins have been scorched from brushing against chimneys, heat has charred the fiberglass-impregnated bushings in both manipulators, and the plastic over the bottom porthole has partially melted.

The currents in the northeast Pacific along the Juan de Fuca Ridge, which plays host to some of the hottest and most vigorous vents known, are notorious. Pilot Will Sellers has spent as long as an hour and a half fighting currents there while trying to approach a black smoker. "How I figured it out," Sellers said, "was to drag our ass in the mud right up to the chimney."

Even in the absence of currents, approaching a smoker can be difficult. "For a long time we didn't go right into the smoke but in the last few years we have driven right through the smoke to get samples of water," pilot Dudley Foster said. "It's always tricky. You go to the bottom of the smoker and visually climb up before you zip across through the smoke. But we still haven't been able to sample some. One smoker is enveloped entirely in smoke. You can't see the smoker through the smoke. The rule is: you can't sample it if you can't see it."

Nor is it always easy to distinguish the source of hot venting fluid. In some places where there are no chimneys, hot fluid will rise from a hole poked in the seafloor. Sometimes the hole may be formed by *Alvin*'s skids making a landing on a seemingly innocuous patch of featureless seafloor. "The vents on the Juan de Fuca Ridge are well known for people bouncing into the bottom and having hot water come out," geologist Bob Embley said. "Several times people have seen that bottom temperature gauge zoom up."

In the late 1970s, a permanent temperature probe was added outside the bottom window to constantly display the temperature on small TV screens inside the sphere. In 1983, the bottom window, which was used infrequently, was shielded with a disk of titanium, making *Alvin* a three-window sub. The metal will not melt in the vent temperatures, but according to Maloof, it could radiate enough heat to melt the plastic window under it. At about 194°F, Plexiglas will soften and begin to deform and extrude into the sphere, he said. At 356°F, it will flow like hot taffy. "Then the core would start melting and probably punch a hole right through. A leak at 4000 meters would act like a knife, like the kind of high-pressure water jets used to cut stone. It could cut your throat or your hand off."

Theoretically, a hole the width of a paper clip could cause the kind of catastrophe Maloof described. But it is highly unlikely, Maloof said, that a hole would form all the way through a penetrator or a three-and-a-half-inch thick view port and allow a decapitating jet spray. At both entry points, water must take a meandering path.

For the most part, as long as the pilots can see, they can avoid hot water, littered cables and wreckage, junked munitions, and tight spots like crevasses and overhangs. But what they can't see they can't avoid. Visibility from *Alvin* is severely limited. The pilot cannot see everything the two passengers see from the side view ports, and the passengers cannot see what the pilot sees at the center window. Despite the two external video cameras hooked to tiny inside monitors, there are blind spots beneath, behind and above the sub. Also, depth perception is altered; things tend to appear slightly bigger than they really are. "Sometimes we'd be going up a canyon wall, sampling, and when we tried to go up, we'd bump the top of the submarine," geologist Dan Fornari said. The First Commandment of diving is: Thou shalt not let anything but water get over the submarine... and when something does bang the top of the sail, get out fast.

Alvin's batteries, manipulators and sampling basket are still droppable. The mercury release was eliminated in the mid-1970s because of environmental concerns. The hull release, once considered the ultimate safety feature, is not entirely trusted.

Alvin's HY100 sphere was released once, but as a test in shallow water. Its titanium sphere has never been released in the ocean. In the navy's pressure chamber, the long bolt in the center of the sphere was turned remotely fifty times, and to nobody's surprise, the greater the pressure, the more difficult it was to turn. Some engineers argue that the hull would release only if *Alvin* were horizontal. In a real emergency, a pilot would first drop the manipulators, batteries and basket. Only if none of these measures helped would the pilot consider releasing the sphere. But by then, the engineers say, the much lighter *Alvin* probably would be nose up or tail down; either angle could put enough stress on the hull release shaft to keep it from turning. If the hull did let go, the ride to the surface in a seven-foot titanium bubble swirling through inner space would not be gentle. The passengers probably would be acutely seasick and bruised from hurtling equipment. But nobody can know for sure unless there is a dire emergency or the release is tested in the field, which is unlikely because of the expense.

The navy eliminated the hull release from *Sea Cliff* and *Turtle* to make structural changes for a three-point launch system. The Alvin Group manager Barrie Walden has considered eliminating it on *Alvin* so braces can be added to the front of the frame, as they are aft. But the pilots say a releaseable hull, even if unproved, is nice to have. "It's like the seat ejection in a plane," Dudley Foster said. "Even though you may never have to use it, it's in the back of your mind that it's there."

Among the most dangerous moments in *Alvin* were six dives in 1983 to the *Thresher*, the nuclear submarine that foundered off Cape Cod in 1963. On reaching bottom, Ralph Hollis found a portion of the mangled *Thresher* once attached to the engine room and reactor. One end was buried in the sediment. Hollis headed toward the gaping end. Careful not to touch bottom and risk getting tangled in debris, he dropped the hovering sub's nose and got to work; he took a core of mud, set out a monitor to measure radioactivity and slurped up some animals.

When he slurped with the left manipulator, *Alvin*'s bow drifted to the right; he used the stern propeller to bring *Alvin* back to center. Taking pictures with the right manipulator edged him to the left. Again Hollis used the aft prop, not realizing that each whir pushed him a little farther into the *Thresher* tomb. As Hollis described it that night in his journal:

> I found that I had moved quite a ways into the hull.... I tried to swing to the right enough to drive the sub out. The sub moved a little and then stopped with a *clunk* as the back part of the sub came up against something solid. I couldn't back up because this would surround me with sediment and I wouldn't be able to see a thing. It was one of those rare times on the bottom that I actually felt fear. If I slid farther into the hull or made the wrong move, then I would be stuck there forever....

Hollis noted his bearing, hoping the magnetic compass was not skewed by the wreckage that now enclosed *Alvin* on all but one side. He wrote:

> I tried backing out... we were engulfed in sediment kicked up by the stern prop.... We couldn't see a thing out of the windows.... I could vaguely see the hull in the sonar. It seemed like hours before there was any change in the sonar picture, indicating that we were moving out of and away from the hull. Finally as we moved up off the bottom and away from our sediment cloud, we could see again. The two observers never knew/realized how serious a situation we were in. It is only with a lot of luck that we are not still there.

On the last dive Shaun Nerolich, a pilot for a year, got trapped in the wreckage. He gently rocked *Alvin* until the stern prop was freed. Hollis finished his journal:

> I asked [Shaun] what it felt like and he replied that he has never had a feeling like that before in his life.
>
> It is hard to describe the feeling. I think that it is the way I would feel just before I was executed. You are not ready for death and you see no way to avoid it.
>
> It is probably more apparent at this dive site because all around you is the evidence of a terrible tragedy.

Nerolich didn't dive again; he left the Alvin Group the following month.

Alvin's passengers, including those who have known scary moments, have special respect for the pilots and an abiding faith in the integrity of the sub. In a quarter century of diving, representing more experience than any other deep-diving craft in the world, *Alvin* has always returned its passengers safely to the surface.

"We took a great deal of pride in bringing the submarine back whole," George Ellis said. "We brought it back dented and scraped, but whole. You know what we were afraid of? Brody. He was a powerful inducement. If you lost anything or even chipped the paint, Brody's response... well, his responses are basically unprintable."

Some pilots and scientists joked with inexperienced passengers by pretending, for instance, that the sub was leaking; as proof they pointed to the condensation which always bathes the sphere. Bob Aller heard creaks on a dive. "Like glass cracking, as if it was the windows," the biogeochemist said. "When I mentioned the cracking sound around the window, Ralph said it shouldn't do that. He would talk about how many dives the sphere had made and wonder how many more it could go. He certainly doesn't go out of his way to make you feel that you're in something that couldn't fail." Biologist Craig Smith recalled getting tangled in fishing line off Panama. "We were hooked on this line," he said, "and George Ellis joked that we were stuck down there. It scared the shit out of me."

No pilot said he quit because of fear, but Hollis, who with eleven years of diving is *Alvin*'s most experienced pilot, said that some pilots left because they were finally too spooked to dive.

There are other reasons for quitting. Keeping enough technicians and pilots has been a problem since the early 1980s, when *Alvin* became a hot item. More business meant more work, longer cruises, and less time with families back home. In 1980 there were only six out of the usual ten men in the crew, including three electronics technicians, two mechanics, and an electrician. Two pilots, Hollis and Ellis, were among them; the budget was never big enough to afford the luxury of paying men just to pilot.

But Ellis was threatening to leave. After more than a year of piloting every other day for weeks at a time, Ellis felt he was "being worked right into the ground; we were all working very long hours and we were just making it. I worked whether I was at sea or in port a minimum of eleven hours a day, seven days a week, forty to fifty days straight. Then I'd take a ten-day vacation and start another one of these stretches. It became harder and harder to come back." He quit in April 1982, leaving the recently hired Bob Brown and Hollis, who dived alone from February through June; then Brown left.

At the same time, Jack Donnelly, who had replaced Larry Shumaker, took another position at WHOI, and *Alvin* engineer Barrie Walden became the new manager, the first with no piloting experience. He couldn't fill in at sea as Rainnie, Shumaker and Donnelly had. Desperate for pilots, Walden went to the *Trieste* base in San Diego to borrow a navy submersible pilot for a few months.

Today every member of the *Alvin* crew is hired as a technician with the option of a pilot track; a college degree is not necessary but the technicians must have electrical, electronic or mechanical skills and be willing to spend at least eight months a year at sea. Of an average five technicians hired annually, two or three do not stay a year.

Alvin has had twenty-two full-time pilots, nine hired before 1982. Attrition is high, Walden said, because "90 percent of the work is routine and boring." About half the responses to the job ads he places come from recent college graduates "looking for adventure"; the others are older and a few have backgrounds in scuba diving, ROVs (remotely operated vehicles), or shallow-water submersibles.

In the beginning, of course, nobody had any experience piloting small deep-sea research submarines. It wasn't until 1971 that the navy drafted a policy on the qualifications of pilots of "manned, noncombatant submersibles." That year Ed Bland became the first *Alvin* pilot to go for questioning, mostly about safety, before a panel which included a naval officer who also accompanied him on his first solo dive and bestowed pilot certification.

Pilot training was "hit or miss," Hollis said. "You'd beg dives as you went." He and Ellis held classroom training sessions for all the technicians on transits to and from dive sites. Today a seat on every fifth dive is reserved for a pilot-in-training, who must pass four oral examinations before separate panels of engineers, *Alvin* and navy submersible pilots and scientists. In addition, every rookie keeps a log of the sub's maintenance, according to his specialty. "We're more qualified than the navy pilots," Hollis said. "*Alvin* pilots spend more time diving because we do many more dives. And we train differently, which can be attributed to experience. We have more experience. Most of our training is self-teaching. We put the guy in the sub, tell him what he's doing right and what he's doing wrong. We can't afford formal training and we don't have the time for it. Over the years I have discovered that you can make anybody a pilot."

Hollis joined the Alvin Group in 1975 because he was bored

with every job he tried after retiring from the air force as an electronics specialist. He had packed groceries, sold real estate, tended bar, and managed a McDonald's on Cape Cod.

Jim Hardiman wasn't sure why he got into the business. He had spent six years in the Coast Guard, four years in the reserves and two years on active duty as a machinery technician, which included classroom and hands-on preventive maintenance work. After graduating from Boston College with a bachelor's degree in physics in January 1982, he had no set ideas for a career, although oceanography was one of several possibilities he followed up on. He had applied a half dozen times for a job as a research assistant in one of the labs at WHOI, but didn't get one.

In the spring of 1982, Donnelly and Walden asked if he was interested in joining their team. Hardiman, twenty-two, with a full black beard and thick wire-rimmed glasses hiding his intelligent grey eyes, figured it could be a foot in the door. He drove to the elbow of Cape Cod for a brief interview. Donnelly and Walden asked him three questions, he said.

"Do you have a valid passport?"

"Yes."

"Do you get seasick?"

"No," he lied.

"Can you be in Costa Rica the day after tomorrow?"

"Sure."

He was hired.

Hardiman had caught a glimpse of *Alvin* in 1980 while making one of his job applications at Woods Hole, but the sub was disassembled in an overhaul. He had never seen *Lulu* before. As he boarded the mother ship at Punta Arenas, he wondered if he had made a mistake. Doubt gripped him a second time when he followed the first mate to his bunk in the Tube of Doom.

"I guess you better meet Ralph," the first mate said.

Hollis was a man of few words, and those were hard to catch. "Can you swim?" he asked Hardiman.

"Yes."

And without another word, Hollis chalked Hardiman's name on the big dive board next to one of the two escort swimmer slots. Then Hardiman threw up.

Those years in the Coast Guard didn't help much. He concentrated on just staying alive. Being an escort swimmer, his major duty for three months, was the most dangerous job he ever had. He deeply feared the sharks. On one recovery, Dudley Foster watched him leap back into the Boston whaler. "Never mind the sharks!" Foster shouted. "Just tie the basket up!"

"I seem to remember being wet all the time," Hardiman said.

"I went through a lot of massive confusion. I would have quit if it weren't for Brody."

Five months after his arrival, Hollis said to Hardiman: "I suppose you want to be a pilot." Hardiman hadn't thought much about it, but he was flattered. Ten months later, he was ready to take *Alvin* down on his own. "Oh, God," he said, standing in the hatch, "please don't let me die at the bottom of the ocean!"

Hardiman's predive supplications were so common that he became known as the Paranoid Pilot. Despite that, Hollis said, he was an excellent pilot who understood the navigation like no other pilot.

"Up until that point in my life I never had to concentrate for eight hours straight as I did as a pilot," Hardiman recalled. So many things to remember. To set *Alvin*'s aircraft altimeter at zero before closing the hatch. There was a right way to close the hatch, gently so it sat evenly all around. Otherwise the mounting pressure would force it to slide into proper position with an eardrum-popping noise. And you had to remember to open the valve to bleed oxygen and turn on the scrubber blower. With too little pressure, you couldn't open the hatch, and you had to have the guys tug on it from the outside. Another ear-popper.

It took practice to make a gentle landing. So many things to think about to avoid smashing into the bottom. Passengers were weighed the night before. Only forty surprise pounds could throw off the calculations for ballast and trim. Too heavy and you would slam onto the bottom. It wasn't dangerous, just embarrassing. Getting the water depth right was just as important. Ellis held a record for shortest dive. Thinking he was in deeper water, he slammed onto the seafloor; all the weights dropped off, and *Alvin* popped back to the surface.

Stopping eighteen tons, even from a walking speed, took finesse with a steering control box of switches, six for the props for forward, reverse, rotate, and so on. "It's like a big sailboat," Hardiman said. "It's easy to oversteer. When you hit a switch for a motor, nothing happens for two seconds."

Hardiman supervised pilot-in-training Will Sellers on a dive with geologist Vernon Asper. When it came time to return, Sellers lightly touched a switch to release a drop weight, and getting a nod from Hardiman, he flicked it. Suddenly there was an explosion. The men, terrified, sat quietly, not daring to move more than their eyelashes. "Well, I guess we're not dead, huh?" Asper said.

"It was the first thought that crossed my mind, too," Hardiman remembered. "It was loud, *so loud*, I expected the hull to come apart. Then I remembered that Skip Marquet had told me not to worry, if anything ever happens to the hull you won't know what hit you. I took comfort in that."

Small pieces of glass drifted by, and they realized that *Alvin*'s main beacon, the thalium iodide light, had imploded. Why, nobody determined. It was a coincidence that the implosion occurred precisely when Sellers released a drop weight.

The green light was mounted a safe two feet above the sphere, a short but critical distance. The shock waves of a nearby implosion could blow out a view port or form a dimple in the passenger sphere. Depth charges work on the same principle.

Because virtually anything not filled with pressure-compensating fluid or gas can implode, all exterior items on *Alvin* are tested to 10,000 psi, equivalent to a depth of about 20,000 feet. By knowing at what depth an item implodes, the critical stand-off distance—how close to the submarine the object can safely be—can be calculated. Brittle materials, such as glass, are unpredictable, even when pressure tested successfully. No glass is allowed in the sampling basket, which at its closest is two feet from the sphere.

The biologist Barbara Hecker had been in tight spots before. On her first dive, a green Ralph Hollis backed the sub into Turner Tower at a deep ocean station, and snagged the aft propeller. On another of her dives, a current pushed *Alvin* against an overhanging cliff, shearing off lights and drop weights. She was in the sub when an electric short caused smoke to enter the sphere, and she was aboard when *Alvin* approached an overhang and her colleague Bill Ryan suggested ramming the ledge to break off a piece. "Is he kidding?" Hollis asked Hecker. "No," she replied, "that's why scientists aren't pilots. We'll do almost anything to get our samples."

They did not bash into the overhang and risk it falling on them. However, at least twice large ledges have fallen on *Alvin*. At the Juan de Fuca Ridge in 1988, a young geophysicist had his heart set on getting a big chunk of sulfide, as big as a Volkswagen bug, he said. Eager to please, the inexperienced pilot tugged at the ledge and *Alvin* dived with it, nose down. Most of the ledge crumbled and fell off the basket on the brief but fast descent, leaving both pilot and geophysicist a little wiser.

When Hardiman screamed for Hecker and Charlie Paull to stop dictating their observations into tape recorders so he could concentrate and free *Alvin* from the opening in the wall of the limestone cliff that held them off Tampa, the scientists immediately obeyed. After waiting what she thought was a safe few moments, Hecker broke the silence.

"Yeah, I knew we were stuck," she lied.

"How did you know?" Hardiman asked. "Shit, Ralph's going to kill me."

"I've got faith in you, Jim," Hecker said. "Whatever you do don't panic. Don't do anything hasty. Don't hit that release."

"Don't worry," he replied. "I wouldn't do *that*."

"We really couldn't move," Paull said. "Jim was trying. We kept hitting something with a thud. He actually sat there and just thought. Then he decided to take one of the arms of *Alvin* and reach out and push back against the wall, and all of a sudden we popped out and rose fifty meters or so."

It took Hardiman about fifteen minutes to free the sub.

More serendipitous pieces of the cold vent puzzle came together in late 1984 when a scientist from Texas A&M called the California physiologist Jim Childress to say he had picked up some strange animals in a trawl about 150 miles off Louisiana in sediment full of hydrocarbons. "These geologists and chemists were doing surveys and catching these animals and throwing them over the side," Childress recalled. "One day, a tube worm that was being thrown over was broken in two and blood went all over this guy's foot. He had just read a paper on hemoglobin in the vent fauna and called me."

The scientists had found another cold vent at a depth of about 2000 feet. More were discovered the following year off Louisiana and Texas where the seafloor held high amounts of hydrogen sulfide and methane.

Vents in the ocean are not heat-dependent; nor are they restricted to the boundaries of tectonic plates or even great depth. A vent needs only a plumbing system. At the warm and hot vents, heat makes the fluid buoyant. At the cold vents or seeps off Tampa, the water, heavy because of its extreme saltiness from the cliffs, oozes from the permeable limestone. The hydrogen sulfide at the cold seep is not produced geothermally but comes from the interaction of bacteria and organic-rich mud. The vents in the Guaymas Basin have both organic-rich sediments and heat.

The summer of 1984, dives off Oregon brought a new twist to venting in the deep sea. Geologists, geophysicists and geochemists dived on the Oregon Slope, a section of the continental slope which happens to be in a subduction zone, where for many millions of years the Juan de Fuca Plate has been sliding beneath the North American Plate. The scientists wanted to determine how the gargantuan stress of subduction—being squeezed and scraped—affects the sediment, which in turn may affect the water chemistry. To their surprise, they discovered clusters of giant clams, tangles of skinny tube worms and *Solemya*, the sewage clam, but in cold water rich in methane and carbon dioxide.

They saw no venting fluid, but suspecting it took down a ball in a tube and watched the ball rise, indicating that invisible fluid rose from the seafloor at about a gallon an hour, carrying organic and inorganic material that nourished the chemosynthetic bacteria, which in turn fed the animals.

"We had been working off Oregon since 1962 and not once did we dredge *one* clam," said geologist Vern Kulm.

Also surprising were the chimneys and slabs of carbonate, which form when carbon dioxide combines with calcium, a common process in shallow water, especially in the tropics where abundant calcium comes from seashells. On the Oregon Slope, the carbonate formed from vent fluid—the methane oxidized and created carbon dioxide, building the three-foot stacks and slabs.

"It was really exciting, not only just that the carbonate was there but how abundant it was," Kulm said. "A lot of people have found this in ancient rocks and called them shallow-water deposits and they're not. It's put an entirely new perspective on some of these deposits. We've added immeasurably to the interpretation of ancient rocks."

On another expedition, the Oregon team tried to determine the source of the cold venting water by using higher resolution seismic profiling. Instead of shooting an air gun and recording the sound reflections with hydrophones towed from a ship, they strapped the underwater microphones to *Alvin*'s sampling basket. The air gun was shot every ten seconds from the *AII*, and hydrophones on *Alvin* recorded the acoustic waves reflected from under and on the seafloor. According to these data, the vent fluids traveled along faults and reached the seafloor right where there was a cluster of animals.

The scientists still don't know precisely from how deep the underground fluid comes, or why the temperature of the water was cold except in a small area where it was several tenths of a degree centigrade above normal or how the venting affects the water chemistry. "At this point, we're just documenting the chemistry," Kulm said. "But we think the animals have the biggest effect on the chemistry."

"As we've learned more, it's gotten more complicated," chemist Karen Von Damm said. "One of the biggest problems with vent chemistry is the variability we see at different vents. The vents have major global control but exactly what, we don't know. We still don't know how to calculate the global flux from the vents. We don't understand what's controlling the vent pH itself, although it seems unequivocal that vents are the main source for the iron and manganese and other alkaline elements in seawater."

Since the first vent discovery in 1977, more sophisticated tools have enabled chemists to measure elements in parts as minute as a thousandth of a trillion, and chemists are returning to the hot springs to collect more hydrothermal fluid and break it down into still minuter parts.

"The processes that go on in the ocean occur on long time scales, longer than the average lifetime of a graduate student," John Edmond said. "And some kinds of chemical reactions are often biologically mediated. When you put all that together, it turns out that it's impossible to mimic oceanic processes in the laboratory. You can't see below the seafloor; the technology doesn't exist to drill in a hot spring. We can't get real samples of actual hot rock and hot water reacting with each other so we have to use the chemistry to understand what's going on at depth."

Alvin returned in 1984 to the vents at Galapagos, Guaymas, and others along the East Pacific Rise, and made its first visit to the Juan de Fuca Ridge off the Northwest United States and Canadian coasts, where black smokers taller than a three-story house gushed black, white and yellow "smoke." And the sub explored other ecosystems unrelated to vents. In late 1985, the *Atlantis II* steamed home after a record season of 161 dives, and *Alvin* went into the high bay at WHOI for a six-month overhaul. The weary pilots and technicians, glad to be coming home for Christmas, sang, to the tune of *Rollin' Down to Old Maui:*

> It's a damn tough life full of toil and strife
> We *Alvin* men undergo
> And we don't give a damn about Ralph and his plans since fifty
>   dives ago
> For it's down in the submarine, me boys, diving every day
> And Barrie Walden's waiting on the dock sayin' there's no
>   raise in pay...
> Oh, rollin' down to old WHOI, me boys
> Rollin' down to old WHOI
> We're homeward bound...
> For two long years we've been out to sea
> Gung-ho all the way
> But it's startin' to fade when I have to wade through the oil
>   every day
> Oh, how I wish for a normal job, nine to five each day
> But I must admit that I sure will miss
> The overtime they pay.

30

Aboard the Atlantis II, *off Bermuda, May 7, 1986.* The *Alvin* team has been tweaking and probing their overhauled sub since leaving Woods Hole in the evening darkness of May Day. Most of the half dozen shore-based engineers and I, a guest, joined the Bermuda cruise. The engineers, pilots and technicians huddle now around *Alvin*. They remind me of interns in a hospital emergency room trying to diagnose. On these trials the ballast pump has failed, a penetrator leaked, mercury leaked and there have been grounds in batteries. Nothing serious, the chief engineer Jim Akens says. Akens wears a black tam covered in dust. He joined the Alvin Group in 1977 after a decade in the rock-and-roll business; he built state-of-the-art sound systems for Joan Baez, Jeff Beck, Sonny Rollins, Steely Dan, Joni Mitchel.

When the penetrator leaked, *Alvin* manager Barrie Walden scraped off the hardened grease with his pocketknife, to the consternation of some pilots who believe that if anything is going to kill them, it will be a failed penetrator. But it stopped the leak.

This cruise caps one of *Alvin*'s most extensive overhauls. And as in all the other tear-downs, the changes made this past winter and spring were limited by only imagination and money. *Alvin*'s annual budget is about two million dollars. Safety continues to predominate and simple is still better than complicated. Despite the advances in computer technology, no major system or part of *Alvin* depends on a microchip.

Most needs were obvious. The Hoover motors worked fine at 6000 feet but at more than double that depth failed regularly. Clinkers clogged the brushes in the armatures or just destroyed them. The motors burned out so often that two spares were carried. Savvy scientists tried to avoid expeditions scheduled directly after a series of very deep dives because they knew they would probably lose a dive day to replace a burned-out motor. Now *Alvin* has brushless motors, a new technology that has rendered clinkers a problem of the past. The motors are mounted in six small thrusters which replaced the wheel-like aft propeller and the two midships props, providing slightly more speed and maneuverability with less power, and much more reliability.

Alvin's titanium still shines like new, although occasional cracks have developed in the frame, usually at welds. The navy's ceramic pump is no longer troubled with bad O-rings and the ceramic no longer flakes off, thanks to new stainless steel pistons sprayed with molten ceramic. Four new battery compartments were built of syntactic foam and fiberglass; the fourth still holds no batteries. It would require too much more buoyancy to compensate for the extra load. The batteries are still lead acid, but instead of golf carts, these are made for forklifts; they're lighter and leaner. The steering box of toggle switches was exchanged for a joystick, the kind of control the original *Alvin* came with. Hoover oil got too expensive. Bray and mineral oils and cherry juice (aircraft hydraulic fluid) are used.

All the penetrators are new, and for the first time all the holes around each porthole are filled with wiring. Jim Hardiman volunteered to handle the tedious chore of grinding the penetrator cones, coating them with blue dye to check for fit and regrinding them. Hardiman, like other pilots, never could understand how a tiny leak at a penetrator could theoretically cause a decapitating spray of water—which is why he volunteered to handle the job of fitting the new penetrators. He went to senior engineer Arnold Sharp and said, "OK, Arnold, tell me about this voodoo engineering with these penetrators." Sharp said there was no voodoo involved.

For the first time, *Alvin*'s sail has no windows. The pilots said nobody ever used them. There's a new pilot seat, identical to the original—a metal box with a cushioned lid that holds emergency tools. The passenger sphere has never been so comfortable. Large foam back pads were affixed with Velcro tabs to the sphere, and the floorboards were dropped four inches so observers don't have to crane their necks so much to see. Jim Akens is most proud of the new cassette stereo and tiny speakers in the sphere. "I convinced Barrie that it would be good for morale," he says. For the first time, people enter the sphere shoeless. The only original part of *Alvin* is the left manipulator, made by General Mills twenty-two years ago.

Walden convinced the navy to stretch the certification period from one to five years. The relationship between the certifiers and the WHOI operators has evolved into one of trust and mutual learning. Sometimes the navy doesn't bother to send a witness to certification dives of the sub or its pilots.

The crew worked six and seven days a week, rushing to meet the schedule, which is crowded with scientists who have waited years for their dives. Unlike the earliest team, these men are paid overtime on salary bases that start around $19,000. A few have made down payments on their first homes. As they worked in the high bay, they sang homemade verses, most composed by pilot Will Sellers, a former navy helicopter technician.

> Pack up all your tools and rags
> Put 'em all in a bag
> Bye, bye, high bay
> Pack up all your handling gear
> and let's go have a beer
> Bye, bye, high bay
> No more submarining for the weekend
> Damn good thing cuz I'd go off the deep end
> So cast the lines and clear the pier
> Let's get the fuck out of here
> High bay, bye, bye...
> (To the tune of "Bye, Bye, Birdie")

And they sing on the *AII*:

> Hosin', hosin', hosin'
> Keep that Bray oil flowin'
> Keep a gallon bucket by your side!
> In any kind of weather
> It'll turn your skin to leather
> Make your hands, brittle, hard 'n' dry
> (To the tune of the theme from the television series *Rawhide*)

The stern is alive with the commotion of pre-launch. Pilots Paul Tibbetts and Will Sellers separate the lines that pull *Alvin* on its rails out of the hangar. Sellers sings: You load 18 tons and what do you get, another day older and deeper in depth.... The outside phone rings, and Tibbetts, whose T-shirt reads INSTITUTE FOR THE SEXUALLY GIFTED, answers: "Thank you for calling AT&T." Pause. "Bazner! Baaazner! No, he's not here."

Bos'n Ken Bazner is with Harris, a budgerigar listed on the ship's passenger list since 1975. The parakeet sleeps with Bazner but spends the day playing in the crew's mess, riding the toaster handles, strutting inside a Pringles Potato Chip can, and chuckling the bos'n's gravelly laugh. "Science sucks," Harris says. "Science sucks, heh, heh, heh." Perched on Bazner's eyeglasses, Harris tours the main lab at day's end and startles scientists with his extensive, if earthy, vocabulary.

The Submarine Blues

In the sphere, Jim Hardiman is trying out the new thruster controls. His voice comes over the radio held by pilot Dudley Foster. "Did I forget anything? Left rudder, right rudder, forward, reverse...."

"Try your mercury," Foster says.

"Left rudder?" Hardiman pops through the sail to see for himself that the port thruster works.

"Right?"

"Yeah," Foster says, but Hardiman pops up again.

Much of the electronic equipment in the sphere was built by George Meier, who has been with the Alvin Group since 1969. "When George does it, it always works," Akens says. "It never fails, it's the best. We tend to assume we cannot buy most of what we need. It requires a good fundamental understanding of what you're trying to do."

"I'm *hiccup* perfectionist," says Meier who has had the hiccups for three days. He says the worst part is trying to drink a beer with people watching. He's upset about the scratches and dirty handprints already on the new panels inside the sphere.

A group of men from Asahi Broadcasting are eager for their dive which the Japanese TV station has paid for. The man diving with Hardiman today is Masa, a famous actor who most recently played the transvestite in *La Cage Aux Folles* in Tokyo. He says he does only love stories and comedy. But he also says Asahi is filming him and *Alvin* as part of a documentary on the Bermuda triangle.

In *Alvin*'s basket the Japanese have placed an egg, cork brick, hard plastic ball, small orange buoy, styrofoam ball and a balloon. It's an old trick. *Alvin*'s crew and passengers have sent hundreds of styrofoam cups and wig holders down with the sub to shrink into perfect miniatures. The raw egg, naturally filled with pressure-compensating fluid, will return to the surface perfectly intact and edible, but salty.

"What are we waiting for, please?" Terri, the translator, asks.

"I have to go to the bathroom," Hardiman says. Terri bows. Hardiman bows and rushes off.

The escort swimmers—*All* sailor Sallye Davis and *Alvin* technician Dave Sanders—lean against the sail for their ride out atop the sub. The strips of bright pink plastic which say REMOVE BEFORE FLIGHT are taken off the drop weights. Military flyers use them. On *Alvin* the strips are attached to small metal pins that keep the drop weights from accidentally falling while the sub is out of the water.

Hardiman returns, and still concerned that *Alvin* has not yet had a perfect dive, tells Akens, "If anything happens, anything, I'm bringing it right back up." Then he gets into the sail and invoking the final words of a convicted killer to his firing squad, Hardiman hollers: "In the words of Gary Gilmore, let's *do* it!" His head bobs down and back up, and with great drama, Hardiman heaves a filthy sneaker, and then another, into the hangar.

The A-frame lifts Hardiman, Masa and another pilot in the eighteen-ton submersible high into the air. We hear Hardiman again when *Alvin* is in the water. "Are my thrusters down?"

There's fresh coffee in the mess. Harris has been listening to the radio; he mimics even the static. Cook Joe "Rib Roast" Ribeiro wields a cleaver—down against a cutting board, and when a sailor walks into the galley, up in the air. "Hey, you! Get out of my passageway!"

At 1:55, surface control says *Alvin* is descending without problems. But in twenty minutes, Hardiman's strained voice comes through. "I think the sub's cracking...."

Akens sighs, puts a hand to his forehead and takes off his dirty beret. "I'm going to take a nap," he tells Foster.

"But what about Hardiman?" I ask.

"He's fine," Akens says and heads for his cabin. I look at Foster, but his droopy blue eyes are calm. He convinces Hardiman to ride out the dive.

At 5:26 *Alvin* and the small Avon boat are at the stern and soon aboard ship. The Japanese cameramen focus on their star as his smiling face appears from the sail. They offer him a cigarette and coffee. Hardiman's blue woolen hat with the little ball on top rises slowly from the sail. The movie star talks with the cigarette in a corner of his mouth and the microphones get closer.

Now it is clear what caused the explosion—the hard plastic ball in the basket had imploded. "That should be a lesson to you," Ralph Hollis tells Hardiman. "Make sure you know what's in the basket before you dive."

Hardiman is most interested in the five days off due him. His bride-to-be awaits him in Bermuda. But there would be no wedding on this island. He cannot get the time off. Hardiman will leave the Alvin Group in a few months and return to school. The wedding will not take place until next year.

After the trouble-free dive, the shore-based engineers and I debark, and a passel of impatient scientists from the National Oceanic and Atmospheric Administration, Duke, WHOI and MIT board the *AII*. Throughout the extra-long Bermuda trials, they worried that their dives would be preempted by the next expedition in the Pacific where *Alvin* is slated to spend most of the next twenty-four months. Now they are relieved, and they can't get out of Bermuda fast enough to begin their journey, to get to what they are almost sure awaits them in the Atlantic.

Their study site is at the slow-spreading Mid-Atlantic Ridge about 1800 miles east of Miami. They dived there in 1982 with *Alvin*, but the expedition was cut short when a scientist injured his groin while trying to board *Lulu* from the small boat. If he hadn't hurt himself, he and his colleagues almost certainly would have discovered the first hot vents in the Atlantic. They didn't know then that *Alvin* had been on the perimeter of one of the biggest, hottest, deepest and orneriest vent fields ever. In 1985, unable to get *Alvin* time, they returned to probe the site from the surface. They recorded slight temperature anomalies and dredged odd-looking shrimp and rocks stained with the telltale colors of polymetallic sulfides. And their video cameras imaged the first black smokers in the Atlantic. They knew what the theory was, that hydrothermal vents existed only at faster-spreading centers. But they didn't believe it.

On May 21, 1986, NOAA geologist Peter Rona and MIT chemist John Edmond made the first dive, number 1675. Almost immediately on reaching bottom, they had much to report.

"I see a long eel-like fish, three of 'em," Edmond said.

"There are at least four or five dozen anemones over here," Jim Hardiman said.

"There're *hundreds* of them over here, and a lot of Galatheid crabs. Beautiful."

"Chimney in front of me!"

"Oh, *wow*, man! It's beautiful!"

"We're right over it, Jim," Rona said, alerting the pilot to high temperatures.

"OK, as long as it's not black smoke."

"No, it's clear, shimmering water.

"Lot of *shrimp* now. Oh, *yeah*...."

"*Holy moly!* There's hundreds of shrimp around here."

"This is a big mother.... This is beautiful, man! This is massive sulfide. *Look* at it! Jesus Christ, man!"

"*All*, this is *Alvin*," Rona said into the mike.

"There's a gigantic fish over there. That guy's about two meters long!"

"Hey, aren't those guys up there supposed to answer?"

"Yeah, try again."

"*All*, this is *Alvin*."

"Three meters? Naw."

"Roger, *Alvin*."

"You guys put us right down on the money.... We have encountered the first hot spring."

"Boy, it's so easy to drive this submarine now."

"*Every*thing is sulfide...."

"I've got clear, shimmering water...."

"Yeah. Shimmering water. Beautiful. Tiny chimneys. Magnificent."

Alvin was in a forest of short, pointed chimneys gently puffing milky smoke that rolled like clouds of ground fog. The colors were not autumnal, but vernal. White crabs, pale anemones. The white fluid turned bluish. The vent field sprawled over a huge mound. The onion-shaped tops of the spires, reminiscent of Russian architecture, led the divers to name the area the Kremlin. *Alvin* continued up slope and at the crest of the mound encountered another forest, this one of seventy-five-foot-high spires belching 350°C (662°F) black smoke. The smokers were puffing so vigorously that *Alvin* could not approach them.

Rona estimated the mound to be the size of the eighteen-story Houston Astrodome, which can hold 45,000 grandstanders. The deep-sea Astrodome, Rona calculated, held some 5 million tons of sulfides, a deposit equivalent to the 2.6-billion-year-old Noranda mine of sulfide ore in Canada. "It's the size, shape and composition of that land deposit," Rona said. "The age of the earth has been estimated at 4.5 billion years. If these deposits are about half the earth's age, that says these seafloor hot springs were active then."

Some 250 miles south of the Kremlin was another enormous hot vent, dubbed the Snake Pit, after the predominant denizens, the eel-like fish. On dive number 1683, with John Edmond and WHOI geochemist Susan Humphris, *Alvin* sailed over yellow-stained rock, and crabs, and the long fish came into view.

"Okay," Humphris said, "I'm starting to see anemones, very small, all over the rocks, large numbers of really tiny ones; they're white, on stalks. We're going up a scarp and there are a lot of crabs still. John, do you see these little mushroomy-shaped white things on the rocks?"

"No," Edmond replied.

"They're all over the rocks. They're tiny little sea anemones, I think. The rocks are covered with some white tubes, probably tube worms, and these small, almost mushroom-like white sea anemones. Some snails on the rocks now.... The water seems to be getting a little more cloudy now. Eel fish.... Depth is 3469 [meters or 11,381 feet]."

Alvin crossed a small gully and slowly climbed a hill, encountering more anemones, crabs and eely fish.

"I can see black smoke all over the place," Edmond said.

"I have in sight two very large chimney structures, three meters and... there's shrimp all over this side," Humphris said.

"There's millions of shrimp now," Edmond said.

"Yeah, they're everywhere."

"What're these?" Humphris asked. "Have you got any of these little sort of smoodgy things on the rocks..."

"Yah."

"...that are little cream-colored..."

"Yah."

"...sort of gelatinous things?"

"Yah."

"Are they anemones?"

"I don't know."

"Wow! These constructional features are fantastic! These are amazing!"

The chimneys were free-form stacks, some with abbreviated branches of sulfide.

"Look at these eels, they're *huge!* Outrageous things. There's shrimp everywhere."

"They're in a perpetual feeding frenzy."

"It's almost like there's a fight for space."

The thumb-sized shrimp, which behaved more like bees than crustaceans, crowded around a stack of sulfide, completely hiding it by their frenzied activity. Dots of bioluminescence on the crustaceans flashed like hundreds of penlights in the darkness. For long moments, the scientists watched, speechless. Finally, Edmond said quietly, "Far out." And Humphris said: "This is amazing."

Edmond likened the hordes of frenzied shrimp to maggots on rotten meat. Then he saw an eelpout, likened to a piece of intestine by a geologist when first spotted in 1979 at the first black smoker vent. "Ugh," Edmond said, "there's these gross amazing fish here, not like the other ones. God, they're revolting bloody animals."

"Oh, it's not ugly."

"It *is*. It's horrible."

Although many attempts were made, no fish were captured at the Atlantic vents.

"These snaky fish were all over the place," Jeff Karson said. "They're one of the weirdest things there. They like to live in the crevasses created by the piling up of collapsed chimneys. We think they're blind, but we think they react to light because sometimes they'd jump when our strobe went on."

On other dives the scientists reported chimneys that flared at the top. "The chimneys look like a frying pan on top, and hot shimmering water is coming up out of this frying pan, and these shrimp are lying in the frying pan," Karson said. "They look like they're getting fried in this vigorous, swirling water. We also saw beehive things of sulfide which may have grown from these frying pans. They're more than a meter across and two meters high, and they have this circular ribbing just like a wasp or bees' nest. They look like they're steaming. Some are covered entirely with shrimp."

The crab and mussels, and perhaps the snail, are new species. A six-sided creature, which the scientists called the "Chinese checkerboard animal," is still a mystery. No one even knows what kind of animal it is, although biologists suspect that it is ancient. The divers were unable to get one of the mushroom-like things to the surface; but biologists confirmed that they were anemones.

The shrimp represent two new species; one, called *Rimicaris exoculata* (Latin for no eyes), has a small reflective patch on its back, which is barely distinguishable after being in formaldehyde. When biologist Cindy Lee Van Dover, who was working on her Ph.D. at WHOI, dissected the shrimp, she found their guts "packed solid" with copper, iron and zinc sulfide particles and not a trace of animal remains, indicating that the shrimp fed on the bacteria covering the chimneys. But how could they get so close to such intense heat?

Van Dover discovered that the patch on the shrimps' back was not a simple coloration like a butterfly's spot, but a substantial organ connected by a large nerve to the brain. She wondered if it was a kind of thermal-sensing eye. Lots of animals in the deep sea are blind; they don't need eyes to see in eternal darkness. *R. exoculata* didn't even have an eyestalk, no vestige whatsoever. At hot vents, wouldn't it be useful to sense when they got too close to heat, where the distance between shrimp scampi and life was a quarter of an inch?

Van Dover did not have the training or equipment to follow her curiosity alone. She walked down Water Street to the Marine Biological Laboratory in Woods Hole to see Ete Szuts, a sensory physiologist. Szuts agreed to analyze the patch for rhodopsin, the light-sensitive pigment in all retinas.

"Ete's technician did the assay for rhodopsin," Van Dover said. "And Ete didn't believe the positive results. They did it two more times, just to be sure. This "eye" was clearly weird. There was no lens so it could not form an image, but it had a lot of pigment, and it was more sensitive than a regular eye."

Next she turned to Steve Chamberlin at the Institute for Sensory Research at Syracuse University, who examined the patch's anatomy and concluded that theoretically the shrimp should be able to "see" thermal light invisible to the human eye. Did the chimneys in the Atlantic glow with heat?

"I wanted to go to the Atlantic in *Alvin* and turn out the lights but I had no idea what kind of detector to use," Van Dover said.

"This was so far out of my field. Ete started talking about rhabdoms and I said, 'What's a rhabdom?' All I knew is where I wanted to get to and I needed help getting there."

Light in the deep sea? How preposterous.

31

You'll never see much in Alvin. *Manned submersibles are doomed.*
   Bob Ballard, *Cape Cod Times,* 1984

When Bob Ballard returned to Cape Cod in 1980 from sabbatical in California, the Woods Hole Oceanographic Institution awarded him tenure. In addition to job security from a lab that liked to say only world-class scientists at the top of their field, that is, the best in the world, get tenure, it was also a collective message of approval from his peers. Ballard could continue his career in research at Woods Hole. But he could not let go his idea to find the *Titanic,* and he announced that he was going to go down the grand staircase of the long lost luxury liner.

Which sounded loony, foremost because nobody knew where the *Titanic* was. And obviously, Ballard could not go in *Alvin.* Penetrating a wreck, no matter how big, would be suicide for humans in a submersible. No, Ballard was not really going down those once magnificent steps. He planned to use tethered cameras which would be his "eyes" inside the ship. He said there was no need for people to be under water; they could stay aboard ship watching banks of TV monitors which showed exactly what the remote explorers saw. This "telepresence" would simulate being down in a craft like *Alvin.* Only better. Vision and time under water would be unlimited. Ballard said it was far preferable to crouching on his hands and knees inside *Alvin* and spending fully half of an eight-hour dive traveling to and from the bottom.

There were already many versions of tethered underwater vehicles; among the oldest were *Mizar*'s sled, Deep-Tow and ANGUS. They consisted of cameras, lights, and other gear on a steel frame towed from a ship. The most sophisticated towed sleds could transmit live data and images. The new small tethered versions incorporated robotics and computers and could be steered remotely. These packages of cameras and lights, used primarily by the offshore oil industry, were called ROVs, or remotely operated vehicles, and they were still in their infancy.

However Ballard was going to find RMS *Titanic* and get down any submerged stairway, he needed money. Federal research dollars were not available just to foot Ballard's dream, which had become an obsession. Looking for the *Titanic* would involve no scientific research. But it certainly would involve engineering. And the Navy was keen on telepresence for such purposes as investigating sunken military hardware. Ballard went for funding to the navy which in 1982 gave him an ROV it had been experimenting with and $2.8 million to assemble a team of engineers to improve it and build better remote tools.

First Ballard's group built Argo. Like the tethered 35 mm camera sled ANGUS, Argo was a large frame of heavy-duty steel piping designed to be towed from a ship. Argo carried video cameras, lights, and sonar, and computer software to continuously transmit black-and-white images live, in real time, to banks of monitors aboard ship. Ballard planned to use the *Titanic* as the target to test and prove the worth of Argo. But first he had to find the sunken luxury liner. He turned to France for help.

A team of French engineers, led by Jean-Louis Michel, agreed to collaborate in this hunt of hunts. They had the coordinates tapped out as the *Titanic* sank about 400 miles southeast of Newfoundland. But a fix in those days of no satellites, not even long-range radio-assisted navigation, was not what it is today. The *Titanic* crew used dead reckoning and celestial navigation—the latter undertaken between sunrise and sunset because it is virtually impossible to get an accurate sight after twilight when the horizon is dark. Furthermore, the *Titanic* was near two massive ocean currents and had certainly drifted—to the south with the Labrador Current and to the east with the Gulf Stream. The ship's position transmitted that terrible night was probably off by several miles—a lot of water.

Based on much homework and intuition, Ballard and the French decided the most likely place for the *Titanic* was within a 150-square-mile area. They planned to locate it with France's SAR, a state-of-the-art sonar system, and then photograph it with Argo.

The summer of 1985, Ballard joined French engineers on the research vessel *Le Suroit*, which towed SAR. But in twenty-two days at sea, the sonar found no trace of *Titanic* and *Le Suroit*'s time was up; it had to go home with SAR. SAR had surveyed 80 percent of the target area. The remainder would have to be covered with the new Argo carried on WHOI's *Knorr,* and within the twelve days on site allotted to the ship.

Ballard and the three Frenchmen who joined the *Knorr* cruise decided to follow up two clues: suspicious depressions SAR had detected and the vague image of a large curved object which Jack Grimm insisted was the *Titanic*'s propeller, captured four years before by camera sleds from Lamont and Scripps. But the so-called prop was only a rock and the craters were left from bigger rocks.

Tedium reigned on *Knorr*. Teams of seven took turns every four hours round the clock in the darkened control van to watch the TV screens linked to Argo. After six days of watching, even Ballard began to doubt. Had they miscalculated the probable area? Should they have used the course data from the *Californian* instead of the *Carpathia*, ships in the area when the *Titanic* steamed into a massive iceberg the clear night of April 14, 1912?

Ballard went to his cabin August 31 shortly before midnight when a fresh crew took over in the control van. He climbed into his bunk with *Yeager*, the biography of Chuck Yeager who had the "right stuff" to break the sound barrier.

In the control van, technicians joked about how they were going to stay awake through another boring shift of popcorn, caffeine and mud, mud, mud in fuzzy black-and-white video.

At 11:49 P.M., a dark cloud appeared in a corner of the monitors.

"Here's somethin'," a technician declared.

The cloud grew bigger.

"It's comin' in."

"Wreckage!"

"Bingo!"

The seasoned lookers knew it was no cloud, but it was much too hazy to be identified. It could have been anything, even a boulder.

But soon, other dark clouds appeared, followed by distinct pieces of something. Then a large piece of twisted steel.

"What is it?"

"I don't know."

"What *is* it?"

"I don't know but it's man-made."

"It ain't no geology."

"Some more stuff comin'."

It was four minutes after midnight. A huge circular image filled the screens.

"Look at it."

"What the hell..."

"Oooo!"

"God, it looks like..."

"The boiler!"

"The boiler!"

"The boi*ler!*"

One refrain was shouted in a French accent. Jean-Louis Michel just happened to have before him one of the many books on the *Titanic* scattered around the control van, and he just happened to have that book opened to a page of a photograph of the *Titanic*'s boilers.

The control van filled with people and shouts. *Knorr*'s cook, John Bartolomei, volunteered to fetch Ballard. Blue jumpsuit over pajamas, Ballard scooted down to the control van and saw for himself. They had done it, they had found the unfindable and unsinkable ship that took 1522 people to their graves because there weren't enough lifeboats to save them. It sat upright in 12,500 feet of water, an Ozymandian monument to hubris.

According to several witnesses on that starry, moonless night in 1912, the *Titanic* broke up at the surface. The underwater pictures from Argo and ANGUS confirmed it. The ship had been wrenched into five sections. The forward section up to the second smokestack had dived at a 45 degree angle. No trace of the midsection with the third and fourth funnels remained. The stern lay more than a mile behind the bow and pointed in the opposite direction; it had pivoted some 180 degrees before smashing into the mud.

Full of improbable victory, *Knorr* returned home to a hero's welcome, and the French engineers, with painful smiles for the hordes of photographers waiting at the Woods Hole dock, seethed with indignity. They thought the Americans, that is, Ballard, were getting all the credit.

Ballard, the French and Jack Grimm determined to return the following summer, this time with a submersible and an ROV to penetrate the grand wreck. The French could not raise the necessary funds. Grimm sent a proposal to the Alvin Review Committee to take *Alvin;* but the committee rejected Grimm's request because there was no science involved.

Next came Ballard's proposal, which didn't mention the *Titanic;* it requested using *Alvin* to test the navy's old ROV, which his group had improved on, at sites the navy classified. His proposal was approved. But when word via the grapevine reached the committee that Ballard really intended to go to the *Titanic* with *Alvin,* "all hell broke loose," as a member of the *Alvin* review panel put it. The committee rejected Ballard's proposal for the same reason it turned down Grimm's—there was no science involved and *Alvin* was not essential to test an ROV. But the committee and peer review didn't matter. They were overruled by the navy which was funding Ballard. Other scientists on the dive schedule were bumped.

Despite the improvements, the navy's old ROV was unreliable, according to Chris Von Alt, one of Ballard's top engineers. In 1985 in the Gulf of California, Ballard's team hooked the ROV's tether into one of *Alvin*'s penetrators, and with the small remote camera package in *Alvin*'s basket, dived. The test run proved that it was possible to operate an ROV from inside *Alvin,* but the navy's ROV was plagued with leaks and the operator, Martin Bowin, thirty-one, was nearly electrocuted. "A couple times I felt an electric tingle run right through me, especially if I leaned against the wet hull," said Bowen, who had piloted ROVs for ten years before Ballard hired him. "When the ROV's motors flooded, I couldn't reach the switch quickly enough to shut off the power. You could have a meltdown before you got to that switch."

In January 1986, six months before the *Alvin* dives to the *Titanic,* Von Alt convinced Ballard to let him and the other engineers in his shop build a better ROV, using some parts from the old version.

The new ROV, called Jason Junior or JJ, was a squarish blue shell of syntactic foam with four brushless motors, the same kind of thrusters *Alvin* used, attached to a bright yellow tether of electrical wires. A titanium cylinder with a glass dome held a tilting color video camera, 35 mm camera and flash.

A week before the scheduled departure to the *Titanic,* the engineers connected JJ's tether to *Alvin*'s penetrators for the first time, and JJ, sitting in the basket aboard ship, worked fine. Then *Alvin* dived with JJ off the WHOI dock. The ROV didn't work at all. The morning of July 9, 1986, sailing day, they tried once more off the dock. The engineers stared hopefully into the water as *Alvin,* carrying JJ and Von Alt, disappeared. Nearby, Ballard told reporters about his plans for going down the grand staircase. Finally, *Alvin*'s sail broke the surface. Ballard bid the reporters adieu and approached Von Alt.

"Can we go?" Ballard asked.

"Yes," Von Alt said.

It was raining when the *AII* crew threw the lines and *Alvin*'s mother ship left the WHOI dock for the *Titanic* burial grounds. Tensions ran high. JJ had worked only once from *Alvin* and had never worked, had never been to, deep water. And there was more than healthy competition between the Alvin Group and Ballard's team. When Ballard christened his shop the Deep Submergence Laboratory in 1982, the Deep Submergence Group changed its name to Submersible Engineering and Operations. The *Alvin* engineers minded their own business at the Smith Building on the waterfront while Ballard's team worked a few blocks away in an aluminum-sided building and trailers. Ballard continued to use *Alvin,* but he had a new dream, he was following a different star. And sometimes that had been hard for the *Alvin* team to take. It was bad enough hearing Ballard say he didn't need *Alvin* any more because he was developing something better, but when he said *Alvin* was no good, he went too far. The Alvin Group had collected newspaper clippings of his remarks; in one, from a July 11, 1984, issue of the *Cape Cod Times,* Ballard said:

You'll never see much in *Alvin.* Manned submersibles are doomed.

and

Once Argo-Jason comes on line, in a few months at sea it will eclipse all the bottom time *Alvin* has had in twenty years.

    The sub and *AII* crews were so angry they tried to embarrass Ballard in front of some navy brass that November by taping up copies of the newspaper paragraph about submersibles being doomed. They were everywhere—in labs, heads, cabins, on mirrors, bunks, toilets and closets. When Ballard boarded and saw his own words in huge print, he angrily ripped off the papers, managing to get rid of the most obvious ones by the time his guest, Assistant Secretary of the Navy Melvyn R. Paisley, arrived. *AII*'s chief steward John Lobo quietly made more blown-up reproductions, which he used as placemats for lunch. Ballard also had missed the six signs taped to the underside of the top bunk in Paisley's cabin.

    Now, as the *AII* steamed through the summer drizzle to rendezvous with Ballard's obsession, the tables were turned, and the *AII* and *Alvin* crews could not resist lobbing a big *touché* at Bob Ballard, the man who had gone out of his way to save *Alvin* from disuse and financial ruin fifteen years before, the man who said he didn't need *Alvin,* who said remote telepresence was better—this man was going down in *Alvin* so he could see the *Titanic* with his own eyeballs.

    Lobo baked a large cake decorated in colorful icing that spelled out "Manned Submersibles Are Doomed." Ballard was a good sport. "OK, OK," he said. "I'll eat my words." And he did.

    In three and a half days the *AII* reached the *Titanic* site. Would they find human bones or safes full of jewels from the first-class passengers? WHOI insisted that all divers would obey the recent bill passed by the United States House of Representatives, which forbade the collection of any items from the wreck. Would there be any wood left on the ship? Would *Alvin* get stuck in the wreckage? There was no question about the danger of *Alvin*'s mission. JJ's operator, Martin Bowen, had run ROVs in underwater wrecks before, but from a barge or on land and in only about 350 feet of water.

Another concern was the potential for a sympathetic implosion. JJ, designed to operate at 20,000 feet, was pressure-tested to insure that no components would implode. The ROV was expendable, that was one of its major advantages, that it could go where no human need risk life to visit. If JJ became entangled in wreckage, *Alvin* was equipped with two means for extricating itself. A hydraulic cutter mounted on *Alvin*'s basket could sever JJ's tether. If that didn't work, the entire basket could be dropped.

Ballard descended with the most experienced pilots, Dudley Foster and Ralph Hollis, to get the lay of the land and assess the dangers. The Alvin Group manager, Barrie Walden, had insisted that if on this first dive either pilot decided the area was too dangerous, there would be no more dives, period.

It was not an auspicious dive. During the two-hour-and-forty-minute plunge by gravity, *Alvin* developed problems that worsened at bottom. The sonar didn't work and a battery tank was leaking. Dependent totally on surface control to steer them right, Ballard, Hollis, and Foster looked for an hour, straining through the twilight snowfall to see an object as tall as an eleven-story building. The pilots debated whether to abort the dive. The sonar kept sweeping and squawking but picked up nothing, and saltwater continued to seep into the battery tank, forcing out more pressure-compensating oil. A short-circuited battery could cause a fire or a hydrogen explosion. But they had come this far. And it was still a little leak.

For ten more minutes, *Alvin* flew through empty ocean. The leak detector dial approached the second of three alert levels. The pilots again discussed returning to the surface. The leaching water was still probably in quarts, not gallons. But there was only one ground detector; while it was on, it could not warn of another ground in another battery tank or elsewhere. The pilots decided to head for the surface. Then suddenly, an hour and twenty-two minutes after touchdown, they faced a towering black wall rooted in the seafloor.

The first direct glimpse of the *Titanic* was brief, perhaps two minutes' worth. Hollis quickly backed *Alvin* away from the swirling sediment, and reasoning that it was unwise to wait for the water to clear, headed for the surface.

For the second dive, Martin Bowen joined Ballard and Hollis. JJ was mounted in *Alvin*'s basket and a duplicate set of the lights and cameras JJ carried were mounted on one of *Alvin*'s arms.

Again, the bright orange sonar scope blinked erratically, but surface control quickly directed them to the *Titanic*. *Alvin* flew along the starboard bow where everyone expected to find a long, deep cut from the iceberg. If there was a gash, it was hidden. The bow was deeply buried in the sediment.

The massive, hulking wreckage was draped in giant globs and slender icicles of rust. "Bleeding rust," Ballard said. "That's what it's doing, just bleeding rust. Well, Ralph, you want to do the other bow? You think that'll be better? *Bleeding* rust, the whole ship is bleeding rust, it's just incredible."

Hollis steered *Alvin* over and down along the bow, paying attention to the words of caution from his passengers, who could see what he could not see. Ballard knew the target well. His sense of direction was almost uncanny, especially when clouds of blinding sediment and rust swirled at the view ports.

"OK, you need to get higher up, Ralph," Ballard said.

"All right."

"Now go forward. You need to get higher up. . . . Overhang. Pivot right."

"We're at 3781 [meters or 12,405 feet] along the bow and there's a large buckle, probably from impact," Ralph told the surface.

"It's buckled, folded right in," said Ballard, looking at the puckered steel hull, thinking that perhaps there was no 300-foot gash, that the impact of hitting the iceberg was powerful enough to split the seams of steel.

The current was almost as fast as *Alvin*. As long as *Alvin*'s thrusters worked, the sub was in control, but if the power failed, the current could easily sweep *Alvin* into the wreckage.

The ship's railing was draped in "rivers of rust" and "rusticles," as Ballard called them. The toppled mast, with the crow's nest still attached, lay across the deck with cables and lines that stretched crosswise.

The deck was gone. What was thought from photographs to be wooden planks turned out to be caulking and the calcareous tubes of wood borers that had ravaged the wooden parts.

Hollis steered *Alvin* further over the forward deck, past a large steel chair, another huge capstan, broken railing, and two more capstans thick with brown ooze by the wheelhouse that no longer existed. All that remained of the ship's wheel was the bronze pedestal. Finally he saw an unobstructed patch of steel subdecking to alight on. At least it looked safe. Besides the voracious wood borers, rust had been eating the ship for seventy-five years. Bowen likened *Alvin*'s landing to the first human footstep on the moon. If the deck gave way under *Alvin*'s weight, they would fall through many more decks and there they would stay for good. But chances were good that they would not fall through. *Alvin*'s primary ballast system was precise enough to vary the sub's weight by ounces and pounds. Hovering in the ocean, the sub weighed only a couple of pounds.

With *Alvin*'s nose into the half-knot current, Hollis carefully adjusted ballast, making the sub just heavy enough to inch *Alvin* down, down, down. The three men listened for the sound of a creaking deck. No creaks.

"OK, here's what I want you to do, Ralph," Ballard said.

"You want to do the impossible," Hollis interrupted.

"No, I want to... lift up and go back to the bow and come in again and land as far forward as you can. Swing left, out, come in again. We can only inch this way, not that way."

"I'll lift up. I'm afraid if..."

"If you pivot right, you can't hit that boom."

"You sure you want to do this?"

"Yes."

Hollis raised *Alvin* and a cloud of particles.

"The entire ship is rust," Ballard said. "And the wood borers have gotten to it. You want to sway your stern to the port side."

"I'm firing stills, Bob," Bowen said.

"You see something on the TV?"

"Yeah, I see railing."

"We're going to be clear of the boom, no sweat," Ballard said. "No, no, don't go back this way, Ralph, go forward. Now you can go back."

"I can see the bottom and the ship," Bowen said.

"Ralph, stay this side of amidship. You got to move to the right, right. OK. You should be clear now."

"I see clear to the bottom," Bowen said.

"OK, you should be able to go down to the bottom, Ralph."

"I see something there," Hollis said.

"That's the boom," Ballard said. "Pivot around, Ralph. I want you to sit down in front of it, but not, you know, too close."

"I just saw something white pass in front..." Bowen said.

"It'll go away," Hollis said. "Patience is a virtue."

"Off to the right, Ralph," Ballard continued, as the sub dropped to the seafloor, raising another cloud. "What you want to do now is go straight up. That's good."

"There's a porthole," Hollis said.

"Straight up. There's a porthole. Stay close. Cable. Back off. OK, straight up."

"Cable's on the starboard side of the bow," Hollis said.

"Keep rising. Another porthole. Keep rising. What's that on the right, a big hole?"

"I don't know."

The Submarine Blues

"Cable coming down, go forward and come right. OK, you're clear."

The bronze tops of the bollards and capstans on the tilted deck had little corrosion; some shined like new.

"There's cables over on top of the anchor. You can sit down, no, let's not sit down yet. Move forward and off to the left of the bridge. We want to clear those cables."

Alvin flew on, past rows of intact portholes, over the immense hole left by the forward smokestack, and alit at the opening to the grand staircase.

Bowen pushed a tiny button on the control panel in his lap to open the gate that kept JJ from sliding off *Alvin*'s basket. But JJ's motors were leaking.

After two and a half hours on the bottom, surface control called *Alvin* home because of mounting seas.

In the morning, *Alvin* approached the *Titanic*'s bow from behind, flying over the artifacts that had spilled from the ship's belly when it ruptured. A man's shoe, china cups and saucers, chamber pots, a chair. And the sub climbed up the side.

"Above the bow," Ballard directed Dudley Foster. "You're clear. The highest part of the ship is below us now. Go forward. The staircase is dead ahead. You just missed it. Spin left, left. You just missed it. . . . Go right, I want to see the compass, the compass should be over here. The ship is now sloping down away from us. The compass used to be right here. . . . look down, that's it, that's it. You have to go north, turn around and come around. There's a landing pad dead ahead."

Foster slowly lowered *Alvin* by the glass ceiling of an elevator. *Alvin*'s shadow grew larger and larger on the deck.

"*Atlantis II*, this is *Alvin*. We have landed at the entrance to the staircase and we'll be deploying JJ down staircase," Ballard said, and began to review with Bowen the possible entanglement areas for JJ. They had no idea what was left inside the ship. The gaping holes in all the photographs showed only as black circles.

Foster was still adjusting ballast. "All right, Dudley, we heavy enough? Huh? OK, Martin, it's all yours."

"Garage door is open," Bowen said.

"Brave new world, Martin, you ready?" Ballard asked.

"Yup."

This time, JJ obeyed. The blue box with two headlights on a bright yellow line disappeared beneath them.

"*Alvin*, this is the *Atlantis II*, Jason is inside the staircase, over," Ballard said calmly, his excitement revealed only by his inverted words.

JJ flew out on eighty feet of yellow tether like a big slow

hummingbird, down three decks and into the room off the grand stairway. There was nothing left of the staircase. Every time JJ's thrusters touched something, an orange cloud billowed.

"We really couldn't get too far in there without getting tangled and kicking up all the dust," Bowen said. "There was only about three feet of clear sailing between the two feet of debris on the floor of each deck and the chandeliers on the ceilings. But now Ballard had fulfilled his promise to the navy to penetrate a wreck. The rest of the dives were gravy. I had never seen him so jubilant as on the ride up from that dive."

Bowen pulled the joystick to make JJ ascend and twice pressed the trigger on the handle to reel in tether, turning JJ to videotape *Alvin* and the faces of the men inside against the portholes. It was weird. They were actually watching themselves. Ballard said it was like a "close encounter."

On the next dive, Ballard and Bowen returned with Will Sellers to the wheelhouse area.

JJ inched out of the basket and flew over the brass lantern and telephone in the crow's nest, landing atop a shiny bronze capstan that said it was made in Glasgow. It flew down to the forecastle deck and past the lifeboat chocks, recorded its reflection in the windows at the crew quarters, peered into the first-class gymnasium, and, back at the grand staircase entrance, photographed brass and crystal chandeliers.

"That's what was hanging my tether up," Bowen said.

"That was the guy we didn't like," Ballard said.

"That guy's a killer," Bowen said.

"Come over to this side, it's cleaner. Stay away from that. No, I don't like that, too much cable."

Orange clouds of rust filled the TV monitors.

"Shit," Bowen said softly.

"Get your optics up," Ballard instructed, meaning tilt JJ's cameras toward the debris at the ceiling.

The *AII* signaled in Morse code and Sellers responded with the letter "B," indicating that *Alvin* was on the bottom and OK.

Another blinding cloud engulfed JJ. "Pivot right," Ballard said. "Pivot right, believe me, go further right and clear."

He was right.

"Oh, look at that shit," Bowen said at the debris.

"That's the nasty stuff."

"Don't go over there. Pivot left, go down, stop, oh, wait a minute, look up."

"Is that red light flashing for a reason?" Sellers asked.

"It means the tape has come to an end," Bowen replied.

"How much further down can you go?" Sellers asked.

"Lots," Ballard said. "Okay, let's go home. I think we need to go to the bow and look for a name."

"Can I land close to the anchor?" Sellers asked.

"Oh, phhht! Ship'll be right in front of you."

"Wonderful," Sellers yawned loudly.

The sea had removed the letters from the bow that spelled RMS *Titanic*.

Strong currents thwarted the fifth dive, twice pushing *Alvin* out of control, and whisking JJ around as if it were a kite, Ballard said. But the current slacked for the sixth dive, and *Alvin* finally crossed the "wine field," about 2000 square feet of scattered artifacts, including many wine bottles, and mangled parts of the ship's midsection that were hardly recognizable, between the bow section and the stern. The field of debris offered haunting glimpses of the human tragedy. A white porcelain doll's head lay on its side. A perfectly intact coffee cup sat upright on a boiler. A toilet bowl and a green wine bottle, together. Many cups and plates, saucers, chamber pots, and lumps of coal. Metal bed frames, a shiny copper kettle, small space heaters and wood stoves, a bathtub almost completely covered in globs of rust, silver goblets, soap dishes, a mirror.

"It's actually like going to a museum," Ballard said. "There are just thousands and thousands of items all over the bottom. If we were ever going to see any human remains, we would have seen them in this area. The closest we saw was a shoe . . . no human remains at all, which was sort of a relief."

At one of several safes, *Alvin*'s pilot grabbed the knob, which like the safe's ornate seal and dial, shined a brilliant gold or bronze. As he turned, the whole safe lifted up; the entire back had rusted away.

If there had been jewels, would the divers have kept their promise not to take them? Ballard swore he would take nothing, no matter how precious or trivial, from the graveyard. When a small length of cable came to the surface on ANGUS, it was immediately thrown overboard.

Only one foray was made to the stern which, Ballard said, was "a little dicey." He reported:

> It's just a tremendous twisted pile of wreckage that is very difficult to maneuver in because it's so irregular and overhangs so much. The stern is just sitting upright on the bottom. The front portion has buried itself deep in the sediments. . . . There is no paint on the exposed part of the hull of the ship anywhere. The propellers are buried but the rudder is quite visible.
>
> We rose up and sat on the stern. At the very end of the stern,

we placed a Titanic Historical Society plaque commemorating those who perished on the ship. We thought the stern was an appropriate place to place it since that is where most of the people died. It was the last part of the ship to go under.

A second plaque from the Explorers' Club had been left at the bow.

JJ did not work for the dive to the stern. It got snagged on wreckage at least twice, and on the twelfth and last dive involved *Alvin*. "We were trying to go on the longest excursion to date from *Alvin*," Bowen said. "On this dive Bob said, 'Go for it. If we lose JJ, this'll be the dive.' We had probably 140 feet of tether out at the bow. The idea was to fly JJ off the boat deck over the port rail and into the promenade deck. We did all that, but while JJ was inside the tether got caught."

When Bowen pressed the joystick trigger twice, instead of tether pulling in, *Alvin* began to slide across the deck. JJ was caught on something and because the ROV was attached to *Alvin*, the sub was caught. As soon as Bowen realized that *Alvin* was moving, he quickly released the trigger to stop them from sliding into what would have been their tomb. *Alvin* had moved only about a foot, Bowen said. In the tiny Sony Watchman on the control panel in his lap, he saw JJ's tether pinched scissor-style between the railing and deck. He wasn't ready to revert to the drastic measures of severing JJ's tether or dropping the basket. Bowen led JJ in circles, slowly uncoiling the tether, and then he carefully moved JJ aft, finally freeing the yellow line.

JJ had worked on only four dives, still considered by its engineers a good score. Most of the trouble had been with leaks in the motors and the tether.

The *Suroit*, *Knorr* and *Atlantis II* expeditions were engineering successes, but it was the *Titanic* lore that captured the world's attention and imagination. WHOI's small public affairs office was overwhelmed. The PA manager, Shelley Lauzon, was used to strange requests regarding *Alvin*. Years before, a cigarette manufacturer wanted to film an ad with a woman diving off *Alvin* and surfacing while smoking a cigarette. A watchmaking company wanted to dive to a giant watch on the seafloor. About a dozen people called every year to ask if they could rent a dive for their wedding anniversary. With about a thousand *Titanic*-related calls a day, WHOI's already strained telephone system went berserk, and one day, when more than fifty people were automatically queued on a line, the PA phone system broke down. In 1989, WHOI got a new and bigger telephone system.

In the summer of 1987, the French science agency leased its new minisub *Nautile,* launched in 1985 and rated for 20,000 feet, to United States and Canadian investors who dived to the *Titanic* and collected many artifacts.

32

The ocean is still the largest and least understood environment on the planet. Minisubs like *Alvin* are myopic and inefficient for certain jobs, such as obtaining measurements over large areas and for long time periods, records critical to understanding the global ocean. But this nearsightedness has led to profound changes in our thinking about the birthing of Planet Earth. As it did for biology and chemistry, deep diving has challenged some of the basic tenets of geology and geophysics, repeatedly teaching the same lesson—that earth's largest environment is far richer and more complex than previously imagined. And the surprises keep coming. As Duke geologist Jeff Karson says, "We know so little about what's going on at the seafloor, you practically have to be a moron not to learn something new every time you dive."

"Every single dive has a chance of something serendipitous happening, and it does at least once on every expedition," geophysicist Tanya Atwater said from her laboratory at the University of California in Santa Barbara. "And fifteen years ago we thought we were done, we thought we had figured out seafloor spreading! There's a wonderful richness coming out of how the whole ridge works. There's a profound thing that happens to every single person who gets in that sphere even if they don't get any samples—they come back a changed scientist. Some jobs cannot be done except with *Alvin:* to explore in fine detail and sample, and to get your eyeball and your gut calibrated. When you look at the seismic record of a twenty-meter scarp, it is barely visible. When you drive up to it in a submarine, it is a sixty-foot-tall vertical wall—and that is a shocking experience. Every scientist who works on ocean-floor data needs to go down in *Alvin* at least a few times."

"*Alvin* has been essential in allowing us to get to the nitty gritty," Lamont geologist Dan Fornari said. "It has been essential for many of our models of how the ocean floor has developed. *Alvin* has allowed us to really quantify our understanding. We couldn't have have come this far without it."

Bill Ryan, another Lamont geologist, recalled the almost elegant poignancy of his first *Alvin* dive. "It was rather spectacular," he said. "The dive was off Cat Island in the Bahamas. We sampled rocks that were 50 million to 120 million years old. I was impressed that in one dive in *Alvin* in the right place, you could cover such a broad range of time history in one sample basket, get hold of pieces of earth laid down in succession, ride up a cliff and with your eyes see 50 million years go by. It would cost maybe $40 million for a commercial well to get that far down."

The earth scientists have seen the contradiction of belief—outcrops and escarpments eroded by currents and animals. They were amazed on *Alvin*'s dives in the late 1960s to see sediment on steep cliffs. They knew the mud held plant and animal remains, but they didn't know that the mud dwellers, virtually ignored by science, could expel enough gooey secretions to hold so much sediment together and keep it on such steep slopes. And they were surprised to learn that animals played a major role in shaping canyons. "We found that one of the biggest processes that erode canyons is biological," said geologist Bob Embley of the National Oceanic and Atmospheric Administration. "They actually cause the movement of materials downslope. Even the corals down there, they get up to ten feet long, they move material down the canyon. Crabs and other organisms dig into the seafloor, and they're able to cause material to break off and move down."

In a most unusual turn of technique, geologists and chemists are using biology to find the targets of their studies. Off Oregon and at other subduction zones off Japan, the Bahamas, and Peru, jumbo clams and tube worms lead to the seafloor source of chemically altered fluid and underground channels of warmth.

One of the biggest lessons the in-person probing and measurements along midocean ridges have taught is that the earth under water has many similarities to that on land—the stretching and faulting relationships on the continents exist on the seafloor. And yet it is different; the seafloor is extraordinarily diverse. "We found as much variability in one dive in FAMOUS as we got in dredging the whole Atlantic," Ryan said.

The seemingly neat boundaries separating tectonic plates are really a complex patchwork of overlapping plates, plate edges that migrate back and forth along ridges, and plates that separate into smaller cells.

Geophysicists diving in 1985 more than six hundred miles west of the Galapagos vents saw for the first time the birth of a new section of a seafloor spreading center. "It was the most tectonically disrupted piece of seafloor we have ever seen," Dick Hey, a geophysicist from the University of Hawaii, said. "We had studied it remotely with just about every kind of instrument a geophysicist could get his hands on. But it wasn't until we could get down with *Alvin* that we could really understand all our previous data sets. There wasn't a single intact pillow. It was like the whole area had been through a ringer. Every single pillow had been broken up. A crack is breaking through, we call it a propagating rift, and behind it a new seafloor spreading center is being born. We found out exactly how this happens. It starts with fissuring and faulting. We could actually see the birth of a seafloor spreading center."

Atwater and Ken Macdonald joined Hey's expedition with their pioneering geophysical tools first used from *Alvin* in 1978 at the Mid-Atlantic Ridge. The data from their magnetic gradiometer indicated that the disrupted portion of seafloor had actually rotated as it was shifted from one plate to another. Precisely what triggers earthquakes to shift a section of earth from one plate to another is not known. Geophysicists have calculated that every million years, a propagating rift lengthens by about thirty miles.

The midocean ridge is not volcanically active everywhere, as geologists once thought. Underground magma chambers are probably not large and long-lived, but small and intermittent. And still mysterious. Magma oozes from not only the plate boundaries at spreading centers, but also from seamounts far away from ridges. And the superhot molten earth is not the same everywhere; indeed geologists have discovered magma with compositions never seen before.

Underwater peaks far away from the midocean ridges seem to be a key in unlocking the mystery of magma variations, indicating that these seamounts are an intimate, and as yet not understood, link in the global lava plumbing. On dives in February 1987 to Loihi, an active volcano about seventeen miles off the big island of Hawaii, scientists found no central crater but two calderas on Loihi's south flank, leading them to believe that the conduits of magma had moved southward. Judging from the lack of sediment on the basalt, they determined that Loihi last erupted a few thousand years or so ago. The seamount, which rises a half mile, is an island in the making, and until *Alvin* dropped in, it had never before been visited.

The scientists expected to find vents at the craters, because surface-towed instruments had detected methane and primordial helium above Loihi, but by the third dive, no vents had been seen. On that dive, 1796, *Alvin* approached the edge of a cliff at 5000 feet.

"Oh, all these big fish, they're all around, weird fish," Scripps geochemist Harmon Craig said. "They're very long and slim.... There's a lot of sediment stirring up all over the place here. We're surrounded by clouds. Anybody see anything that could be hot water?"

"Just white sediment," replied Cambridge University geologist Dave Hilton.

"Look at this white sediment."

"I can't see a thing."

Pilot Dudley Foster worked pincers around a clump of pale earth but another cloud formed and the rock disintegrated in *Alvin*'s hand.

"Boy, you can't see much when you sit around here, the sediment gets so stirred up," Craig said. "Let's go up and see what we can find."

Alvin rose up the caldera wall.

"Boy, what a sheer wall this is," Craig said. "Lot of these fish here. Where the hell are the vents?"

Seeing nothing but the needlelike fish, they descended again and spotted small holes in little piles of earth on the sandy white floor of the caldera.

"There are a lot of little holes," Craig said. "It's like there are a million vents all over the place. It's so porous. Maybe it's like the land volcanoes. It doesn't have to concentrate into streams and make chimneys. There's got to be a better name than chimney for those little things."

Then Hilton saw basalt and Craig noticed the telltale coloring of sulfides. Globs of pale bacterial material floated around *Alvin*.

"We've got to be in a hydrothermal vent area," Craig said. "Don't you think, Dave? There's the stains, there's the chimneys ...there's the bacteria, and you've got basalts over there. Everything."

Alvin stopped at a so-called chimney, a craggy lump of what looked like hardened sand.

"Can you knock it over?" Craig asked Foster. "This one's pointing toward me and it's open, it's got an oval opening. Look, you have a hole, beautiful, and nothing's coming out. Nothing. It could be that these holes have a little bit coming out, but so faintly that we can't see them, do you think?"

The three men strained to see rising, shimmering fluid.

"No water at all, huh?"

"Nope."

"Jesus."

On the fifth dive, John Welhan of Memorial University in Newfoundland accompanied Craig back inside the caldera.

"White stuff," Craig reported from his view port. "Oh, it's all reddish now, loaded with particulates."

Welhan, from his side of the sphere, could hardly believe what he saw. "Do you see shimmering water going downhill?"

"What?"

"Shimmering water, it's flowing downhill. *Look* at that."

"I can't see," Craig said. "Well, let's go to the top. You do see shimmering water running downhill or is it that we're just running past it, John?"

"Oh, yeah, well..."

"I don't think it's running downhill, but it is shimmering, I agree with you."

Finding nothing interesting like shimmering water at the top of the caldera, they descended again.

"There's some more shimmering water going downhill," Welhan said.

"Really?"

"Yes!"

"Downslope of us?"

Again Craig saw little holes in small piles of earth. "Here's a little white mouth but there's nothing coming out of it," he said. "Oh... here's shimmering water, I got a lot of it right here. It's definitely flowing downhill, John. Well, I'm not so sure it is.... It *is* flowing downslope. Everywhere. It really is."

The 30° C (86° F) water was saturated with carbon dioxide, helium and iron. Every liter of water had seven liters of carbon-dioxide—the highest concentration ever found in seawater. The water tubes exploded at the surface like champagne bottles. It seemed to explain why the venting fluid ran downslope—it was so full of CO_2 that it was denser than the surrounding water. But chemists later determined that the water was actually less dense. Many scientists now believe the downhill shimmering was an optical illusion.

Craig wired his lab:

HAVE MAPPED A LARGE PART OF THE CALDERA AND DISCOVERED A SINGLE HYDROTHERMAL VENT SITE ON RIM OF SUMMIT CRATER. ONLY ONE ACTIVE SITE BUT THAT ONE IS TERRIFIC AND NOTHING LIKE WHAT WE EXPECTED. VENT WATER IS 30 DEGREES C ISSUING AS SHIMMERING FLOW OVER RED TERRACES WITH WHITE OUTLINES AROUND SMALL ORIFICES. THE WATER CONTAINS 7000 CCSTP [cubic centimeters of gas measured at standard temperature and pressure] OF CO2 PER LITRE, MORE THAN 140 TIMES TOTAL CO2 OF AMBIENT SEAWATER AND FULLY 4000 TIMES CONCENTRATION OF FREE CO2 IN SEAWATER. NO SULFIDES, NO H2S, NO FAUNA NEAR VENTS. LACK OF ECOSYSTEM EXCEPT POSSIBLE BACTERIA MAY BE DUE TO TOXICITY OF CO2 RICH WATERS. CO2 SOURCE IS HIGHLY VESICULATED BASALTS AT SITE DEPTH OF 930

M. ROCKS FIZZ WHEN BROUGHT TO SURFACE.... STUDIES CONTINUING IN TOTAL FASCINATION WITH INSIDES OF AN ACTIVE VOLCANO.

In addition to the long, thin fish and furry red carpets of bacteria, the divers saw another creature. Until it moved, pilot Skip Gleason thought it was a rock. "You gotta see this," Gleason said, moving away from the center view port for his passengers.

"We saw this very strange thing looking right back at us," Gary McMurtry, a geochemist from the University of Hawaii, said. "It wasn't afraid of the sub at all. It had huge scales, kind of looked like a frog, like it could crawl around. It was a kind of whitish purplish red thing with a huge mouth. You could easily put your fist in its mouth. It was so ugly. You know, it's funny, it had to be the world's ugliest fish, but it had the most beautiful blue eyes."

The divers brought up two twenty-five-pound Big Mouths, which Scripps biologist Dick Rosenblatt identified as a rare species of goose fish, a kind of angler or monk fish. Anglers ambush their prey, and they need big mouths. They don't bother chewing; they inhale dinner. Protruding from their head is a long, narrow appendage which ends in a tag, their reusable bait. Depending on the kind of angler, the tag may resemble a tiny shrimp or worm, or for deep-sea anglers like Big Mouth, a spot of bioluminescence. "They're not swimmers, they're walkers," Rosenblatt said. "Their pectoral fins are extra long and become like an arm with a wrist and elbow which they use to push themselves around on the bottom. This musculature is probably used when they're getting lunch." Rosenblatt is uncertain if the Loihi Big Mouths are a new species. They appear to most resemble a specimen which has resided at the British Museum since its capture in 1906 from the Indian Ocean.

Before leaving Loihi, the scientists signed their names in bright yellow paint on an aluminum plaque and left it on the floor of the caldera. On a return visit, they retrieved the plaque to measure the thickness of iron oxide precipitates, and, for posterity, replaced it.

The water chemistry around Loihi is still a puzzle. "We've been twice and returned with different water chemistry," McMurtry said. "That says the sampling wasn't representative or something happened in the six months between dives to cause the chemistry to change. It's very difficult to sample these waters even though the temperatures are relatively low, because the water starts to de-gas before we get it to the surface. Over Loihi there's a plume of tremendous quantities of helium-3 and methane. It's like the smog in LA. Exactly what are the processes that go on on these volcanoes? What kinds of rocks are being altered? At what temperatures? We don't know."

From Loihi, the *Atlantis II* brought *Alvin* for the first time to the western equatorial Pacific for five months of dives by six international teams of scientists, with port calls at Piti, Guam; Tanapag, Saipan; and Tokyo.

Alvin explored inside the Kusugas, volcanoes near the Mariana Islands, 1500 miles east of the Philippines. Thick deposits of white sulfur covered domes on top of the volcanoes, which emitted warm fluid and hosted white crabs, worms and bacterial globs.

Other seamounts by the Marianias intrigued geologists because rocks dredged from these volcanoes indicated that the solitary peaks, rising 9000 feet above the seafloor, are not built of erupted volcanic rock like the nearby islands but of the slick greenish rock serpentinite. The dredged rocks also contained the mineral aragonite which is made of calcium carbonate. On dives to Pacman, a seamount shaped like the video game creature, geologists found little of interest. But nearby on Conical, they discovered the source of the serpentinite; it had oozed like toothpaste from cracks in the seafloor and formed odd twisted green shapes, some the consistency of cream cheese. And rising from this emerald carpet were dozens of white stacks the divers described as "eerie, twisted tombstones." It was a cold-water vent, subsequently called the Graveyard. The three-foot-high chimneys got their coloring from the aragonite crystals and bacterial mats that covered them. Only when the pilots pushed *Alvin* against a chimney and broke the top did the divers see cold fluid issue forth.

Janet Haggerty, a geologist from the University of Tulsa, had predicted the presence of chimneys formed in cold water in the southeast Pacific. She also found something unexpected—an unknown mineral on the white stacks.

Nickel-sized limpets with bright red flesh were the sole inhabitants of the Graveyard. Biologists are looking for symbiotic bacteria in the hemoglobin-rich blood. "Some biologist was smart enough to leave formalin and instructions on the *AII*," Haggerty said. "But these things weren't well preserved."

The "Rambo" style of sampling, as the pilots called it, came in handy to determine if a lifeless chimney really was dead. But after finding the Graveyard, Haggerty asked her pilot not to ram any more chimneys. "What is this, a Save the Chimney dive?" he asked. She wanted to return and see what nature, not humans, had done.

Other expeditions to the southwest Pacific explored the deep trough or back arc basin formed by the tectonic plate carrying the

Marianas as it grinds beneath another plate. Over the millenia, the subduction also formed the 35,800-foot-deep Mariana Trench (the planet's deepest depression) and the volcanic island chain the Mariana Island Arc, which includes the Marianas, Saipan and Guam. Expectations of what *Alvin* would find in the trough varied. Some scientists anticipated finding vents because towed instruments had detected tiny temperature anomalies and a huge plume of methane. However, no primordial helium, a sure sign of hot vents, had been found, and no ventlike animals had been captured in the several hundred photographs taken before *Alvin*'s dives.

The Scripps geochemist Harmon Craig and Canadian geologist John Welhan made the first dive in the Mariana Trough April 7, 1987. According to the log for dive 1822, they climbed into the sub at 7:50 A.M. In five minutes, Ralph Hollis shut the hatch and the escort swimmers' small Avon boat (operated by an *AII* female sailor called the "Avon Lady") was lowered. Three minutes later the A-frame lifted *Alvin* and in four minutes had the sub in the water. On *Lulu*, launch took from twenty to thirty minutes.

For this dive *Alvin*'s left manipulator, the last original part of the sub, was taken off for good and replaced with a $100,000 prototype—a computer-controlled hydraulic manipulator made by the California firm Schilling Engineering. The pilots called the new arm "Son of Schilling"; the scientists dubbed it "Strangelove" after the spastic character in the movie of the same name. The night before, when the new arm was attached, it suddenly shot straight out, narrowly missing the Scripps machinist Helmut Kueker. Then it reached over and ripped the irreplaceable flag of the Explorers Club.

Craig had convinced the club to let him take the cherished banner, which had accompanied Sir Hubert Wilkins to the Arctic in 1931 and Auguste Piccard to the stratosphere in 1946. Craig and Kueker had just mounted it in *Alvin*'s basket, intending to photograph it for posterity in the deep sea, when Strangelove attacked. "That arm's a real killer," Craig said. "It nearly took my machinist's head off. Someday it's going to be just perfect, I know, but it had a bug in the control panel somewhere, and it went after anything that went near it. Like a cobra. We were horrified. We were trying to think of excuses to tell the Explorers Club. What could we say—a shark bit it?" *AII* sailor Big Ed Brodrick grabbed an old Mexican flag (green, white, and red) from the collection of foreign flags on the *AII*'s bridge to patch up the torn red, white, and blue banner.

Alvin carried another special item that depended on Strangelove's cooperation: a half-gallon Old Forester jug that held, in addition to bourbon, the cremated remains of Bud Burke. In the course of his career as a salesman for various marine instrumentation companies, Burke had befriended lots of people in the underwater business, including the early Alvin Group. A few weeks before Burke died of lung cancer, WHOI agreed to his request to let *Alvin* place his ashes in the deep sea.

1100 [*Alvin* to *AII*] Depth 3585 m... right on target.
1139 [*Alvin* to *AII*] Get good fix this position.
1208 [*AII* to *Alvin*] Still in same position?
   [*Alvin* to *AII*] Yes. Sampling.
1338 On top of ridge. Bud Burke deployed here.

Then *Alvin* entered a vent field where walls of clear water shimmered in a bloom of springtime hues. "It was just bursting with life," Craig said. "There were millions and millions of anemones, pink and white, just beautiful."

FROM: ATLANTIS II 07APRIL87
TO: SCRIPPS
ATTN: VALERIE CRAIG/ EV HERNANDEZ   URGENT

LOCATED AND SAMPLED TWO HYDROTHERMAL VENT FIELDS STARTING ONE HOUR AFTER OUR FIRST DIVE REACHED BOTTOM IN 3,600 METERS OF WATER. 'CHOCOLATE GARDENS' IN MASSIVE DARK PILLOW BASALTS, AND 'SNAIL PITS' IN BRECCIATED SILICA-STAINED BLACK BASALTS ARE TEEMING WITH INCREDIBLE CONCENTRATIONS OF VENT FAUNA. PRESENT ARE VERY LARGE ANEMONES WITH DENSITIES RANGING FROM ONE TO 100 PER SQUARE METER, HAIRY GASTROPODS WITH BI-MODAL HAIR LENGTHS, BOTH KINDS OF USUAL CRABS, LARGE MUSSELS WITH RED GILLS AND FULL OF H2S, NUMEROUS SMALL BARNACLES, SHRIMP, AND OLD FRIENDS POMPEII WORMS. NOT PRESENT ARE TUBE WORMS AND CLAMS. WE DO HAVE VERY LARGE VENT FISH UP TO FIVE FEET LONG WITH HEADS EIGHT INCHES VERTICAL. SWIM LIKE EELS AND HARD TO CATCH AT LEAST FIRST DAY. 'SNAIL PITS' LOOK LIKE WOKS COMPLETELY FULL OF WRITHING SNAILS, SHRIMP AND CRABS. TEMPERATURES SO FAR TO 23 DEG. C. ALVIN WORKING BEAUTIFULLY AT NEAR-MAXIMUM DEPTHS WITH IN-SUB REAL TIME NAVIGATION. ATLANTIS II RUNNING SUPERBLY AS USUAL. OUR FOUR-YEAR WAIT TO GET DOWN TO EARTH ON THIS BACK ARC TERRAIN HAS FINALLY PAID OFF WITH RICH RESULTS.
                                                                CRAIG

07 APRIL 87
VIA COMSAT
H. CRAIG

EXCITING MESSAGE RECEIVED. BRAVO... WHAT WILL YOU DO FOR AN ENCORE?

VALERIE

FROM: ATLANTIS II 10APRIL87
TO VAL: WHAT I WILL DO FOR AN ENCORE FOLLOWS:

FOURTH DIVE LOCATED MAGNIFICENT VENT AREA NAMED 'TOPLESS TOWERS' FOR VAST FIELDS OF LARGE CHIMNEYS AND HYDROTHERMAL MOUNDS. VENTS ARE OXYMORON TYPES DUBBED 'CLEAR SMOKERS' EMITTING TRANSPARENT WATERS RESEMBLING HOT GLYCERIN AT TEMPERATURES OF 268 DEG. C., FROM CHOKED ORIFICES OF SPHALERITE, CHALCOPYRITE, ETC. CRABS ASSEMBLE IN NEAT ROWS IN COLOSSAL AMPHITHEATER TO WATCH THE SHOW: LOCAL WOKS WITH HOT SHRIMPS, SNAILS, POLYCHAETES, ETC. ENTIRE AREA LOOKS LIKE ONE GIANT SULFIDE MOUND. BECAUSE PLUMES HERE HAVE NO HELIUM WE SUPPOSE THIS TREMENDOUS HYDROTHERMAL SYSTEM MAY BE CIRCULATING IN THICK SULFIDE DEPOSITS STRIPPED OF HELIUM BUT GENERATING METHANE. MAY BE ANSWER TO PARADOX OF OUR FINDING METHANE WITHOUT HELIUM IN SOME BACK ARC BASINS. METHANE CARBON ISOTOPES WILL TELL IF WE ARE RIGHT AND MAY BE THE KEY SIGNATURE FOR EXTENSIVE SULFIDE DEPOSITS. EXCITEMENT RUNNING HIGH AS WE CONTINUE TO EXPLORE THE RUGGED AND FASCINATING KRAZY KAT COUNTRY (FOR MASSIVE OCHRES AND UMBERS PLUS CHIMNEYS) OF A BACK-ARC BASIN SPREADING CENTER. WE HOPE TO UNDERSTAND THIS INCREDIBLE AREA IN A SMALL WAY.

CRAIG

The methane analysis did not solve the mystery of why there was no helium in the water when it was present in the basalt. The scientists later realized that the methane plumes found before the dives did not come from the vents. "It was some kind of serendipity," Craig said. "We still don't understand it."

In contrast to the Snail Pits, there was little life at Topless Towers and the other vent fields covered with acres of chimneys.

FROM: ATLANTIS II 16 APRIL87
TO: SCRIPPS

OUR FINAL DIVES LOCATED SEVERAL MORE HYDROTHERMAL VENT FIELDS WITH UNIQUE FAUNA, ETC. ONE OF A METROPOLIS TEEMING WITH LARGE GASTROPODS APPROPRIATELY NAMED 'WHELKS CLUB.' TEMPERATURES UP TO 285 DEG. C. MEASURED, ALL ON 'CLEAR SMOKERS.' GOOD SAMPLES OF FAUNA AND SPHALERITE CHIMNEYS. EXOTIC BEHAVIOR ON CHIMNEYS INCLUDES HERDING OF GASTROPODS TO WARM WATER BY SHRIMP AND CRABS TESTING WATER TEMPERATURE WITH CLAWS. FINAL DIVE LOCATED MAX TEMPERATURE VENTS AND WAS CELEBRATED BY COOKOUT INCLUDING GRILLED GASTROPODS FROM CAPTURED CHIMNEY. EXCELLENT ESCARGOT A MATCH FOR BEST PARISIAN SNAILS. OUR TEN DIVES HAVE YIELDED AN AMAZING AMOUNT OF NEW DATA, IDEAS, THEORIES, AND SHOWN ONCE AGAIN THE TREMENDOUS POWER OF ALVIN FOR RAPID ACQUISITION OF KNOWLEDGE IN NEW TERRAIN. NOW HEADING FOR SAIPAN ON GORGEOUS MOONLIT TROPICAL SEAS WITH GREAT SENSE OF FULFILLMENT ON ONE OF THE BEST CRUISES EVER, THANKS TO SKILL AND DEDICATION OF THE ENTIRE ALVIN/ATLANTIS II GROUP. A TOTALLY PROFESSIONAL OPERATION IN EVERY WAY.
                                                                CRAIG

The chimneys were remarkable. Rather than a single large opening, they had many little holes which precipitated sulfates, and they were covered with fuzzy white splotches that resembled lichens on a tree. The white matter was barite or barium sulfate, a scavenger of radium.

When Craig and Dick Hey ate four golf-ball-sized snails, they did not know their escargot came from a radioactive vent. Neither did the half dozen Explorers Club members who sampled the snails Craig brought them when he returned the battered flag. "We boiled the hell out of them and loaded them with garlic," said the Explorers Club president, John Bruno. "They were pasty and for some reason they had a metallic taste."

At about the time of the Mariana dives, a scientist at the Geological Survey of Canada in Ottawa happened to walk into a room where sulfides from the Juan de Fuca Ridge were stored. The scientist, who handled uranium ores, wore a radioactivity counter. Near the vent sulfides the counter went completely off scale. He immediately sought Jim Franklin who handled the vent sulfide mineral analysis.

"I told him no, no, no, it must be a mistake," Franklin said, "because these sulfides came up from the bottom of the sea in areas that were almost zero years old."

Naturally occuring radioactivity is generated from within the earth by the decay of uranium, thorium, and potassium, but it takes about a million years for these elements to become radioactive. So it made no sense that at hydrothermal vents, which last only decades and perhaps less, the sulfides could be naturally radioactive.

But Franklin went to check out the sulfides anyway.

"Ah, sure enough," said he. "They were extremely radioactive. They were in excess of two thousand times background levels of radiation. I was just astonished. The whole thing was a bit of luck. We discovered that those with the mineral barite were the most radioactive. It's an extremely effective mineral for absorbing radium. Our radiation people here determined it was radium 226, which has a short half-life of a few thousand years."

Radium emits the gas radon, a carcinogen.

Franklin immediately called Steve Hammond, the director of NOAA's vent program at Oregon State University, and the word spread.

Not all vent sulfides are radioactive, but those with barite usually contain radium, which in a poorly ventilated room will build up harmful levels of radon. Scientists were advised to wash their hands after handling sulfides, and technicians who routinely sawed through the samples were cautioned about the danger of inhaling flying barite particles. Folks who kept sulfide paperweights on their desks and mantels got rid of them and stopped prying off pieces to give away as souvenirs; and the Pacific Geoscience Centre in Sidney, British Columbia, moved the large sulfide chimney that had been on display in its front hall for several years quietly building up high levels of radon. (It had also been displayed at Expo '86 in Vancouver.)

What made the vent sulfides so radioactive? No one knows. But the high radioactivity allowed the sulfide to be dated precisely. The Mariana vent chimneys were only a year old.

33

Hydrothermal vents probably exist throughout the world ocean wherever seafloor is spreading and also on active volcanoes or seamounts. Cold seeps may exist wherever organic-rich sediment exists, and on seamounts that rise near subducting plates. The connection between the cold-seep and hot-vent fauna is still being debated.

"There are strong similarities, from the taxonomic to the biochemical level," Scripps biochemist George Somero said. "Tube worms, mussels and clams are at cold and hot vents. The symbiosis is driven by sulfur or methane. Things are set up pretty much the same. But it seems like each animal species has a different kind of sulfur symbiont even within the same vent. There is quite a lot of biochemical differentiation among the bacteria. Some prefer thiosulfate or hydrogen sulfide or methane. It has developed into a very complicated story, one that isn't over yet."

Verena Tunnicliffe, a biologist at the University of Victoria in British Columbia, said hot and warm vent animals, no matter where they live, clearly seem to be related to each other. "But the animals at the cold seeps seem to be evolutionarily more removed from the hydrothermal vent animals. I think something different is going on there."

Exactly what, is not clear; nor do scientists know how the animals get from one ephemeral oasis to another, or why there are always similarities *and* differences, even between two nearby vents. Geneticists like Judy Grassle, from the Marine Biological Laboratory in Woods Hole, found that mussels from the East Pacific Rise at 13 ° north looked just like the Galapagos vent mussels, but they were genetically quite different. This seems to indicate that these molluscs did not migrate from one vent to another. Some suspect that at least for some animals the cold seeps may be stepping-stones to the hot springs.

"I think all of us now agree that there is not just one strategy for dispersal," Tunnicliffe said. "Really the best way for getting from site to site is by moving larvae. But that idea hasn't worked out real well. For instance, we haven't found any tube worm larvae. Some vent animals have long-lived larvae and some have short-lived larvae and some have no larvae at all, so they've got to crawl." If they can. Sessile creatures like tube worms stay rooted to the same spot for an entire lifetime.

On a return visit to the Galapagos vents in 1988, few tube worms were left at the Rose Garden; the mussels, which had grown to jumbo size in the three years since the last visit, were the dominant creature. The spaghetti worms seen in 1985 were gone. But the water chemistry had not changed. "It's very strange," biologist Bob Hessler said. "Why?"

No one knows why most of the Pacific vent creatures are sessile while free-swimmers are common at the two Atlantic vent sites, or why certain animals are restricted to certain vents. No mussels have been found on the East Pacific Rise at 21° north latitude, and no giant clams at the vents between 11° and 13° north. A species of large crabs from the 13° north vents has been found nowhere else in the world. Likewise, the small feathery palm worm is known only from the Juan de Fuca vents. Nor do biologists know why different animals dominate vents—shrimp at the Mid-Atlantic Ridge, hairy snails at the Mariana vents, tube worms and clams at the Galapagos.

Because the Mariana vents are not on a section of midocean ridge, but in a faraway trough formed by a subducting plate, Hessler reasoned that the fauna there may have evolved independently. "There was every reason to expect that the life in the Mariana vents would be completely different from other vent animals," said Hessler, who joined the third leg of the Mariana expedition. "That was pretty exciting, but what was more exciting was finding similar animals. How did they get there? We haven't the faintest idea. Our knowledge of the global pattern of vent animals is so poor that we don't even have the right to speculate at this point."

The Mariana vent shrimp are new species but related to one type of shrimp found at the Mid-Atlantic Ridge vents; both belong to the same genus. "The differences between them are picayune; in fact, they almost look like twins," Hessler said. "The wonder of it is that no one would have expected these animals that live halfway around the world from each other to look almost like twins." The white crabs and limpets resemble other Pacific vent versions.

Other animals at the Mariana vents have no equivalent elsewhere. The huge snails, whose shells are covered in dark fuzz, harbor symbiotic sulfur-oxidizing bacteria. Apparently the shrimp feed off the snail's hair, which traps bacteria and other food. And the barnacles were extraordinary. "When I saw all these barnacles," Hessler recalled, "I sent a telex back to Bill Newman at Scripps. When Bill got that telex, he practically fainted with excitement."

These are heady times for cirripedologists.

A barnacle, as the English biologist Thomas Huxley once declared, is a shrimp standing on its head kicking food into its mouth with its feet. He neglected to add that the small crustacean does its acrobatics from inside a shell. Most barnacles live in shallow water and most are sessile; they spend their entire lifetime cemented head-down to the same home—the back of a whale, a turtle, a rock, pilings, even a floating coconut. Their permanent shells leave a good fossil record. The earliest barnacles had a stalk, a kind of foot, which allowed them to move. They lived during the Cambrian period, when the major phyla appeared. Four hundred million years later, during the Cretaceous period when dinosaurs roamed, sessile barnacles first appeared.

Alvin returned in 1979 from the vents on the East Pacific Rise at 21° north with a stalked barnacle never seen before, not even in the fossil record. *Neolepas,* which was captured live but died quickly at the surface, evolved 225 million years ago in the Cretaceous period. It was like netting a *Tyrannosaurus rex.* The next most primitive barnacle found was the sessile creature from the Mariana vent; it was thought to have died out 145 million years ago.

"What we got was another living fossil," Newman said. "It looks like it's from a group that went extinct in the Miocene. It's a missing link. Fossils like it had been described, but when you study it, you see it was not like those fossils, or maybe some of the fossils we saw aren't what we thought they were. These are very ancient critters. When you start looking you can't believe your eyes. They look so crazy, you're seeing something you've never seen before. It's flabbergasting."

Unlike other barnacles, those at vents have smaller and more numerous plates in their shells and the most sophisticated feeding parts ever seen on a barnacle. But while these filter-feeders eat bacteria, they apparently are not chemosynthetic. If not, how could they endure an environment so loaded with poisonous gases and metals that some scientists refer to the vents as heavy metal heaven? "Barnacles can handle a lot of stress," Newman said. "They're tough little devils. They live in the Antarctic, and under the burning sun of the tropics, and in the highest intertidal. The fact is, I don't think we've exhausted this. How could we be so lucky? I'm expecting much greater finds. I think there's a possibility of finding something so astonishing that we don't even have any idea of it. There had to be a barnacle before stalked barnacles, something not stalked or sessile. It would be marvelous to get one."

Newman believes that a vent, like an isolated island, is a refuge for ancient animals. Most vent animals are new to science and many represent higher taxa; in general, the higher the taxon, the longer the animal has endured through geologic time.

By 1985, scientists had counted fifty-eight species at the eastern Pacific hydrothermal vents; of that number, fifty-three belong to different genera, Newman said. Of this group 33 or 62 percent are found only at vents; 45 percent represent new family groups. And six species of tube worms belong to higher taxa, representing new classes, orders and a phylum, the highest category. In 1985, Meredith Jones created the new phylum Vestimentifera for tube worms from hydrothermal vents and cold seeps. At least six kinds of tube worms are still undescribed. And today, Newman's count of species has more than doubled. "It's just exploding," he said. "More and more information keeps pouring in."

The most ancient vent barnacle, *Neolepas*, represents a new species, genus and subfamily. Like barnacles, brachyuran crabs, which are ubiquitous at many vents, were thought to be rare in the deep sea. The vent crabs represent several new species, genera and a superfamily which is estimated (no fossil record for it has been found) to have been around for 85 million years. Twenty-three species of vent limpets represent four new families dating to the Mesozoic era, 200 million years ago. Leptostraca, the shrimp-like crustacean with teeth on its eyestalks, belongs to a primitive group which lived as far back as the Paleozoic era, *before* the dinosaurs. The spaghetti worms and tube worms, some theorize, may have first appeared 600 million years ago when the Cambrian period dawned.

The amount of primitive taxa at vents, Newman said, is no less than astonishing. "Hydrothermal activity must have existed since the earth cooled sufficiently for the oceans to form," Newman concludes. "Therefore, hydrothermal organisms could have existed since the origin of life; indeed it has been proposed that hydrothermal activity may have been involved in the origin of life itself."

A few scientists have hypothesized that life really did begin in a place like the Garden of Eden, the Galapagos vent Jack Corliss named in 1977. And it began with chemosynthetic bacteria, the basis of the vent food chain.

Among the vent microbes are the archaebacteria, the most ancient life form, present three to four billion years ago when life first appeared on earth. The archaebacteria are generally not symbiotic, although many species are chemosynthetic. In this group fall all the methane-producing microbes, among the most ubiquitous life forms—life that needs no sun. While it appears that most vent bacteria oxidize sulfur, others use a half dozen other reduced compounds, including ammonia, iron, manganese and methane.

In the modern vents, this microbial life, from the microscopic to the jumbo *Beggiatoa*, lives in massive profusion—on the seafloor in angel-hair mats and plush carpets, on rocks, and on the shells and backs of animals and inside some of them. So far, only the tube worms, clams, mussels, and the Mariana vent snail have been shown to harbor symbiotic bacteria. Copepods have been seen nibbling at the bacterial globs floating in the water, and fish grazing on the bacterial carpets.

Microbiologists John Baross and Jody Deming of the University of Washington, and geologists Jack Corliss and Sarah Hoffman, both formerly at Oregon State, have dared to take this further. They argue that hydrothermal activity linked to seafloor volcanism commenced simultaneously with the creation of the primeval oceans soon after our planet formed. Drifting plates formed a basin for a nascent ocean which, like the planet, was warm or hot and provided every imaginable gradient and ingredient to produce the raw elements of the first life—the single-celled organisms.

Based on experiments with bacteria collected from the first hot vents in 1979, Baross and Deming proposed that life, that is, vent bacteria, can reproduce at the outrageous temperature of 250°C or 482°F. "It has caused such a huge stir in the community that it has come close to driving us out of science," said Deming. "Conventional wisdom says when you get beyond the boiling point of water you can't have life. We said, if you have pressure keeping water from boiling and vaporizing, why not? We thought it was a very logical thing."

Their work was not perfect; all the organisms died before they could repeat the experiments. Thermophilic bacteria are notoriously fickle and fragile; in the lab, they sometimes die for unknown reasons. In 1985 Baross and Deming collected more fluid samples from black smokers for new experiments. The results did not confirm their original findings, but indicated that bacteria grew at 120°C or 248°F. This is still considered too high by many of their peers. "If you discard our 250°C, the 120°C holds the record for highest temperature for the activity of life," Deming said. "So, we're on record for being ridiculous. But what has happened as a result of our last paper is that more people working on thermophilic bacteria have begun to say, well, maybe the outer limit of life is 150 degrees (302°F)."

Holger Jannasch and Karl Stetter of the University of Regensburg in West Germany have isolated from the the Guaymas vents bacteria which seem to live at 110°C or 230°F. Chemists have found in the methane-producing vent microbe *Methanococcus jannaschii* peculiar lipids lining the cell walls; perhaps, they theorize, the lipids have something to do with making the bacterium heat-tolerant.

Most vent animals live in the deep sea's average near-freezing temperatures, but at least one, the Pompeii worm, endures 45°C or 113°F. The grotesque little *Alvinella pompejana,* which is covered with hairlike appendages, builds tubes directly on 350°C or 662°F smokers. It has been seen leaving, not entering, the tubes, presumably when the heat pushes above 45°C. Most scientists find it hard to believe that *pompejana* can live in those scorching tubes. Likewise, it is assumed that the *exoculata* shrimp which cluster in a frenzy around chimneys at the Mid-Atlantic Ridge vents are grazing on bacteria while keeping a safe distance from the intense heat. How *pompejana* endures even 45° is still a mystery. Scientists have been unable to bring up a live specimen, necessary for biochemical analyses.

"I personally believe they're experiencing temperatures a lot higher than 45°C," Verena Tunnicliffe said. "The argument is always: but is the worm really feeling that 350°C on the chimneys? Until we get some very fancy equipment we'll never be sure. I'm sure it's not 100°C (212°F), but if it's over just 50°C (122°F), the DNA molecule is not stable. So they may be doing something to protect their molecular structure. Nobody has looked at the molecular structure of these worms yet."

Those who accept the notion of the hydrothermal origin of life wonder if animals and bacteria live on other planets, or whether, if we were to lose our sun or if a nuclear holocaust decimated life on earth, vent animals would survive. Still others think all this is crazy.

"It's an unfortunate set of criticisms because it is just dumb," Baross said of his detractors. "The earliest organisms were thermophilic archaebacteria; the archaebacteria utilize volcanic energy sources; and the early fossil evidence—it just goes on and on; it all points to a deep or shallow vent environment for the beginning of life."

"I am *convinced* that life begins at the vents," Corliss said. "The more I look at it, the more sense it makes. What amazes me most is that it makes so much sense and people don't believe it. These hot springs are beautifully designed flow reactors produced by magma. Five or ten years from now everybody is going to know how life began on earth. It's amazing how resistant science can be to an idea."

Following the Mariana dives in August 1987, the *Atlantis II* stopped in Japan to play host for an open house. Crown Prince Akahito climbed into *Alvin* for a firsthand comparison with Japan's own 2000 meter (6560 feet)-depth *Shinkai* (from *shin*, deep and *kai*, ocean), launched in 1968. The government was building a second *Shinkai*, designed for 6500 meters (21,300 feet).

Then the *Atlantis II* crossed the Pacific to Astoria, Oregon, for dives to the vents on the Oregon Slope and the Juan de Fuca Ridge, and in late October, worked its way south to the Santa Catalina Basin off California, a deep ocean station visited repeatedly since 1984.

It was a good but uneventful cruise, led by Craig Smith, a biologist from the University of Hawaii. A couple of graduate students—geologists Robert Wheatcroft from the University of Washington, and Robin Pope from North Carolina State University—made the fifteenth and last dive, 1949, on November 10. They were in about the last hour of their dive, enough time for a final plankton tow, when they realized that *Alvin* had gone off track. Before returning to the bottom station coordinates, they spotted a long row of whitish blocks on the flat sandy bottom. Pilot Paul Tibbetts steered toward it.

"What the hell is that?"

"I dunno," Tibbetts said and moved *Alvin* closer.

"It looks man-made."

"It's unbelievable."

"A fish?" Tibbetts offered.

"It's a *fish*? That big long thing?"

"That's no fish."

"There's *clams* here!" Tibbetts shouted. "*Calyptogena!*"

"Limestone deposits?"

"*Calyptogena?*"

"What is that?"

It was much bigger than *Alvin*.

"I think that's a whale."

"Oh, you're kidding, a *whale?*" Tibbetts said.

"Yeah, those are whale bones. I think . . . whale vertebrae."

"*Whale vertebrae?*" Tibbetts didn't believe it.

"Yeah."

"Can we pick up some of it?" Tibbetts asked.

"Yeah, if you think it's safe to pick up."

"Why are those clams here, that's what I'd like to know," said Tibbetts, who began to maneuver a manipulator.

"Well, it's just a huge enrichment. Ah, we think we found a dead whale. [Chuckle.] Ah, it's really large . . ."

"Oh, come *on*, no cloud!" Tibbetts cursed as sediment swirled at the viewports. "That's why I hate driving around here... *shit!*" Tibbetts grabbed his first piece of whatever it was but lost it. "I can't see anything out here. We have a local cloud."

"Wait for a little bit. We don't have a whole lot more to do. I'll give you some time as long as I get some of the whale. I'm interested to see who's bored into it."

"See what?" Tibbetts asked.

"Animals bore into it.... The big pieces are probably its skull."

"Wow," Tibbetts said. "You want to get a big piece?"

"I'd rather get a vertebra. We have a better chance of telling what it is."

"*Clams!*" the pilot shouted again. "I don't believe it, all over the place."

"Yeah, when you said that I thought we found a new vent." [Chuckle, chuckle.] "A seep, a cold-water seep. We could be in *Science.* Craig [Smith] would get a large kick out of this. This is what he did his thesis dissertation on... animals falling down on the bottom. And he actually tried to sink a whale once." [Chuckle.]

"How?" Tibbetts asked.

"Dynamite. Actually, they put 1200 pounds of ballast on it and it still wouldn't sink."

"Where do you go to get a dead whale?" Tibbetts asked.

"He found one floating on the water."

"It's a pretty big-sized whale, huh?" Tibbetts said.

"Well, you know, grey whales come through here all the time ...I mean those things get up to sixty feet, don't they?"

"Wow," Tibbetts said. "This thing's huge. It's dinosaur bones."

The scientists laughed.

"Why couldn't it be a dinosaur bone?" Tibbetts demanded.

"Because these basins didn't exist when the dinosaurs died out."

"All right, I'll buy that," Tibbetts said and began to try again with the manipulator. After long moments, he announced success. "It looks like big tusks. That's what reminded me of a dinosaur."

"We can try for one of those big curved ones. Basket's going to be really heavy."

"Okay, here we go," Tibbetts said and moved *Alvin* to the head of the mound.

"Holy *shit!* That is massive! Those are huge! Don't put one of those in the basket, they're too big! They're, like, six feet long."

"Holy shit, it's huge."

"Holy smokes."

"OK, let's go back to science. We want to do this plankton tow...."

The skeleton was about sixty feet long and belonged to an

adult blue or fin whale which weighed about fifty tons. While some flesh may have been buried in the sediment, none was seen. The whale was at least five years old and had lain on the seafloor for about the lifetime of a hydrothermal vent, a few decades. The bones reeked of hydrogen sulfide. When the biologists sawed into a vertebra, blood, still fresh, like the baloney sandwiches that sat in the deep sea for ten months in *Alvin*, oozed out.

As far as Craig Smith knew, it was the first time anyone had come upon a whale carcass or skeleton in the deep sea. It was also the first indication that vent animals might be nourished not from geothermal energy as are the warm- and hot-vent communities, or from organic-rich sediment that feed the cold-seep creatures, but from the flesh and marrow of a dead animal. Another oasis with another twist; here, dinner came from the putrefaction of a whale carcass. The carcass provided another plausible explanation for how vent animals moved around—by hitchhiking on whale falls, microcosms of nourishment.

On a return cruise to the whale fall the following year, biologists collected a small vent mussel, the clam species that lives in the hot mud at the Guaymas Basin, and three unknown species of limpets and snails. The clams and mussel housed chemosynthetic bacteria. Microbial mats reminiscent of those at the Guaymas Basin and the cold seeps in the Gulf of Mexico covered parts of the whale skeleton. "The dominant clam here has only been collected at hydrothermal vents," Smith said. "We haven't had a chance to sort through all the samples, but it's very likely that we may find other vent animals."

"Some people were sort of upset that whale bones rather than geology may serve as stepping-stones for vents," Smith said. "Whales are distributed throughout the world ocean, so they could die and sink virtually anywhere. If you do some reasonable calculations, you come up with some surprising numbers. One is that before the whaling industry, there were probably 200,000 sulfide-rich whale sites on the bottom of the ocean. That's a large number, so it's not at all unreasonable to think that whale falls play a role in the dispersal of vent animals from one locality to another. In the wildest speculation, one could argue that until the advent of whales, vent animals may have been much more restricted in their distribution."

"Since the discovery, people have gone back to look at the bones of whales that have come up in trawls and they found the same kind of animals we found," he continued. "So, this is not unique after all. It looks in fact as if whale skeletons are pretty widely distributed, and many others may have this characteristic fauna linked to vents."

Alvin spent almost all of 1988 at hydrothermal vents in the

Pacific. On the East Pacific Rise, a plaque was dropped to mark *Alvin*'s 2000th dive, which was noted in the federal Congressional Register.

At Guaymas, the oil-refinery vent in the Gulf of California, oceanographers tried to identify the small wisps of hot "smoke" from finger-sized holes in the bottom mud. Using radioisotopes as tracers, they determined that the fluid percolated up through the mud in the same way groundwater rises on land. "The flow there is remarkable," WHOI chemist Fred Sayles said. "The velocities were measured in meters per year which is extremely high. It's got to be flowing in some sort of microchannels. We've never seen them and don't know how far they extend. I suspect that maybe the cracks get bigger as you go down, but I really don't know. They change as soon as you disrupt the system. As soon as you take a core [of sediment], there is no more flow."

They also found an astronomical amount of acetate, indicating that it is almost certainly an energy source for bacteria. When heated, acetate loses a molecule and transforms into methane. Two other acids—propionic and buteric—were detected, but in lesser quantities. Chris Martens, an organic chemist from the University of North Carolina, thinks there are even more kinds of acids, all used by the robust microbes at vents. He speaks of *Beggiatoa*'s diet as one of Italian salad dressing with goat cheese on the side. (Acetic acid is vinegar; propionic and buteric acid are present in cheese.) "If offered a choice of Italian salad dressing, that is, pre-existing organic molecules, versus sulfide or methane, *Beggiatoa* would pick the salad dressing, because all they have to do is eat it," Martens said. "On the other hand, if they chose the other substances, they have to use the complicated process of chemosynthesis to convert the energy to organic carbon."

It's still a theory. To prove it, Martens must return to take many more delicate and complicated measurements. "It's very difficult to do this *in situ*," he said. "It's like trying to measure how fast a tomato soup molecule is being broken down. We don't know at what rate the acids are being used. We can barely measure these rates in the laboratory. But the information is fundamental to understanding not just how these communities are structured but the way natural gas is formed."

Only a fraction, perhaps 1 percent, of the 40,000 miles of midocean ridge has been visited. Any one scientist has spent less than fifty hours total at the best-studied vents. Only about a dozen hydrothermal vents and cold seeps have been visited.

Today there are more than eighty published papers on just the Galapagos vents. From the 1979 cruise alone, 65,000 creatures, mostly microscopic, have yet to be described; the samples have been sifted, however, and biologists expect no surprises. Maybe. A white buttonlike animal seen in pictures taken on the 1979 dives has never been seen again. What biologists thought was a brachiopod, spotted in the same photographs, was collected in 1985 at the Galapagos vents; it was an unknown species of scallop. No one knows what kind of animal the eggs collected in 1979 came from. Biologists have just begun to scrutinize in depth the most recently discovered life from the Mariana, Juan de Fuca and Atlantic vents.

Some vent animals remain elusive. For almost a decade, biologists yearned to catch the pink, blue-eyed fish seen only at the Galapagos vents. These fish did their drunken dance in and out of the warm shimmering water when scientists returned to the site in 1988. "Everybody wanted it because it was such an enigma," Bob Hessler said. "What was it doing there? It wasn't a predator. It wasn't behaving like a scavenger or a carnivore. It was the only fish we knew of that loved being in the vent water, as opposed to being just nearby."

"The first thing on our dive plan was to go down to the Mussel Bed and try to catch one," Hessler continued. "There were dozens and dozens of them. Ralph [Hollis] took a net and by god, he did it on the first swing, it was absolutely incredible. Then he said, 'I got the fish but I dropped the net.' I had my heart in my mouth because I wanted it so badly."

But Hollis kept at it and soon pronounced victory. One vent fish—they could not have known that it was a mother filled with life—was in a coffin in *Alvin*'s basket, heading for the surface. On the *AII* deck, the biologists saw that the fish was a live-bearer. It had exploded with the pressure change during the ascent, spilling the babies growing inside.

Hessler immediately cabled Dick Rosenblatt at Scripps, who immediately called the Natural History Museum of Los Angeles County to tell Dan Cohen, the ichthyologist who tried everything short of dynamite in 1979 to catch a Galapagos vent fish. "We got the fish!" Rosenblatt said. "Where is it?" Cohen asked. "On the boat; it'll be here soon." But the pink vent fish almost didn't make it.

From the Galapagos, the *AII* steamed to its next port of call to face one of the strangest and most unpleasant experiences the crew ever had with customs, but this time it wasn't in a foreign country. When the ship docked in San Diego on May 18, it was boarded by U.S. Customs agents and a dog, enforcing the Reagan Administration's nine-week-old "zero-tolerance" program designed to get rid of drugs. The agents found two homemade pipes and a roach (the end of a marijuana cigarette too small to smoke) in a desk drawer in a sailor's cabin. The sailor said the items weren't his. The *AII* crew frequently switched cabins, according to vacations, changing watches and promotions. He had been in his current cabin for only three weeks. But he was led off the *AII* in handcuffs and taken to jail. And the *Atlantis II*, including the vent fish and everything else aboard, was held under "constructive seizure."

Rosenblatt was permitted to board the *AII* but forbidden to remove anything. He asked a customs agent what they did with "constructively seized" goods, like one-of-a-kind animals collected with great difficulty. Oh, the agent said, they would probably give them to the Smithsonian. At least the fish would not be discarded, Rosenblatt thought.

Thousands of miles away in the Black Sea, press reports about the drug bust reached the *Knorr*, whose skipper and crew took umbrage with some erroneous headlines, and telegraphed WHOI:

> NEWS REPORTS ... INDICATE THAT THE USA MEDIA IDENTIFIES THE A-II AS THE SHIP THAT FOUND THE TITANIC.
> THOSE OF US WHO WERE PART OF THE CREW OF THE RV KNORR DURING THE ORIGINAL DISCOVERY OF THE TITANIC REQUEST THAT [YOU] MAKE EVERY EFFORT TO CORRECT THE OBVIOUS MISCONCEPTION HELD BY THE USA MEDIA AND SET THE RECORD STRAIGHT....
> IT IS A TOUCHY SUBJECT WITH US AND WE EXPECT CREDIT WHERE CREDIT IS DUE....

("Listen," skipper Richard Bowen later said, "I still get stuff in the mail addressed to Admiral Ballard, Captain of the RV *Knorr*.")

In a few days, customs allowed the scientists to unload their gear and the precious vent fish finally made it into a shore laboratory. The creature, a new species of bythitid and the deepest living member of its genus, remains an enigma. "We still don't know how it feeds," Rosenblatt said. "We didn't find anything that would explain why this fish lives in such a peculiar place of low oxygen, high hydrogen sulfide and warm water. We found nothing anatomical that would induce this fish to live in this place, so we deduced it must be physiological."

"The thing that surprises me is that there are always so few fish at vents compared to invertebrates," Rosenblatt said. "A lot of the deep-sea fauna that you would expect to come in for dinner, we don't find. Squid and octopus aren't there either. I'm surprised that over geological time more fish haven't learned to deal with sulfide. It's a mystery. Many of the vent invertebrates have proven to be phylogenetically old, primitive, representing early stages of evolution. But the fish there are quite modern."

Compared with other deep-sea fish, the pink vent fish has an extremely high metabolic rate, as do the other fish, all eelpouts, found at vents. New species of eelpout, representing a new genus, have been found at Pacific and Atlantic hydrothermal vents and the cold seeps off Tampa. The pale, homely eelpout, which looks like an unbaked French baguette loaf, has no scales or pelvic fins. It lives in deep and shallow water throughout the ocean.

While dissecting an eelpout from the East Pacific Rise at 21° north, Rosenblatt found parts of a tube worm trophosome, which seemed to indicate that the eelpout can cope with the poisonous sulfur. But biologists are unsure. "There is no indication that the fish have chemosynthetic bacteria," Rosenblatt said. "But that fish definitely had sulfur in its stomach. My fingers smelled as if I had been playing with matches. Now, maybe that eelpout would have died with that lunch, but it certainly didn't hesitate to eat that stinky trophosome."

Still under "constructive seizure," the *AII* left San Diego on May 31 for the vents on the midocean ridge off the coasts of Oregon and Washington, where *Alvin* would spend most of the last six months of the 1988 season before returning to Woods Hole and a badly needed overhaul. Two months would pass before customs dropped forfeiture proceedings against the *AII* because, the U.S. Treasury Department said, WHOI was an "innocent party" unaware of the trace of marijuana aboard.

34

Aboard the Atlantis II. *Northeast Pacific, September 19, 1988.* Ten hours steaming out of Seattle, 6:30 A.M. We're in Canadian waters, headed for Endeavor, the north end of the underwater mountain range called the Juan de Fuca Ridge, the edge of the Juan de Fuca plate, which extends along the coasts of British Columbia, Washington, Oregon, and northern California. It took us two hours to snake our way from the University of Washington dock to the strait of Juan de Fuca and the open ocean. I have lost my coffee—I wear it now; and breakfast, an orange, has rolled away. But seas are only moderate.

Chief pilot Ralph Hollis heads below deck for the pre-dive.

"Morning, Ralph."

"Yes, sir, how are you this morning," he answers me. I wonder if he knows about the weather that awaits us at the first dive site two hundred miles away.

"Beautiful, like this, only twenty-foot seas."

He may be serious; it's hard to know with Hollis.

The chief scientist is geochemist Mike Mottl, thirty-nine, from the University of Hawaii. This morning he happens to look like death. Seasickness is nondiscriminatory. The other principal scientists on this two-week expedition are geologists Dick von Herzen, fifty-eight, of WHOI; Debbie Stakes, thirty-seven, of the University of South Carolina; and Stephanie Ross, twenty-eight, of the U.S. Geological Survey at Menlo Park, California. Mottl's technicians from Hawaii will handle the water chemistry while von Herzen's team from Woods Hole tends to the sampling gear designed to measure the "smoke" in the towering chimneys along the ridge.

The *AII* and *Alvin* have been here in the northeast Pacific exploring the vents since May. The previous cruise, led by John Delaney of the University of Washington, found pagaodalike edifices of smoking sulfides thirty to fifty feet high, inhabited by colonies of slender tube worms. The fluid issuing from these towers was 662°F or 350°C.

One of the instruments Delaney's team left at Endeavor for

six weeks recorded heat flux, a combination of velocity and temperature. "We measured one and a half megawatts per square meter," geophysicist Adam Schultz said. "That means this vent field is putting out as much heat as a good-sized nuclear reactor. That means the top surface of the pagoda we were sitting on was putting out something like fifty megawatts. Phenomenal!"

Actually, it was even more powerful because the instrument made its recording away from the smoker where the flow was diffuse.

But what made *The New York Times* and *Los Angeles Times* from the Delaney expedition was WHOI grad student Cindy Lee Van Dover's experiment. With *Alvin*'s lights extinguished, an electronic camera aimed at a black smoker recorded on magnetic tape a distinct halo, the brightest glow at the chimney opening. Exactly what this means is still being debated, although it points to the possibility of geothermally driven, instead of solar-powered, photosynthesis. The camera was sensitive to light wavelengths ranging from 400 to 1000 nanometers. The heat-sensitive infrared begins at 750 nanometers, about when light becomes invisible to the human eye. Purple bacteria, found in environments of low oxygen and high sulfur, such as swamps, photosynthesize between 850 and 1000 nanometers. Theoretically then, there is enough light at the deep-sea hot vents for photosynthesis to occur.

Van Dover wants to do her postdoctoral work as an *Alvin* pilot. What better way to see the ocean? But Barrie Walden is skeptical. She has no mechanical, electrical or electronic skills; and he worries that scientists may consider her competition. "When scientists are out there utilizing a tool like *Alvin,* they expect that tool to be for their use only," Walden said. "If a big discovery is made, what happens if Cindy is there? The other scientist may say, 'Now in another two minutes, I would have made that discovery.' Now suddenly Cindy, a scientist, knows the significance, and it registered with her first. What do you do?"

With funding and dive time increasingly dear, the competition has hardly lessened. The kind of scenario Walden described has happened before. Scientists have been known to write papers based on material "borrowed" from someone else's *Alvin* dives. The Alvin Review Committee's policy governing dive samples and photographs has been tightened. Duplicate dive tapes for the chief scientist are now made on the *AII,* and the original video goes straight to WHOI, where it is locked in a vault. For a year, no one may see the film without the permission of the expedition's chief scientist; after that, the material is public information.

Before Walden became the Alvin Group's manager, another woman tried to apply for a job as pilot but was told that she didn't have the credentials. Nobody can remember another scientist wanting to be a pilot. But Van Dover is determined and Walden didn't exactly say no.

For three weeks before our expedition, she did everything the pilots did except drive the sub. She was on the stern every morning at 5:30 for the pre-dive to clean the windows and grease the release for the drop weights; she built a sampling basket, learned how to "crunch the net" (calculate the coordinates for navigation), did all the pre-dive science briefings, and escorted the sub on and off the mother ship. "I figured, if you really wanted to be a pilot," she said, "you'd be hanging around the submarine, working. Basically I did not enjoy swimming; it was not my idea of a good time. When I first saw those guys up there with the sub waving back and forth, I thought if there is one thing that will keep me from being a pilot it will be having to swim. There are some wild times out there."

"It is very difficult to throw a wet line from the Avon," she said. "With a wet suit on, you have little freedom of arm movement. I threw the line three times, and three times it hit the side of the ship. Finally one of the guys on the ship told me to aim for his face. The responsibility is scary. I was always relieved to hear that the weights came off. The first time I put too much grease on the hatch. I worried about getting a hair in the hatch and it wouldn't seal. I was incredibly conscious of everything I did on that submarine."

One day a wrench from her back pocket fell into the superstructure behind the sphere. The pilots told her not to worry, they always discovered a half dozen wrenches among the syntactic foam compartments during overhauls. But she worried. That night she tried to find the wrench with a flashlight. Next day she removed a fiberglass skin and with great contortion pulled out a wrench, but it was not her wrench.

One pilot complained that she was in the way; another "went berserk" when she volunteered to be an escort swimmer. They told her that if she really wanted to be a pilot, she'd be in the Pilots' Lounge throwing iron. The Pilots' Lounge is a euphemism for the ship's filthy hold, where *Alvin*'s drop weights are stored and periodically must be stacked on pallets and hoisted to the stern. "I went down there and I kept right up with them," Van Dover said. "I was orange brown with rust every night and I got a blood blister, but you better believe I wasn't going to say anything. I didn't even say ouch."

Of all her hard work, it was hauling iron that most impressed the pilots. Hollis gave her a pair of dolphins (the pin submariners wear) and declared her an honorary pilot. "Knowing Ralph, I thought it was a joke," she said. "I still don't know if it was a joke."

Ralph, I ask, is this for real? "Yes, sir. She did all the science briefings and she put the weights together, which is a lousy dirty sweaty job. Each of those steel wafers is forty-five pounds."

All five tables in the *AII*'s main lab are full. The white dive board lists twelve requisite players, from pilot to the Avon's coxswain. Overhead a wooden plaque says: PB4UGO.

At 7:25 Hollis gives the order to slide *Alvin* out of its hangar. Launch only looks easy. Bos'n Wayne Bailey, thirty-seven, the senior A-frame operator, says every send-off has a different feel, depending on the sea state and the wind and what his stomach tells him. "You have to get the feel of the boat, it's a real feel," he says.

There's a feel from the deck, too. It takes my breath away to watch *Alvin* dangle from a main line thirty feet above the rolling stern as the waves lap angrily at the *AII*. Twice the sub has hit the *AII*'s stern, knocking a camera and light off, and twice the mother ship has struck *Alvin* in glancing amidships blows, but the damage was not serious. The problem of getting too close to the stern was solved with a small parachute the swimmers attach to *Alvin*'s stern when it surfaces. Like the parachute that some jet fighters inflate on landing, it increases drag, slowing the sub's forward motion toward *AII*'s stern.

Before thrusters replaced *Alvin*'s propellers, a line wrapped itself around an amidship prop, and an escort swimmer got out of the way just before the line shot a hundred feet into the air with the propeller. "That was one of the hairiest times," Bailey says. "That line could have cut off anyone's head."

Otherwise the record is uneventful. "We don't have to work so hard any more," Hollis says. "We just go out there and stand around and drop it in the water."

Those of us not on the stern by *Alvin* are at the starboard side where the Avon is being launched; or in the doghouse at the A-frame controls; or on the bridge; or in the pump room below deck where a ship's engineer watches the pressure gauges that monitor the A-frame's hydraulic systems. With a steep drop in pressure, indicating a failed hydraulic line, the engineer would switch immediately to manual and alert the outside crew on his walkie-talkie. He has been needed only once, which is why he's there now.

The orders come quickly over the two-way radios. "Telly leg up," Hollis says, and the A-frame's telescoping leg lifts the fat main line attached to *Alvin* and drops the sub in the blue-black swells.

For the next eight hours, the activity moves to the top lab and the equipment for keeping track of *Alvin*—a plotter, depth recorder, computer, satellite navigation readout, radio and underwater telephone. When *Alvin* is down, there is always at least one person here. Except for the port arm, the dive to Endeavor with Mike Mottl and Debbie Stakes is going well.

"No, no, no, no, don't do that!" pilot Tom Theodore ("T^3") Tengdin says to himself—or perhaps he is talking to Strangelove.

"Oh, no, don't drop it!" Stakes says, but the chunk of sulfide falls. The pilot starts again and gets his rock.

Tengdin's background is electronics; he left his job as a computer specialist at Purdue last year because an old friend, *Alvin* Engineer Ned Forrester, convinced him to work for the Alvin Group.

"Anyone else see that weird instrument over there?" Mottl asks, referring to a bright orange plastic highway cone. Delaney saw a baby's crib here last week.

"Holy shit," Stakes says. "There are pillows [of basalt] down here covered with malachite. Hold on, folks! This is amazing...."

"Malachite?" Mottl is skeptical that there could be enough copper to produce the green ore.

"What *is* this place? Mike! There's malachite staining everywhere and there...giant crabs, oh my God...Mike? Mike, do you hear me? I said *malachite!*"

Alvin stops at a ridge that resembles Mount Rushmore, except for one profile which is a dead ringer for Sammy Davis, Jr.

After the dive, the collected sulfides, covered in fuzzy white radioactive barite, are laid out in the main lab for curating. Stakes tries but fails to find a Geiger counter. Stephanie Ross says that on dives to vents in a nearby trough in June, she and her colleagues had their urine analyzed for radioactivity. The tests were negative. "We concluded that it was OK to handle these rocks," she said, "but we shouldn't eat our lunch in the main lab and we should wash our hands after handling them."

Mottl and Stakes also brought back black spiders that look like miniature tarantulas. These represent two new species, one named after Verena Tunnicliffe, who has been studying the life along this ridge for several years. "If you pick up those spiders and turn them over, very often you'll find about six eggs," Tunnicliffe said. "What happens is the spiders get together, they copulate, and she hands the eggs to the male who attaches them to his stomach. He wanders around for a year or so while she beetles off and has a good time and he's stuck with the children."

September 21. Dive 2111. Ross, von Herzen and pilot Gary "Count" Rajcula are far beneath the whitecaps whipped by twenty-five-knot winds, searching for a lifelike plywood mannequin of pilot Dudley Foster dressed in a pair of his old jeans and an *Alvin* T-shirt. In addition to being a marker, the plywood pilot is meant to give a sense of scale. It was John Delaney's idea.

Alvin flies toward a huge jagged edifice of sulfide studded with several chimneys. It takes a half hour to circle it. No sign of the wooden Dudley.

At 11:00 A.M. Rajcula's voice comes over the receiver in the top lab. "Praise the Lord!" They have found Dudley, finger pointing to nothing at the bottom of nowhere.

Rajcula had no background or special interest in the marine world. Before coming to Woods Hole in 1986, he made machines that put printed circuits on wafers.

Pilot Paul Tibbetts is at surface control in the top lab. The rule is to make contact at least once every thirty minutes, otherwise communications are kept to a minimum to conserve *Alvin*'s battery power. "Hey, Gary, how's it goin'?" Tibbetts' voice travels through two miles of water and echoes in the seven-foot ball of titanium, now bathed in condensation. "Very slowly," the pilot replies. Strangelove is misbehaving again.

Tibbetts' hair, the color of wet straw, is tied back in a small knob at the back of his neck. He wears a blue bandanna around his head. At six feet five and a half inches, he has to bend his head beneath the messroom ceiling. "Did they put you in the Green Room?" he asks.

"No, I'm in the Nunnery."

Women are usually assigned to the Nunnery, a four-bunk cabin two decks above the main level next door to a head. Almost everyone else sleeps in the Snake Pit below the main deck where sixteen people in six rooms share one head. Three of the six women aboard are assigned to the Snake Pit.

"Yeah, the Nunnery," Tibbetts continues. "That's the Green Room, that's where you get green from being seasick. But you can stay with me in my room if you like."

The plotter creeps another tiny movement across the paper. Tibbetts tells Hollis he's worried about the starboard view port; it leaked a tablespoon of water on his dive. If he's that scared, Hollis says, he shouldn't dive. And to me, Hollis says we're all going to die someday.

The starboard view port has been a concern for several months. Hollis has already examined it with a magnifier and found a few tiny scratches and occlusions inside the plastic cone. There is talk of what may be a tiny crack. Also, the pilots have heard what they think are penetrators "popping," indicating that they may be galling, that is, fusing themselves to the titanium sphere.

Date: Thu, 22 Sept 88
From: ralph
To: office

Good morning All
Still making progress out here.

The One Armed Monster, Son of Schilling, did not make today's dive. Count Rajcula has it scheduled for surgery. You know why Count Rajcula likes to dive? Because it is always night-time down there.

Barrie
I decided not to leave the window in the starboard side. I found all kinds of documentation on that window but no reference to what we see in the window now.

We installed a 1981 spare that we had that I have good test data on and no records of anything wrong with it.

Science was very pleased with today's dive, lots of data....

September 25. On their dive today, Mottl and chemist Christine Andrews discovered a new vent of ten-foot high delicate spires spouting clear hot water and sulfide mushrooms seventy feet high festooned with strawlike tube worms. "It was just fantastic," Mottl said. "Like a fairyland."

Tonight we are steaming south about 250 miles west of Astoria, Oregon, over the middle and highest point of the ridge, marked by Axial Volcano, a monster. It is almost five miles long and rises about a mile from the sea floor. The 572°F or 300°C vents are clustered at the southwest corner of the two-mile wide caldera. At one side are a clutch of thirteen-foot-high chimneys which are really more like buildings than stacks. They are called Hell, Inferno, Miserable, Demon and the Magic Mushroom, and they precipitate colorful sulfides. About 100 feet away is the Virgin, made entirely of anhydrate, the white mineral produced when seawater is cooked.

On another recent cruise, oceanographers left a camera before the Virgin and got the first continuous 365-day photographic record of a smoker. The Virgin shot up four feet in a day. "It's been going like gangbusters for four years," geologist Steve Hammond said. "We knock it down and come back the next day and it has built up again."

The clear shimmering fluid from Axial's chimneys has the highest dissolved gold content of any ever measured in the ocean. Why, nobody knows. There is dissolved gold in all seawater, but in minute amounts difficult and expensive to extract.

No one has ever seen boiling water in the deep sea, but the Axial team has found evidence for it. The boiling point of water is dependent on pressure as well as temperature. In the near absence of pressure on land, water boils at 100°C or 212°F. The temperature has to be considerably higher in the deep ocean to rouse the molecules into a boil. But at Axial, which at a depth of 5052 feet is the shallowest active hydrothermal vent known, the water needs to reach only 300°C or 572°F to boil.

When water boils, it forms a vapor (steam) and a brine (the salt left by the dissipated steam), what chemists call a two-phase separation. At Axial, the vapor phase built the Virgin; and a hundred feet away, the brine phase formed Hell and Inferno. This seems to contradict the laws of physics. It would be like heating seawater in a kettle and while the steam whistled out the kettle spout, the minerals in the water would create a stack of anhydrate across the street. "We don't have an explanation for it," Jim Franklin of the Canadian Geological Survey said. "Astonishing. One would never think that you could ever find the two phases coming out separately thirty meters apart."

The first vent discoverers in 1977 called their 20°C (68°F) water "hot," thinking that was about as high as the temperature could get in the deep sea. Two years later, 350°C (662°F) was recorded at the black smoker fields on the East Pacific Rise, and until recently this was considered the limit. Actually, the highest temperatures taken in 1979 on the Rise were about 400°C (752°F), but the scientists, uncertain if those data represented truth or a flaw in their instruments, stuck with the more conservative number. In 1984 Delaney got a reading of 400°C at an Endeavor vent, but he was unable to relocate the spot for another reading. Before this cruise, his remote instruments recorded 400°C for a full thirty minutes at Endeavor. "I'm beginning to believe it myself," he said. But many of his peers don't believe it.

On this trip, we're passing over Axial for the cleft at the ridge's south end, where another team of scientists in 1986 found a huge plume dubbed "Megaplume." The particulates, minerals and gases from smokers usually rise from 300 to 600 feet. The Megaplume was more than 2000 feet tall and 12 miles across. It was so big that the scientists at first thought there was something wrong with their instruments. Perhaps, some of them theorized, Megaplume was the result of a cataclysmic burp from within the earth—a deep-sea volcanic eruption, something nobody has ever witnessed. Yet.

Now the physical oceanographers who have had virtually nothing to do with *Alvin* are intrigued. Such an enormous plume of gases and heat must affect the movement of water. How much do the vents influence circulation? Do the plumes follow the prevailing currents? Or do they guide the currents?

Hydrothermal vents are cooling-off spots for the fire and brimstone of earth's birthing. Thus, scientists reason, the vents play a major role in controlling the planet's heat budget; perhaps as much as 25 percent of the total global heat is cooled at vents. But nobody knows for sure.

Mottl and von Herzen would love to find Megaplume. Their main goal is to measure all the plumes at several vent fields with an instrument built expressly for this purpose by WHOI engineer Jerry Dean. Dean's array carries temperature sensors and current meters. Attached to *Alvin* on a 164-foot cable, it records data in real time on a lap-top computer in the sphere. Launches have been tricky because the array must go down over the starboard side just as *Alvin* is hitting the water at the stern. And of course, the array is a potential hazard for *Alvin* to get hung up on.

September 28. Dawn. The submarine is being put through its 2118th pre-dive check. *Alvin* will make an unprecedented two hundred dives this year. It's not cute any more. Nobody calls it the "little submarine," as some scientists did, even in their published papers. It isn't little, unless you compare it with a nuclear attack submarine. This research submersible means business. *Alvin*'s last overhaul was in 1986. It has never gone so long between teardowns, and it looks it. The plastic quivers in the basket are charred and partially melted. The small aluminum bonnets that envelop the lights are dented, and paint has been seared off the tip of a thruster. There are patches of melted fiberglass skin, a chunk of skin missing altogether, and a crack about a foot long. The tip of the ski under the basket is bent inward, as if from a collision.

"Ralph?"

"Sir."

"What happened to the ski?"

"Nothing. That's our bumper."

Because of more suspicious weeping, two more view ports were removed, and like the starboard window, the grease was so dry that it flaked off or had to be pried off. The pilots, relieved to discover the cause of the weeping, regreased the view ports. Some of us still wonder why the starboard viewport came out of a box labeled "NFG" (No Fucking Good).

The moon is still visible as Stakes and Ross, both in sweatpants and jackets, climb into *Alvin* for their dive to the cleft, a notch at the southern end of the ridge. Scientists from the National Oceanic and Atmospheric Administration left six new transponders in the cleft but only one now responds to electronic interrogation. Once down, *Alvin* can also follow the buoyed squares of numbered syntactic foam left by other previous visitors. Today the women divers are taking a new square which says PORT STOP, tied to a port bottle filled with salad oil. The other side of the square reads: Stephanie Ross was here FINALLY!

On an earlier expedition this year with *Sea Cliff*, the navy forbade Ross and other female geologists from diving because they were women; the men were allowed to dive. So Ross is feeling more than just scientific satisfaction.

AII chief engineer Charlie Hall and Paul Tibbetts are in the doghouse at the A-frame controls. "Got any groans out of it this morning, Paul?" the chief asks.

"Nope," Tibbetts says. "Not even a grumble."

"It's pretty foolproof, although it can take some finesse," the chief says, sucking on a Pall Mall.

"You know what we need in here, Chief?"

"No, what?"

"A stereo..."

"Jeeezuz! This is one of my sanctuaries, this and the engine room..."

With Hollis' terse command "Telly leg up," the doghouse immediately quiets.

Date: Wed 28 Sep 88
From: ralph
To: office

Good Morning Marge [Stern]
All is going well out here but as usual, not easy. This morning we had to recover the sub and launch it a second time in rough seas because the disconnect for the von Herzen array came apart. This is the second morning that we have had to do two launches.

The weather out here has been poor the whole time. We have only seen the sun a couple times and it is always windy, rough and cold....

Ross, Stakes and Rajcula are in the small notch in the narrow valley far below. In 1984 U.S. Geological Survey divers photographed red and yellow spires, thatches of thin tube worms, clamshells, pillows of basalt with a beer can and "fluffy materials of uncertain nature," including soapsuds, lemon drops and cotton balls (all probably bacteria, Tunnicliffe said). Today there is mostly death in the cleft.

"It looks for all the world like a burned-out town," Stakes says at thatches of slender tube worms covered in sulfide soot.

"Clear smoke coming out of this chimney."

"How close?" the pilot asks.

"About eighteen inches."

"Don't let it get any closer. Tell me if it is."

"We're really close, Gary," Stakes says. "We're going to hit that chimney!"

"Ooo!"

"Gary, you have just wiped out the last life in this community!" Stakes views the vent areas as national park land, places to be treasured and preserved.

The water fills with red particles which Ross calls blood dust. "It's acting very strange for dust."

"Yeah."

"A bloodbath."

They don't know what it is. Bacteria? Sulfide particles?

Thud. Alvin bashes against the east wall of the cleft.

"Well, here's the east wall," Rajcula says.

"Timber!"

Another chimney topples, stirring up more blood soup. When the particles finally settle and they can see, the Port Stop marker is dropped.

The next divers to the cleft report a similar finding: little if any life. Few of the markers left years before are found. Those that are seen are covered in some kind of biology, as the geologists say, making them difficult or impossible to read, or the nylon lines have melted and parted.

October 1. In the before-dawn darkness, we gather at the stern for dive 2121. Dick von Herzen runs to the mess to grab breakfast, and I listen to advice on how to crawl through *Alvin*'s oiled hatch without getting grease on my back and feet.

Rajcula is in first. Von Herzen and I follow up the ladder; we remove our shoes and step onto the submarine, where Hollis stands, radio at his mouth. "You've got a hole in your sock," he tells me. This time he's not kidding.

Pilot in training Tim Conners takes a last look through the hatch. "You watch those nasty currents down there," he says.

"Yes, Mom," Rajcula replies.

The hatch with the tiny clear Plexiglas center comes down and we are sealed like vacuum-packed coffee beans; three big beans in a seven-foot ball. I have interviewed more than three hundred people, about half of them *Alvin* users, but it hasn't fully prepared me for my first dive. I have been in the submarine before, but on the mother ship. Now it is fully loaded with people and gear. At five feet five inches, I am the little one. Von Herzen, who is six feet, three and a half inches, must really feel like a coffee bean. I'm too busy breathing deeply to ask him. I write:

> 6:35 A.M. *We're in. Something so final about that hatch. I think I might sweat. Or get out. What am I doing here? I don't have to be here.*

I have chief scientist Mottl to thank for being here. He expects me to behave like a scientist—take pictures, program the lap-top computer that gathers data from Dean's array, and recite my observations into a tape recorder. I even have a deadline for submitting a typed transcription of my tape.

I took ten pages of notes during my night-before briefing with him.

> *Time is critical. I can't stress it more. Any time you let 35 min. pass and haven't turned on an outside camera, you have wasted a big chunk of time....*
>
> *Some pilots are really good; they'll ask: shall I do a video survey? That's great, but some pilots won't ask. If you see a forest of chimneys, ASK....*
>
> *You hv 3 lights on yr side of the sphere. If you leave the lights on, you'll get yelled at for wasting battery power. You can see better as you approach the bottom with yr lights OFF....*
>
> *Windows melt at 60°C....*
>
> *Get a temperature measurement BEFORE a water sample. Soon as pilot sticks temperature probe in, read off temperatures into tape recorder. Make sure tape is running....*
>
> *Please don't ask pilot any dumb questions. Scrunch up and keep yr mouth shut until yr down....*

Mottl finally paused. "That's all I can think of. Have a good dive."

I got another piece of advice last night from one of Mottl's aides, who jiggled her HERE and said with not a trace of humor: "Better practice."

Thirty-three minutes after the hatch shut, and the *AII* is still trying to get itself into the correct patch of ocean. Rajcula plays George Winston on piano.

"Telly leg up," Hollis finally says. And we, too, rise. In a moment we're in the water. All I can see out the starboard porthole is water, but I know the submarine is sitting buoyant at the surface as the swimmers detach the lines.

"Hatch is shut, oxygen is normal, CO_2 scrubber's on," Rajcula tells the surface. "I need altitude and permission to dive."

He gets it, and down we go in a froth of bubbles. There is no sensation of any motion until I look out the window and see that we are drifting through galaxies, by hundreds and thousands of stars, pin points of light that are the life and death of marine snow, which appears to be falling up. This is truly inner space. At times the stars grow into tiny carousels and helixes.

There's a nook for everything, reachable without getting up (which I don't think is possible anyway). To my left is a reading light with a red and white bulb. By my knee is a holster filled with a flashlight. A tape recorder hangs on a short chain from the tiny TV screen at my tiny window. Now the stereo cassette recorder plays, *Swing low, sweet chariot, comin' for to carry me home....*

The bulky net holding lunch dangles from the hatch. A kitchen timer dangles by my flashlight. At von Herzen's signal, I crank it to sixty minutes to remind us that the videotape has run out and we should change cassettes. There are only a half dozen commands for the lap-top computer, but I practiced last night. At $20,000 a dive, I don't want to screw up the experiment.

The computer, which will collect the data from the array attached to us, is on the science rack, a floor-to-ceiling shelf unit at the back of the sphere. The space beneath the lowest shelf was originally intended as stretching space for cramped legs, but it is always filled with gear. A tiny monitor hooked to the video camera out front is on another shelf.

> *7:44 A.M. at 639 m, my little TV set says. My pants are wet. I'm sitting in a pool of water. The whole ball is very wet. Dick says a penetrator on his side is leaking. He runs a finger up the drip. Yup, salt water. I have a drippy penetrator too. Gary tastes. He says my drip is condensation. He is noncommital about Dick's drip. I can't blame him. All dribbles taste salty to me. I am sitting next to the NFG viewport.*

The temperature underneath us, which is displayed on our monitors at "Wdu Prb" (window probe), says 02.310. "If that ever changes, let me know," Rajcula says. I promise.

We wear layers of clothing. Some pilots wear long underwear. Hollis wears a pair of fleece-lined L. L. Bean booties in the sphere. It's cold, but still not as cold as the steel sphere was and still not as cold as it was before foam back pads were installed in 1986. Titanium is warmer than steel.

At 8:42 A.M., depth 2088 meters or 6847 feet, Rajcula starts the scanning sonar. Wa WA wawa WAWA wowo wowo WOWO wawa WAWA. The sonar is above my head on a shelf that I will bang my skull against many times on this dive. The junglelike noise seems appropriate for a science-fiction movie. But what could be weirder than this?

I keep looking through the darkness for the lights of the city we are to land at. It reminds me of flying over the desert at night. We are 100 m from the bottom. Gary has slowed our descent. The marine snow no longer soars up; it just sits there suspended, as we are. I'm not going to think about the pressure on us.

Lights on. We are in the barren and desolate cleft. Hardened black lava stretches in long folds like a wrinkled road dusted with yellow and green vent particulates. The same colors stain a stump of earth crowned with delicate spires outside my window. I stare at it for long moments before realizing that the water is shimmering.

"And there's a lot of yellow something down there," I observe. "Looks like pollen...."

A huge ray glides by.

"Oooo."

"Pretty nice aquarium out here, huh?" von Herzen says.

"Sure is. I see brittlestars, I think. They look like anorexic starfish. A lot of yellow and kind of green sediment. Black glassy basalt. Can't see the top of it. Looks charred, as if from a fire."

"Depth is twenty-two twenty-two-point-two," Rajcula tells surface control.

"Roger."

Or 7289 feet. There are 3237 pounds of force pressing in on every square inch of the titanium ball that surrounds us. My car weighs less than that. On every square inch. Wawa wa wa WAWA wowo wowo WOWO wawa wawa WAWA With experience, Rajcula says, you can tell from the sonar's screech and whine how close the sub is to something solid.

We head at a walking pace down the valley, rising to climb boulders in our path; but mostly the cleft is straight and unobstructed. It reminds me of a sand-swept asphalt beach road.

"How you doing?" von Herzen looks at me and the computer.

Terrible headache. Pages of my journal are wet. My jacket is wet.

"So far so good."

"You're doing a great job of not stirring it up, Gary," von Herzen says.

"I see what looks like, perhaps tubes. . . . Ow! Isn't that interesting, there's a hole. . . ."

wowoWO WO

"Ah, I've got a cramp."

I make room for von Herzen's knotted leg. Geologist Jeff Fox, who is six feet six, told me that diving meant "a long and painful day" for him. Bob Hessler said he suffered from disobedient limbs, and Fred Grassle was good enough to let him ascend with his ankles perched on Grassle's shoulders. The record dive for linear feet inside *Alvin* was probably set by pilot Don Collasius, six feet four; chemist Fred Sayles, six feet seven; and Grassle, six feet three. I cannot imagine how those three men could fit in here for eight hours. "It was overcrowded," Sayles deadpanned. But where did they put their legs? Sayles couldn't say, except to reason they were wrapped around someone or something.

Rajcula works for more than a half hour to get a water sample and temperature from a tall chimney flanked by two other columns belching superhot water.

"Shimmering water . . ." I report.

"It's not going against the window, is it?" our pilot asks, and not waiting for an answer, he is suddenly lying across my lap to see for himself.

"It's getting kind of cloudy down below," von Herzen says. "Can't see much. Can you see?"

"Yeah," Rajcula says, working the joystick. "I can see."

"I'm kind of amazed that we haven't done it already, but there's no danger of burning off the array, is there?" von Herzen asks. "I worry about burning off that cable. Three hundred degrees [Celsius] could burn it pretty easily."

Rajcula concentrates on getting the long temperature probe inside the rushing column of black water.

"Good shot!" von Herzen says. We watch the centigrade temperature readout on our monitors—100, 150, 180, 200, 245, 290, 300, 350.

POP weewee wo WOP ka POP weewee wo

From marker 0/, we continue over the dusty beach road of wrinkled black lava to 1/, which is supposed to be near another block painted with a mustache and wine glass, in honor of geologist Bill Normark, who sports a handlebar mustache and has his own vineyard in California.

"Can you see it, Victoria?"

"What?"

Rajcula moves so that I can look out the center view port to gasp at the monstrous sulfide blocks and spires chugging black smoke. I have worried that such words as "incredible" and "spectacular," words scientists rarely use but words they have used repeatedly to describe their dives to me, will lose their meaning in this story. But they are right. I'm speechless. Von Herzen asks me if I'm OK.

"How high up does it go?"

"Maybe ten meters," he says. "You can't see the top of it." Von Herzen thinks he sees another chimney.

As we approach the tall smoking yellow block, the temperature probe at the bottom porthole rises for the first time, from 2.8° C (37° F) to 3° C. In a moment it drops to 2.9.

Von Herzen says it is a young chimney. Its smoke sparkles with green sulfide particulates.

Wawa wawa WAWA wowo wowo WOWO wawa wawa wawa WAWA WAWA

We can't find the buoy painted with the mustache and wine goblet, but drop one of our own, a square painted with a Roman nose, after von Herzen's distinguished schnoz. Nose Job floats, incongruous and too tiny. No wonder Delaney left the life-sized Dudley.

Several building-sized columns capture our attention. They happen to be on my side.

"Do you see it, Gary?"

"*Now* it looks like we got smoke!"

"Sure do."

"Which one you wanna go for?" Rajcula asks.

"Let's go for the fat one."

"I can knock it over with the basket."

"Yeah, we could try that."

"That is *really* belching!"

"Visibility OK?" von Herzen asks.

"Great."

Our pilot works Strangelove's pincers around a bit of chimney.

The stacks are now out of my view, but I see all kinds of other things. "Why do these fish swim nose-down, anyone know?" I ask. They don't know. "Is that a crab?"

"Yeah."

"*That's* a crab?"

"Yup. Big spider crab."

Cousin of the Alaska king crab, the sea spider crab, whose leg span is as wide as three feet, is rare and not part of the vent community. Verena Tunnicliffe dissected several from this area and found their stomachs full of the strawlike tube worms. "These crabs have realized there is food on the hoof, or on the tube, as it were," she explained. "They wander around on the periphery of these vents, dash in and dash out again and eat the tube worms like spaghetti. They are a major predator on the vents, which makes them an interesting liaison between vent production and the normal deep sea. They're also a mechanism for vent animals to wander from one vent to another. We have seen vent polychaets (worms) on their legs."

Another giant crab, standing motionless, like a sentry.

"I see white, um, white stuff."

"Sediment?"

"Could be, Dick, I really don't know."

Rajcula looks and declares, "It's fish turds."

"I think I see a starfish. What's it doing out here all alone?"

"Oh, just making waves."

"I need to cut my head off and put it in my lap so I can see better."

"Yeah, those windows are weird," says Rajcula, who has the best seat in the house. It's hard to imagine how divers could avoid stiff necks before the floor in the sphere was dropped in 1986.

"Another big fish and a white thing in front of it. What is it?"

Rajcula is in my lap again. "I don't know, a piece of garbage?"

Von Herzen can't see. "White? It wasn't a smoker, was it?"

"It looks like a big white pillow."

"Yeah, like a rag or something," Rajcula says.

"A rag? It's a long way for garbage, isn't it?"

As the pilots well know, nowhere in the ocean is too deep or remote for our garbage.

We have reached a field of boulders stained in autumn colors. The orange is so bright, it looks as if it has been painted on. Cornmeal dusts clusters of straw, dead tube worms. White petals gently flap near a clump of sulfide.

"Dick, is this alive, these petals, these white things?"

"It's probably some biology of some kind. I don't know what it is."

If only Tunnicliffe were here. I tell her later about the petals. "It's funny, people develop their own nomenclature," she said. "Like puffballs. It took me ages to find out what puffballs were. They're the dandelions from the Galapagos. Geologists once came back talking about jelly bag creatures. The *Alvin* pilots did a magnificent job of getting these, but we still haven't figured out what they are. You drive through some of these fissures and the walls are like walls of Jell-o. My guess is it is mucus from bacteria, a manifestation of bacteria going mad."

Perhaps what I had seen were jelly bag animals? No, they were not like Jell-o, I insisted, they were *petals*.

Rajcula spots marker 7/. He has to pick it up and shake it to read the number. We are in a forest of tall spires.

"This looks like Bryce Canyon, *incredible,*" von Herzen says.

"This is neat." Rajcula.

"WOW!" me.

"That's what I mean by Bryce Canyon."

"Where's Bryce Canyon?"

"Utah."

"It reminds me of Gaudi."

"Who's that?"

The Spanish architect who designed surrealistic buildings and spires that remind me of stacks of partially melted marbles.

We see clusters of small tube worms and the white petals which may or may not be life, but they are waving. Is it the current that is causing them to flutter? But there's no current here. How can big petals be life? I recall the dandelion.

Rajcula flies *Alvin* thirty feet up to the top of a column. We're going to collide with a spire. It falls noiselessly. "Ah!" he says. "Just knocked it over."

"Were you trying to?" I ask.

"No! I didn't mean to."

"You've got some more spires on the right. Very close."

We're hemmed in on three sides by smoking spires. Rajcula manages to insert the temperature probe into the top of one and von Herzen reads the numbers: "215, 239, 271, 280...210, 46, 90, 136, 50, 23...23, 42, 131, 239, 249, 258, 279 (nearly shouting) 299, 319, 324, 325, 326, 350 degrees. Incredible, wow, Old Faithful."

"You know, there's a worm here, looks just like a feather."

"There is something called a feather-duster worm that has been found around the vents in the Galapagos."

"This is a little guy, about as big as my little finger, and it's pink, red, and covered with this white feathery stuff, looks like a hairy shrimp."

Wee wa WA WA woo woo wee WA WA wa

Tunnicliffe said they were palm worms, named by a geologist who reported from a 1983 dive: "I see spaghetti all over the place and little red palm trees."

Our basket is full of sulfide chunks. We have traveled about four miles along the cleft and it is time to go home. Rajcula deploys our last marker, trips the last water sampler, and releases the array and a 250-pound side weight. He also changes the carbon dioxide scrubber—a relief to me.

"All weights away and the array is away.... On our way up and we had a great dive," he tells the surface. Now he can relax. He puts on a tape. *Give me one more night, give me just one more night....*

"Well, very few people have been down there to see that," von Herzen says.

In fact, 1112 people, including pilots and mother ship crew members, have made the 2150 dives of *Alvin*'s first quarter century. We wonder if we are the only three people in the Pacific—perhaps in all the world—who are at this moment in the deep ocean.

My window still weeps and my head still throbs.

"That was nice driving up that cleft, Gary," von Herzen says.

"Sonar makes it easy," our pilot demurs.

"Yeah, but you got us up there in good time."

"Without really banging into anything either," Rajcula says. "That was amazing." He turns with his back to the center view port and stretches his legs, feet propped on the science rack.

The big sack hanging from the middle of the ceiling comes down with the meatloaf and peanut butter and jelly sandwiches (each in an airtight plastic container), candy bars, fruit, granola bars, and a thermos of coffee. The *AII*'s galley made a sandwich just for me—of grape jelly and whole sardines on white bread. The Delaney divers got a rubber spider in a sandwich and a wine bottle full of water.

According to my monitor, we are ascending at 38 meters (125 feet) a minute.

"I would have loved to have gotten up a little farther north if we had time," von Herzen says. "It's unexplored."

The sulfides we collected are only one or two years old. The underground plumbing seems to be migrating north toward the new vent Mottl likened to a fairyland. The highest temperature we measured is 375° C (700° F), above the magic 350° C (662° F).

I pour coffee. Rajcula warns that it will be like mud, but this is good and hot, and my obedient bladder can take it. Lucky me.

"Well, we really did pretty well this trip," von Herzen says. "We got a lot of data compared with the last trip in 1984 when we were at 13° north [on the East Pacific Rise]. The weather was great but nothing worked. The submarine's data logger didn't work, our array didn't work. I think this data will give us a much better handle on the heat and chemical outputs of the smokers. It may be difficult to ever get it all. We'll see. Hopefully."

"What's your depth, Gary?" surface asks.

"Seven seventy-five, and I'm positive that the array's away." He turns up the music (new song, same singer) *It's in the way that you use it, it comes and it goes....*

"Yeah, it's up here," surface says.

"We're driving west, by the way."

"Roger, we want you to go south now," surface says.

"Couple of meat sandwiches here if anybody wants them," von Herzen says.

It's in the way that you use it....

"So where you folks go next?"

"Monterey Canyon and after that somewhere off San Diego." And after that home, two days before Christmas.

"Nobody wants an apple?"

At 485 meters (1500 feet), the battery indicator shows that we are almost out of power. But we don't need any power now; *Alvin* is ascending by its natural buoyancy.

How odd. We left in the dark and it is still dark.

"198 meters."

"Is that light up there we're seeing?"

"100 meters... 15 meters one-five...."

"We're here, back into the real world," von Herzen says. "Hang on, gets a little choppy. This is where you hope that they pick us up in a reasonable time." He remembers his dive last week with Ross when they nearly threw up. Ross told herself she could throw up when she counted to a thousand. She had both feet on the *AII* long before reaching a hundred and staggered to her bunk in the Snake Pit.

Escort swimmer Craig "Mad Dog" Dickson waves under water at my view port. The lines are being attached to *Alvin*.

"What is that, the propeller?"

"Main line down, bring it up."

"Did I see a propeller?"

"Yup."

"They keep those off?"

"Hopefully."

We're in the air. Hollis, radio in hand, faces us far below on the stern. "OK, Mike, bring it in."

We fly in a brilliant cloudless sky, swinging gently from side to side, and slowly down, down... BANG!

"God!"

"Hang on, Mike, I got to get a little further forward."

The A-frame operator pulls up and begins to lower us again.

"Hang on," I hear Hollis.

"Little more?"

"Yeah."

"Got to center that little better on the sled."

This time we land on the rails.

"That looks good enough. Sub secure on deck," Hollis says.

"Gary, thanks for a great ride. Can you open the hatch?"

Not until the hatch is hosed down with fresh water.

"You feel OK?"

"Fine."

"OK, Gary," Hollis says.

The hatch pops, with our ears. I take deep breaths, feeling the wind against me.

Epilogue

Woods Hole, June 5, 1989. A quiet hum pervades the high bay as men and a woman work, succumbing to the inexorable trek of time, which has stolen their weekends so *Alvin* can make its 2151st dive this summer. Pilot Gary Rajcula, on his back beneath the passenger sphere, pulls out a penetrator coated in blue dye. "Got to be lapped some more," he declares. Cindy Lee Van Dover and another pilot, huddle atop the sub with a fistful of wires. They say they are threading noodles through small holes.

Barrie Walden has agreed to let Van Dover become *Alvin*'s first female pilot and the first pilot who is a scientist, operating that *damn toy.*

Alvinella pompejana, Alvinocaris lusca, Opisthotrochopodus alvinus, Oasisia alvinae, Nereimyra alvinae, Paralvinella grasslei....

This summer Japan's new *Shinkai-6500* reached its depth limit of 21,320 feet, becoming the world's deepest-diving sub. With the bathyscaphs long gone, the *Shinkai* and four other subs rated for 20,000 feet are the world's deepest divers: the Finnish-built Russian subs *Mir-1* and *Mir-2,* launched in 1987; France's *Nautile,* née 1985; and the navy's *Sea Cliff,* whose HY100 hull was replaced with a titanium sphere in 1984. The other *Alvin* look-alike, *Turtle,* was fitted in 1980 with *Alvin*'s first passenger sphere, the steel ball certified for 6000 feet, and allowed to dive to 10,000 feet.

There is still confusion over what to call these vessels. Until Germany recently built a minisub to carry torpedoes, the big and small submarines were often differentiated by their intended purpose: combat submarines were meant for combat; the small subs were designed for science and exploration, inspection and retrieval. Frank Busby, who has been tracking underwater boats for decades, calls the noncombat submarines "manned submersibles." But he also includes in this heading tethered subs, diver lockout capsules, diving bells and diving suits—all with an umbilical cord to the surface and much shallower diving. As further distinction, he calls *Alvin*-like subs "one-atmosphere, untethered submersibles." But according to his definition, so are combat submarines. It's not easy. Busby asked six sources, including the United States Navy and the

American Bureau of Shipping, to define manned submersible; he got six different definitions. And his publisher pressed to excise the "manned" part of the title.

Of course, there are great differences between the big and little underwater boats; the modern nuclear submarine has vastly more power and room, and no windows.

Today *Alvin* was declared a "historical landmark" by the American Society of Metals International, joining such company as the Statue of Liberty and sixty-seven other "unique engineering achievements [of] advanced materials." Also, the Elmer A. Sperry Award, for "advancing the art of transportation," was announced; and the National Science Foundation bestowed a "distinguished public service award to commemorate twenty-five years of providing a unique and responsive service to the ocean science community." In the award's processing through NSF's bureaucracy, it was initially made out to "Dr. Alvin Group."

That *pregnant guppy*, that *washing machine*, that *chewed off cigar with a helmet*. How could anyone have known it would so profoundly change the way we think of our world?

The Canadian Geological Survey in Ottawa has just completed the analyses of last year's sulfides from the Axial vent on the Juan de Fuca Ridge. They were full of gold. "Astonishing," Jim Franklin mutters. "Three to four parts per million, or to put that in perspective, it's at the bottom end of what would be considered economical to mine. Six to seven parts per million is about average. This has a lot of implications for exploration back on land. Now we know that if you want to find a good gold deposit you look for environments that were relatively shallow."

That is, ancient geology that once rested in the ocean at shallow enough depths for seawater to boil, as it apparently does at the Axial vents.

The Mid-Atlantic vent team has also found gold in the sulfide chimneys—*23 ppm* of gold, five times the amount of gold in the average land deposit, almost an ounce per ton. The Atlantic vents are at 3700 meters (12,136 feet), too deep for boiling. At least, no one has found any evidence of boiling. Apparently, boiling is unnecessary. "That's something new we've learned," Franklin says. "We're learning something new all the time."

Some people think there may be something far more precious than gold at the vents: cold fusion. "Cold" because the temperature inside the earth reaches only thousands of degrees while that on the sun where hot fusion occurs is somewhere in the unimaginable millions of degrees. Unlike the separating atoms of nuclear fission, the atoms merge in fusion, producing helium 3 (a nonradioactive isotope) and tritium, which decays during a half-life of about twelve years to helium 3.

In March 1989, B. Stanley Pons and Martin Fleischmann claimed that they had produced cold fusion at room temperature in a small glass tube. The details of their experiment, which remains highly controversial, have not been published. Using a similar setup at Brigham Young University, physicist Steven Jones made the same discovery at about the same time. But his claim was more conservative—the fusion signal was *small*, but *consistently reproduced*—and was published in the scientific literature.

Jones's experiment led him to wonder if helium 3—found at volcanoes and deep-sea hydrothermal vents—was not left over from the Big Bang, but produced by cold fusion. And he wondered if cold fusion is under way in the core of Jupiter; it may explain how Jupiter can radiate twice as much heat as it gets from the sun. And it may explain the inexplicable pulse of tritium in the air recorded during an eruption in the 1970s over the Hawaiian volcano Kilauea.

Jones has teamed up with *Alvin* user Gary McMurtry, a geochemist at the University of Hawaii, to ask the Department of Energy for funding to look for a sign of cold fusion at the hot vents on Loihi, the newly forming Hawaiian volcano. They hope to find tritium. Volcanoes are not supposed to burp tritium. Helium 3 is not supposed to occur naturally in the earth, except from the residue of our planet's birth. "We're in the steep part of the learning curve now," McMurtry says. "If we find unequivocal evidence of tritium coming out, then it'll really set geochemistry on its ear. It'll blow it away."

A Note About This Book

My salary to research and write this book was paid for by a grant from the National Science Foundation; contributions also came from the Office of Naval Research and the National Oceanic and Atmospheric Administration. I am grateful for their support.

While *Water Baby* was never intended for a technical audience, my proposal for its funding went through the same anonymous peer review process that any proposal for scientific research must submit to—and pass. To my reviewers, whoever you are, thank you.

Many people, too many to list here, contributed to this book; they gave of their time, their documentation and, as you have read, their hearts. The Woods Hole Oceanographic Institution allowed me all the access I requested to its archives. Colleen Hurter tirelessly chased down information and more. And I was lucky to have Joyce Berry as my editor at Oxford.

V. Kaharl

Index

Akens, Jim, 201, 239, 273, 274, 276, 277
Alexander's Acres, 101–2, 111, 154
Alldredge, Alice, 233–34, 236, 244
Aller, Bob, 234, 236, 265
Allsopp, Steve, 219–20
Aluminaut: design, 16–21, 153; alternatives to, 47; cost, 48; role in H-bomb search, 66–67, 68, 74, 76–77, 79; 84, 117; role in *Alvin's* salvage, 122–23; retired, 128.
Alvin: Awards, 89 (H-bomb search), 346; Commissioning, 45–48; Construction, 41–42; contract, 26, 29–30, 35; Cost, original, 30, 39, 40, 61; modern, 273, 335 (per dive); *see also* Funding; Design, 16, 38–39; *see also* Seapup; Dimensions, xi, 41, 42; Fanmail, 103–4, 130–31; Name, origin of, 32–33, 45; Overhauls and structural changes, 51, 61–62, 92–93; after salvage, 125–27; 146–47, 182, 225, 272–74, 321, 345; Record numbers of dives, 68, 161, 176, 187, 224, 225, 318, 331, 341; Special events: first deep dive, 59–60; swordfish attack, 95–96; accidental sinking, 114–116, 130; Special expeditions: H-bomb search, 62; FAMOUS, 156; warm vents, 173; first geophysical, 198; hot vents, 200; cold vents, 256
Alvin the Chipmunk, 32–33, 45
Alvin Review Committee, 163, 169–70, 207, 209, 210, 211, 223–24, 286, 324
Alvinella pompejana. See Pompeii worm
AMAR. *See Alvin*, Special expeditions, first geophysical
ANGUS, 155, 166, 171, 172, 173, 199–200, 226, 283, 284, 286, 294
Anhydrite, 202
Archimede, 46, 148, 154, 156, 159–60
Argo, 284–85, 286
Arm of *Alvin. See* Manipulator
ARPA. *See* Department of Defense

ARTEMIS, 24, 57, 60, 62, 85
Asper, Vernon, 235–36, 268
Atlantic vents, 278–81, 310, 314, 346
Atlantis II, 254–255; seized by U.S. Customs, 319–20
Atwater, Tanya, 189–90, 192, 194, 196, 197, 198–99, 297, 299
Axial Volcano/vents, 329–30, 346

Backus, Dick, 101–2
Bacteria: 135, 233, 234; abundance at vents, 188, 228; archaebacteria, 312–13; *Beggiatoa*, 226, 313, 318; chemosynthetic, 175, 204, 209, 227, 257, 271, 317; methane-producer, 226, 313, 314; symbiotic, 228, 230, 235, 310; 300, 302, 303, 332 *(soapsuds)*, 333 *(blood dust)*, 339 *(going mad)*; source of nourishment, 228, 271, 309, 313, 317, 318; *see also* Chemosynthesis, Lunch, Origin of life
Bailey, Wayne, 326
Ballard, Robert, 109, 128–30, 131, 138, 140, 141, 142–43, 153–54, 155, 156, 158, 165–66, 170, 171, 172, 183, 212–14, 218, 283–95.
Ballast, 38–39, 41, 86, 93, 125, 126, 146, 86, 290–91
Baloney sandwich. *See* Lunch
Barite, 204, 307, 327
Barnacle, 305, 310–12
Baross, John, 209, 313, 314
Bartlett, Art, 72–76
Barton, Bob, 243–44, 248, 249–50
Barton, Otis. *See* Beebe, William
Basalt, 158, 159, 166, 173, 174, 176
Bascom, Willard, 12–14
Bathocyroe fosteri. See Ctenophore
Bathyscaph, 8, 11, 42, 160; *see also Trieste, Archimede*
Batiza, Rodey, 227
Batteries, 41; droppable, 42, 263; flooded, 66; 125, 142, 182, 274; grounds in, 273, 289
Bazner, Ken, 275
Beebe, William, 7

349

Bernard the Rat, 243–44
Big Mouth, 302
Black smoker, 200–202, 262, 272, 278, 330; see also Chimney
Bland, Ed, 102, 114–16, 139, 147
Boiling. See Seawater
Boston whaler, 241–42, 267
Bowen, Martin, 287, 288–89, 291–93, 295
Brittle star, 99, 185, 336
Broderson, George, 58, 60, 68, 83–84, 88, 110, 116, 142, 148–49, 184; 217, 218, 220–21, 245–46, 253, 265
Brodrick, Ed, 108, 119, 137, 219, 304
Brown, Bob, 241, 242
Bruce, John, 71
Bryan, Bill, 154, 155, 158–59
Bunce, Betty, 104, 190, 191
Bureau of Ships (BuShips), 17, 24, 29–31, 37, 51; audit, 52–53; see also Certification
Burke, Bud, 304–5
Busby, Frank, 345–46
Butman, Cheryl Ann, 240–42
Bythites hollisi. See Pink vent fish

Calyptogena magnifica. See Clam
Calcium carbonate, 125, 126; in sediment, 234; 271, 303. See also Lithoherm, Coral
Cameras on *Alvin*, 92, 166–67, 263
Caminite, 202
Canary, Bob, 123
Canyons, 87, 88, 154; Alvin Canyon, 103; 230, 232
Carbon dioxide: scrubber, 74, 76, 260–61, 268, 335, 340; poisoning, 76, 260, 336
Cavanaugh, Colleen, 228
Cayman Trough, 165–67
Center of gravity, 31, 51, 255
Cephalocarida, 97
Certification, 52–53, 56, 60, 147, 149, 182, 274
Chemosynthesis, 175, 188, 204, 226, 229
Childress, Jim, 182, 225, 226, 230, 270
Chimney, 199, 200, 201, 202, 203, 210, 211, 226, 227, 271, 278–81, 300, 303, 306, 307, 308, 323–24, 327, 329–30, 333, 336–40; see also Black smokers
Chinese checkerboard animal, 281
Clams: shells of, 172, 199, 200, 210, 256, 332; size, 174, 178; growth rate, 235; hemoglobin-rich, 176, 230; oldest, 234; 181, 185, 201, 203, 270, 298, 305, 309, 310, 315
Clambake, 172–73, 174, 182–83, 185

Clark, Dan, 54–56, 123–24, 254
Cleft, 330, 331, 332–33, 336–41
Coefficient of friction, 148
Cohen, Dan, 187, 212, 319
Cold vents, 256–57, 270, 271, 303, 309; see also Oregon Slope
Collapsed lava lake, 183–84
Collasius, Don, 337
Competition among scientists, 207–13, 324
Conning tower. See Sail
Cooks, 92, 94, 139, 217, 219, 220; see also Brodrick
Copepods, 112 (Alvinae); 154, 233, 313
Coral, 255–56, 298
Corell, Bob, 207, 209
Corliss, Jack, 171, 172–75, 176, 177–78, 179, 189, 201, 208–10, 312, 314
Costa, Kenny, 220
Craig, Harmon, 299–302, 304–7
Crane, Kathy, 171, 172, 184, 189, 192, 207, 209
Crew, 198, 199, 217–19, 220–22, 244–46, 272, 273
Crustacean with teeth on eye stalks, 185, 312
Ctenophore, 230–31
Customs: U.S., 179, 181, 319–20, 321; foreign, 246–47
Cyana, 154, 155, 159, 199, 203, 226, 254

Dahlella caldariensis. See Crustacean with teeth
Dandelion, *xii*, 173, 174, 176; Dandelion Patch, 178; 179, 181, 184–85, 230–31, 339, 340
Dangers of deep sea diving: attacks by animals, 95–96, 139; currents, 261, 269, 290, 294; fire, 156, 148, 259, 260, 269, 289; fissures, 158–59, 166–67, 257, 269–70; H-bomb search, 69, 73, 75, 78–79; heat, 261–62; hypothermia, 261; implosion, 268–69, 289; overhangs, 269; penetrators, 148–49, 155–56; power loss, 259–60; smoking in sub, 76, 260; wreckage and munitions, 60, 94, 147, 262, 263–64, 295; See also Dives, fear of; Implosion, Leaks, Safety provisions, *Thresher, Titanic*
Delaney, John, 323–24, 327, 328
Deep Ocean Stations, 136–38, 167, 156, 182, 230, 231–37, 240, 269, 315
Deep Ocean Work Boat (DOWB), 117–18, 128, 130
Deep Quest, 117, 128, 150

Deep sea life: abundance of, 94, 98, 236, 237; early perceptions of, 91–92; 138; like a desert, 173, 185; diversity of, 97; how to quantify, 111; like a rain forest, 237; life span, 234–35
Deep sea sediment, 226, 227, 232, 234, 236, 298
Deep Star, 127, 229
Deep-Tow, 170, 172, 179, 197, 199, 203, 214–15, 283
Deming, Jody, 313
Department of Defense, Advanced Projects Agency (ARPA), 128–29, 138, 161–62
Depth limit: 43 (of 6000 ft.), 47; 147 (of 1200 ft.); 149; 150 (of 8000 ft.); 151 (of 10,000 ft.); 166 (of 12,000 ft.); 182 (of 13,120 ft. or 4000 meters); excursions beyond, 88, 99, 145
Dietz, Robert, 12
Dillon, Bill, 242, 261
Dinsmore, Bob, 253–54
Dives: cold, 87, 335; duration of, 61, 175; entertainment during, 225, 235, 274; fear of, *xi,* 17, 93–94, 98, 102, 104–5, 139–40, 151, 265, 276–77, 334; lack of interest in, *xii,* 10, 11, 12–15, 17, 21, 47, 48, 88–89, 108, 154; sleeping during, 236–37; training, 155–56, 165–66; *see also* Dangers, HERE, Lunch, Urinating, Women
Diving Saucer, 41, 254
Donnelly, Jack, 122, 128, 130, 155, 158–59, 173, 218, 259, 266, 267
Drop weights, 86, 155, 185 (as landmarks), 247 (impounded), 259 (stuck), 268, 276, 325–26
Druffel, Ellen, 255–56
Drugstore, 21, 107, 128
Dyer, Bob, 167
Dymond, Jack 174, 177, 179

East Pacific Rise at 21° north, 199, 200–205, 210, 212, 225–26, 228, 272, 310, 311, 330
Edmond, John, 174–75, 176, 179, 205, 240, 257, 272, 278–80
Eelpout, 200, 203, 280, 321
Electrical wiring. *See* Penetrators
Electric Boat, 18–21, 25, 26
Ellis, George, 187, 225, 242, 244, 246, 265–66, 268
Embley, Bob, 262, 298
Emergency systems. *See* Safety provisions
Emery, K.O., 96, 103, 153
Endeavor, 323–24

Environmental Protection Agency. *See* Radioactive waste
Escort swimmers, 126, 267, 276, 325, 334, 342
Ewing, Maurice, 13, 154
Experiments *in situ,* 135, 188; *see also* Deep Ocean Stations
Explorer's Club, 295, 304, 307

FAMOUS, 150, 153–60, 161, 162, 166, 200, 150, 165, 298
Fiberglass skins, 61, 331
Flegenheimer, Dick, 247
Food chain, 234
Formaldehyde, 177, 186 (forgotten), 203 (with Borax), 303; *see also* Sample curation
Fornari, Dan, 227, 263, 297
Foster, Dudley, 167, 176, 186, 195, 200–201, 231, 239, 262, 263, 267, 277, 276, 289, 292, 300, 328
Fox, Jeff. 165–66, 227–28, 239, 246, 337
Frame of *Alvin*: aluminum, 41, 125, 127 (replaced); titanium, 182, 274 (cracks)
Francheteau, Jean, 199, 200
Franklin, Jim, 308, 330, 346
Froehlich, Bud, 16, 23–28, 30–31, 33, 37, 39, 41, 46, 51, 61, 109
Frosch, Bob, 26, 89
Funding: of *Alvin,* as a national facility, 162–63; charging for dives, 128–29; 24, 108, 122, 125, 127–29, 150, 154, 157, 161–63, 223–24, 225, 273; of scientists, 12, 138, 181, 211–12, 284; *see also* Competition
Fye, Paul, 17–21, 26, 29, 30; 45–46 (*Alvin's* name); 52–53 (commissioning); 77; 85 (*Lulu's* name); 107, 115, 121, 122, 129, 153, 145–48 (titanium passenger sphere); 149, 150, 161–63 (funding crisis), 166, 177 (claims vent biology for WHOI), 182, (retires)

Galapagos vents, discovery, xi–xii, 172–79; second expedition, 181, 182–88; 204–8, 209, 211–12, 213, 225, 229, 272, 310, 318, 319
Gallagher, Bill, 112, 126
Garbage in the deep sea, 135, 136, 327, 339
Garden of Eden, 176, 183, 187, 200, 312
Garner, Sue, 195
General Mills, 24, 25, 27–28, 29, 30, 274
Geophysical measurements: gravity, 199, 200; magnetism, 141, 197–98, 199, 299

Gleason, Skip, 302
Gold, 329, 346
Grassle, Fred, 129, 136, 138, 181, 183, 184, 188, 195, 211–12, 226, 232, 235, 236–37, 239, 240–42, 244, 246, 337
Grassle, Judy, 309
Graveyard, 303
Grice, George, 112
Grimm, Forest, 41–42
Grimm, Jack, 214–15, 284, 286
Guaymas Basin, 226, 270, 272, 317, 318
Guest, William, 66, 68, 71, 76, 77, 78, 79, 80
Gulf Stream, 138, 213, 261, 284

Haggerty, Janet, 303
Hahn & Clay, 31; quality control, 35–36, 37–38; *see also* Passenger sphere
Hammond, Steve, 308, 329
Hampson, George, 97, 99–100, 112
Harbison, Rich, 230–31
Hardiman, Jim, 257, 259, 267–68, 269–70, 274, 276–77, 278
Harris, 275, 277
Haymon, Rachel, 202, 203, 210–11
Hays, Earl, 21, 24, 31–33, 51, 53, 56 (naming *Lulu*), 58–60 (at *Alvin's* first deep dive), 61 (calls for help to find *Alvin*), 62–63, 65, 67–68, 70, 84, 89 (receives award), 101, 130
H-bomb search, 65–79
Hecker, Barbara, 194, 196, 256–57, 269–70
Heirtzler, Jim, 154, 259
Helium, 202, 299, 302, 304, 306, 346
Hellcat, 109
HERE (Human Element Range Extender), 93, 195, 334; *see also* Urinating
Hessler, Robert, 85–86, 97, 98–100, 112, 169–170, 183–84, 185, 186, 187, 310–11, 319, 337
Hey, Dick, 298–99
Hilton, Dave, 300
Hoffman, Sarah, 313
Holland, John, 7
Hollis, Ralph, 222, 198, 225, 227, 239, 240–41, 245, 256, 260, 263–64, 265–66, 267, 268, 269, 277, 289–91, 304, 319, 323, 325–26, 328, 329, 331, 332, 333, 334, 342
Horn, Henry, 54–55
Hot water, 169, 171, 175, 202–3, 226; *see also* Seawater chemistry
Hull release. *See* Safety provisions
Humphris, Susan, 279–80
Hutchinson, Louise, 104–5
Hydrogen bomb. *See* H-bomb

Hydrogen sulfide, 175, 176, 178, 202–203, 204, 226, 228, 229, 230, 257, 270, 309
Hydrothermal fluid, 226, 271, 318
Hydrothermalism, 157, 159, 168–79, 202–5, 227
Hydrothermal vents: antiquity of, 279; biomass, 228–29; distribution of, 309; extinct, 179, 188; toxicity of, 176, 229, 301; *see also* Atlantic vents, Cold vents, East Pacific Rise, Galapagos, Juan de Fuca, Loihi, Kusugas, Mariana vents, Origin of life
HY100 steel: experimental, 30, 31–32; welding, 35–36; machining, 37–38; purity, 38, 145; *see also* Passenger sphere

Implosion, 147, 269, 277

Jannasch, Holger, 135–36, 228, 313–14
Jason, Jr., (JJ), 287, 288–89, 292–95
Jones, Meredith, 209, 228, 229, 312
Joystick, 69, 92, 274, 337
Juan de Fuca Ridge vents, 226, 261, 262, 269, 272, 310, 315, 323–24, 327–28, 329–30, 332–33, 336–41, 346
Juteau, Thierry, 200

Karl, David, 182
Karson, Jeff, 259, 280–81, 297
Keller, George, 154, 155
Kristof, Emory, 156, 166, 167, 181, 184, 186, 213
Kulm, Vern, 271
Kusugas, 303

Lake, Simon, 7
Landry, Dave, 247–48
Lantern fish. *See* Alexander's Acres
LaPrairie, Yves, 153
Launch and retrieval, xi, 58, 94–95; cradle accident, 114–16, 130; with hatch closed, 137; 148, 218–19, 240–42; from *Atlantis II*, 254–55, 277, 304, 334–35, 342; dangers of, 67, 241–42, 326, 331, 332
Lava, composition of, 227–28; *see also* Magma
Leaks, 95, 96, 273, 289, 295, 328, 331, 335, 341
Le Pichon, Xavier, 153, 160
Levin, Lisa, 232–33, 260
Light in the deep sea, 281–82, 324
Lights of *Alvin*, 109 (thallium iodide), 139, 331; *see also* Implosion
Limestone cliffs, 256–57, 269–70
Lithoherms, 138–39, 255–56
Litton Systems, Inc., 38, 39, 40, 46, 61

352 *Index*

Lobo, John, 288
Loihi, 299–303, 347
Lonsdale, Peter, 226, 227
Look-alike subs, 60, 127, 130, 224–25, 263
Lukens Steel, 31–32, 146
Lulu: attempt to steal, 218, 219; cost, 55, 223, 249; cradle, 113–14, 115, 116, 130, 137, 211, 246; design and construction, 53–56; escort ships, 85, 116, 156, 189, 192, 223, 249; funding of, 53–54, 107, 223, 249; lack of privacy, 137; mascots, 244; naming of, 56, 85, 107; 92, 105, 108; replaced, 253–54; safety of, 119, 220, 221, 222, 242, 248; speed, 85; structural changes to, 84–85, 107, 137, 243; swim call, 245; Tube of Doom, 84, 107, 119, 220, 222, 243; 239–50; *see also* Women
Lunch, 100, 108–9, 335, 341, 342; ten months old, 123–24, 135–36
Luyendyke, Bruce, 200, 240

Macdonald, Ken, 197, 199, 210, 299
Madin, Larry, 230–31
Magma chamber, 226, 227, 299; *see also* Lava
Maloof, Roger, 260, 261, 262
Manheim, Frank, 94
Manipulator, need for, 25; 42, 87–88; used with three passengers aboard, 92; lost, 101; found, 103; rebuilt, 109; 125; weight limit, 167; second added, 181; 185, 263, 274; Son of Schilling/Strangelove, 304, 327, 328, 329, 338
Mariana Trench, 15, 141
Mariana Trough vents, 303–4, 305–6, 308, 310, 311
Marine snow, 91–92, 94, 105, 138, 174, 204, 233–34, 237, 335, 336
Markel, Art, 68, 76–77
Marquet, Skip, 57–58, 62, 68, 89 (receives award), 92–93, 103, 109, 121, 142, 145, 151, 157, 223–24, 268–69
Martens, Chris, 318
Mavor, Jim, 20, 36–37, 42, 148, 150, 151
Maxwell, Art, 162
McCamis, Marvin, 57, 59–60, 61, 67; H-bomb search: 68–70, 72–77, 78–79; 88 (exceeds depth limit), 89 (receives award), 91–92, 93–94, 96, 98–99, 101–2, 103 (retrieving *Alvin's* manipulator), 110, 113–14, 118, 122–23 (aboard *Aluminaut*), 130 (quits)
McLellan, Tracy, 138, 181, 192–93, 230

McMurtry, Gary, 302, 303, 347
Megaplume, 330, 331
Megow, Larry, 31, 35, 37–38, 80
Meier, George, 276
Mercury release, 42–43, 263
Mercury trim, 41–42, 62, 127, 273
Methane, 202, 270, 271, 302, 304, 306, 309
Methanococcus jannaschii. *See* Bacteria
Michel, Jean-Louis, 284, 285–86
Mid-Atlantic Ridge, 150, 151, 153, 161; *see also* FAMOUS, Atlantic vents
Midocean ridge, 140–41, 151, 153, 198–99, 202, 205, 226, 297, 298, 299, 318; *see also* Spreading centers
Mile Down Club, 196
Mir-1 and *-2*, 345
Mizar: role in H-bomb search, 67, 68, 77–78; search for *Alvin*, 121–23; in FAMOUS, 155, 283
Momsen, Charles, Jr., 16–18, 20–21, 23–24, 26, 29–30, 46–47, 63; accompanying DOWB, 118–19
Mooney, Brad, 71, 74, 79
Moore, Jim, 154, 156, 158–59
Motors, 42; clinkers in, 60, 62; brushless, 273–74
Mottl, MIke, 323, 327, 329, 331, 334
Mud dwellers, 298; *see also* Deep Ocean Stations
Mullineaux, Lauren, 236
Mussel Bed, 183, 185
Muzzey, Charlotte, 80–81, 89

National Geographic, 166, 167, 171, 185, 212–14, 247; *see also* Publicity
National Oceanic and Atmospheric Administration (NOAA), 129, 161–63, 223–24
National Science Foundation (NSF), 138, 161–63, 166, 168, 170, 171, 177–78, 181, 182, 209, 213, 223–24, 254, 346
Nautile, 296, 345
Navigation: 86, 171, 172, 142, 157 (FAMOUS), 183, (lost), 278, 305, 331, 327; searches for, H-bomb, 66–67; Hellcat, 109; *Alvin*, 116–17; DOS No. 1, 137; lithoherms, 256; *Titanic*, 284; markers, 176 (beer bottles), 185 (drop weights), 332 (blocks)
Nerolich, Shaun, 264–65
Neumann, Conrad, 138–39
Newman, Bill, 311–12
Normark, Bill, 200, 337
North American Aviation, 25–26, 126
Nuclear fusion, 346–47

Octopus, 98, 166

Index 353

Office of Naval Research (ONR), 12, 15, 16–18, 23–24, 28–29, 53; funding problems, 107, 125, 128, 145, 161, 223
Omohundro, Frank, 53, 115
Oregon Slope vents, 270–71, 315
Origin of life, 312–14

Page, Bill, 119, 126, 193–94, 218–19, 222
Palm worm, 310, 340
Panama Basin, 232, 234, 236
Passenger sphere, steel: 41, 125; construction of, 32, 35–36, 37, 38; hull release, 57, 263; in *Turtle*, 345; titanium; condensation, 155, 335, 336; construction of, 145–46; cost, 146, 150; crowded, 333–34, 337; hull release, 263; interior of, 124 (gear), 335; *see also* Dives, Leaks, Pressure tests, Safety provisions
Paull, Charlie, 257, 269–70
Peer review, 181, 223, 224, 234, 240, 286
Peer-reviewed publications, 154, 161, 225, 318
Penetrators, 41, 43, 95–96, 148–51, 155, 156, 157, 273, 274, 328, 335
Petrecca, Rose, 225, 245
Petrone, Pete, 182
Photophores, 102
Piccard, Auguste, 3, 8, 12, 37, 304
Piccard, Jacques, 8, 12, 15
Pilots, 20–21, 57, 94, 102, 113, 127, 130, 142, 151, 155, 157, 159, 196, 200, 251, 265, 266 (certification), 275 (salary); women: 324–25, 345
Pink vent fish, 176, 178, 187; captured 319; 320–21
Plankton, 112, 154, 230–31, 232–34, 313
Plate tectonics, 140–41, 157–58, 161, 298–99
Plutonic rock, 166
Plutonium, 65, 67, 69
Pogonophora, 229; *see also* Tubeworm
Pompeii worm, 203, 211, 305, 314
Pope, Robin, 315–16
Porembski, Chuck, 220
Porteous, John, 177, 178–79, 249
Porthole. *See* Window
Pressure, effects of, 35, 41, 62, 59, 96, 125, 148–49, 268, 276, 336
Pressure tests, 29, 269, 289; of steel hull, 39–40, 43; of titanium hull, 147, 150
Propellers, 41, 60 (stop below 3000 feet), 62, 125–26, 127 (replaced), 121, 167, 182 (titanium added to), 273–74 (thrusters); 326

Protozoa, 232–33
Publicity: 70 (H-bomb search), 77, 78, 80, 89, 92, 128, 130 *(boob of the year)*, 131 (New England Aquarium); FAMOUS, 156, 161, 166; Cayman Trough, 167, 171; 188, 212–13, 276, 288, 324
Pump, 41–42; ceramic, 146, 273, 274

Radioactive waste, drums found, 167–68, 182
Rainnie, Bill, 20–21, 40, 43, 51, 59–60, 61; H-bomb search: 68, 69–70, 78, 79–80; 89 (receives award), 113–14, 121, 125, 128 (hires Ballard), 130, 149–50 (quits).
Rajcula, Gary, 327–28, 329, 332–33, 334–43, 345
Rechnitzer, Andreas, 15–16, 25, 128
Red Sea, 169
Reese, May, 21, 77
Remotely Operated Vehicle (ROV), 266, 283–84, 286–87, 289; *see also* Jason, Jr.
Reynolds, J. Louis, 16–18, 20, 21, 24, 47, 76
Ribeiro, Joe, 220, 277
Rimicaris exoculata. See Shrimp
Rona, Peter, 278–79
Rose Garden, 184, 187, 310
Rosenblatt, Dick, 302, 319, 320–21
Ross, Stephanie, 323, 331–33, 342
ROV. *See* Remotely Operated Vehicle
Rowe, Gilbert, 136–37
Russian trawlers, 66, 70, 117, 167
Ryan, Bill, 142, 214–15, 269, 297–98

Safety factor, 43, 146
Safety provisions: hull release, 25, 52, 75, 263; saw, 43; drill, 93; 109, 126–27, 166–67; rebreathers, 127, 260; 42–43, 166, 260, 262, 289
Sail, 38, 41; destroyed, 123, 125; rebuilt 126, 255; painted red, 172; no windows, 274
Salvage of *Alvin*, 117–23; 122, 123 (Fickle Finger)
Sample basket, 101 (lost), 109, 173, 112 (with ski)
Samples: curation of, 177, 203, 208, 257; thefts of, 208, 210, 212; as souvenirs, 210, 308; ownership disputed, 177–78, 208–12, 324; impounded by Customs, 247; *see also* Sulfides, radioactive
Sampling from *Alvin*: 98–99; 101 (without manipulator); 109, 141–42, 154 (classic geology); 138, 256 (lithoherms); 227 (kamikaze), 271 (seismics), 303 (Rambo-style)
Sampling tools, 87 (pick), 97 (six

shooter); 109–11; 156 (plumber's helpers); 182, 186 (slurper); 184 (dandelion catcher), 226–27 (coffin), 224 (bottom smotherer); 331, 335, 337 (Dean's array)
Sampling from surface ships, difficulty of, 9, 85, 47, 92, 157, 169, 232
Sanders, Howard, 97–98, 99–100, 124, 133, 135, 183–84
Santa Catalina Basin, 233, 215
Sayles, Fred, 318, 337
Scheltema, Rudy, 97, 99–100
Schevill, Bill, 11, 18–19, 111, 136
Schlee, John, 86–88, 112
Schultz, Adam, 324
Scripps Institution of Oceanography, 17
Sea Cliff, 127, 345; *see also* Look-alike subs
SeaMarc, 214–15
Seamounts, 108, 155, 156, 227–28, 232, 299, 303, 309
Sea Probe, 213–14
Seapup, 23, 24–25, 33, 128
Seasick, 245, 267, 323, 342
Sea spider, 100, 327
Sea spider crab, 338–39
Sea urchins, 98–99
Seawater chemistry, 202–3, 204–5, 206–7, 255, 271–72; boiling, 202, 313, 330, 346; 300, 302–3 (Loihi); 306 (Mariana vents); 346 (Atlantic vents); 329 (Juan de Fuca vents)
Seismics, 10, 142, 271
Self-portraits of *Alvin* underwater, 167 (Cayman Trough), 185 (Galapagos vents)
Sellers Will, 251, 261–62, 268–69, 275, 293–94
Serpentenite, 303
Sewage clam, 230, 270
Sharks, 58, 218, 245, 267
Sharp, Arnold, 151, 274
Sheet flows, 198
Shinkai-6500, 315, 345 (world's deepest diving sub)
Shrimp with "eyes" on their backs, 278–82, 310, 314
Shumaker, Larry, 16, 150–51, 161, 167, 208, 223–24, 266
Siphonophore, 184, 230, 231
Ski, 112, 331
Silver, Mary, 233
Smith, Craig, 195, 236, 265, 315, 316, 317
Smith, Ken, 187–88, 195, 221, 239
Smithsonian, 177–78, 208–9, 211, 320
Snail, hairy, 305, 306, 307, 310
Solemya. See Sewage clam
Somero, George, 309

Sonar, 101, 103, 111, 116, 117, 168, 289, 335–38, 340–41
Southwest Research Institute, 29, 38; supertank, 39–40
Spaghetti worm, *xii*, 174, 178, 182–83, 188, 203, 310, 312, 340
Spiess, Fred, 26, 199, 210, 214–15
Spreading center, 140–41, 168 (rate), 169 (Red Sea), 297, 298–99, 309; *see also* Midocean ridge, Plate tectonics
Stakes, Debbie, 173, 177, 189–90, 193–94, 198, 323, 327, 331–33
Steele, John, 246, 247
Stern, Marge, 49, 107, 332
Stimson, Paul, 114–16
Strangelove. *See* Manipulator
Stroup, Janet, 194
Subduction zone, 270, 298, 304, 309, 310
Submarines, early history, 5–8
Sulfide, 159 (from FAMOUS), 173, 174, 176, 177, 199, 200, 201–2, 203, 226, 227, 269, 278, 279, 281, 280, 300, 306; radioactive, 307–8; 323, 327, 328, 329, 333; 337, 339, 340, 341
Summer, Jim, 28–30, 31
Swordfish attack, 95–96
Symbiosis. *See* Bacteria
Syntactic foam, 41, 51, 92 (brow added)

T-shirts, 198, 243, 245, 249–50, 253
Temperature anomalies, 157 (expected); 169 (skeptical of); 171–76, 184 (Galapagos vents); 199, 200–201, 202, 330, (21° N); 226 (Guaymas), 227 (Red Seamount); 278, 279 (Atlantic vents); 301 (Loihi); 303 (Kusugas); 304, 305, 306, 307 (Mariana vents); 323, 329, 330, 337, 338, 340, 341 (Juan de Fuca vents)
Temperature, limit for life, 313–14
Temperature probe, 173, 200–201 (melted), 226 (base charcoaled)
Tengdin, Tom, 327
Thermocline, 10–11, 53, 61
Thiosulfate, 230, 309
Thompson, Geoff, 245, 259
Thresher, 47, 52, 116, 127, 263–64
Thrusters, 276, 273–74
Tibbetts, Paul, 275, 315–16, 332
Titanes, project, 146
Titanic, 13, 128, 213–15, 283–96, 320; *see also* Grimm, Jack
Titanium, 145–46 (lightweight), 335 (warmer than steel), effects of: seawater on, 125, 274; ink on, 126

Index 355

Tongue of the Ocean, 56, 87–88
Toye, Sandra, 163, 223, 224
Trieste, 8, 12, 15–16; alternatives to, 17, 23; 28, 37, 47, 127, 150
Trophosome, 228, 321
Tube of Doom, 107, 119, 137, 142, 190, 239, 267
Tubeworm, *xii* (largest sampled), 176, 177, 178, 179, 183, 184 (largest seen), 187, 188, 203, 210, 225; at Guaymas, 226; opening found, 228; not all new, 229; at cold vents, 256, 257, 270, 271; at subduction zones, 298; not found, 305; no larvae, 309–10; new phylum, 312; portions of found in eelpout, 321; straw-like, 329, 332–33, 338, 340
Tunnicliffe, Verena, 309–10, 314, 327, 332, 338, 339, 340
Turner, Ruth, 129, 131, 137–38, 192–93, 195, 212, 221, 235, 242, 244, 245
Turner Tower, 235, 244, 269
Turtle, 5–6, 41, 127, 345; *see also* Look-alike subs

Urinating during a dive, 93, 129, 195–96, 196–97

Van Andel, Tjeerd, 154, 157, 165, 168–70, 171, 172–74, 178, 189, 212, 218
Van Dover, Cindy Lee, 281–82, 324–26, 345
Van Leer, John, 97
Vent fauna, at: Galapagos: 173–76, 178–79, 181, 182–83, 184–85, 187–88, 310, 319, 340; 21° N: 201, 203, 310; at 13° N, 310: cold vents: 256, 270; Atlantic vents: 278–81, 310; whalefall, 315–17; Mariana Trough: 305, 306, 307, 310–11; Juan de Fuca: 310, 323, 327, 329, 332–33, 336, 338–40
Vent fauna, ability to detoxify, 229–30; ability to withstand heat, 281, 314; evolution, 310–11; most unknown, 181, 185, 312; hemoglobin-rich, 176, 229–30, 270, 303; means of dispersion, 309–10, 317, 339; primitive taxa, 188, 312; short-lived 188; *see also* Bacteria, sources of nourishment, Hydrothermal vents
Vestimentifera, 312; *see also* Tubeworm
Vine, Adelaide, 46, 47–48 (at commisioning), 124 (flag)
Vine, Allyn, 9–15; proposal to build a sub, 17, 89; role in *Aluminaut* design, 18–19; search for pilots, 20; 32, 36, 45, 51, 86, 53, 56, 85, 154
Vine, Lulu, 20, 56, 85, 243
Vine-Matthews Hypothesis, 141.
Virgin, 329, 330
Volcano. *See* Seamount
Von Alt, Chris, 286–87
Von Damm, Karen, 205, 271–72
Von Herzen, Dick, 168, 323, 327, 331, 333, 334, 335, 336–42

Wakelin, Jim, 45, 46–47
Walden, Barrie, 126, 151, 247, 255, 263, 266, 267, 273, 274, 324, 325, 329, 345
Walsh, Don, 15, 16
Walsh, Joe, 20, 32, 36, 39–40, 45
Watkins, Bill, 111
Weaver, Roger, 114–16
Weeks, Bobby, 36, 118, 123–24
Wegener, Alfred, 140, 141
Welhan, John, 300–301, 304
Wenk, Ed, 16–17, 26, 29, 153
Whales, 111
Whalefall, 315–17
Wheatcroft, Robin, 315–16
Wirsen, Carl, 135–36
Wishner, Karen, 233, 234
Whitlatch, Bob, 232, 234, 236–37
Wilson, Val, 62; in H-bomb search, 68–69, 72–77, 78–79; 86–87, 89 (receives award), 130, 147, 148, 151
Winget, Clifford, 110–11, 122, 124, 125–26, 127, 138, 156–57, 167–68, 182, 186
Windows: of bathyscaph, 8; of *Aluminaut*, 18–19; of *Alvin* specifications for, 25; 37, 39, 41; pressure-tested, 43; 96, 111, 139; leaks/condensation, 328–29, 331, 335, 341; 337
Women: first to dive, 104; allowed to dive, 137–38; aboard *Lulu*, 189–90, 192–94, 196; bad luck, 191, 193; discrimination of, 129, 190, 191–92, 332; attacked, 193, 194; in Nunnery, 328; *see also* HERE, Urinating
Wood borers, 129, 137, 235, 290, 291
Woods Hole Oceanographic Institution (WHOI), 10, 13; trustees of, 18, 161–62; first departments, 21; public affairs, 17, 295; 190–92; 50th anniversary, 225

Xenophyophora. *See* Protozoa
Xylophaginae. *See* Wood borers

Young, Earl, 143, 214

Zarudsky, Rudy, 94, 95
Zooplankton, 233